Wilfried Rähse

**Industrial Product Design
of Solids and Liquids**

Related Titles

Bröckel, U., Meier, W., Wagner, G. (eds.)

Product Design and Engineering

Formulation of Gels and Pastes

2013
Print ISBN: 978-3-527-33220-5

Norton, I. (ed.)

Practical Food Rheology - An Interpretive Approach

2011
Print ISBN: 978-1-405-19978-0

Tadros, T.F.

Rheology of Dispersions

Principles and Applications

2010
Print ISBN: 978-3-527-32003-5

Jameel, F., Hershenson, S. (eds.)

Formulation and Process Development Strategies for Manufacturing Biopharmaceuticals

2010
Print ISBN: 978-1-118-12473-4

Seider, W.D., Seader, J.D., Lewin, D.R., Widagdo, S.

Product and Process Design Principles

Synthesis, Analysis and Design

Third Edition
2009
Print ISBN: 978-0-470-04895-5

Tadros, T.F. (ed.)

Emulsion Science and Technology

2009
Print ISBN: 978-3-527-32525-2

Wilfried Rähse

Industrial Product Design
of Solids and Liquids

A Practical Guide

WILEY-VCH

Verlag GmbH & Co. KGaA

The Author

Dr. Wilfried Rähse
Bahlenstr. 168
40589 Düsseldorf
Germany
Raehsel@t-online.de

Library of Congress Card No.: applied for

British Library Cataloguing-in-Publication Data
A catalogue record for this book is available from the British Library.

Bibliographic information published by the Deutsche Nationalbibliothek
The Deutsche Nationalbibliothek lists this publication in the Deutsche Nationalbibliografie; detailed bibliographic data are available on the Internet at <http://dnb.d-nb.de>.

© 2014 Wiley-VCH Verlag GmbH & Co. KGaA, Boschstr. 12, 69469 Weinheim, Germany

Print ISBN: 978-3-527-33335-6
ePDF ISBN: 978-3-527-66762-8
ePub ISBN: 978-3-527-66761-1
Mobi ISBN: 978-3-527-66760-4
oBook ISBN: 978-3-527-66759-8

Cover Design Formgeber, Mannheim, Germany
Typesetting Laserwords Private Limited, Chennai, India
Printing and Binding Markono Print Media Pte Ltd., Singapore

Printed on acid-free paper

Contents

Preface

Explanation of the Product Design

Chemical product design is a novel and comprehensive approach in product development, bringing close together the product and process developer with the marketing people. This new field deals with the interdisciplinary development of innovative nature- and chemistry-based products in the industry. Through involvement of the customer, there is a high probability that the developed products shall fulfill their needs. *Target is both an accelerated development and an improvement of the sales.*

On the application of the customer, tailored products pay off for both sides and represent an interesting "win–win" situation. A notable product idea that can solve a customer's problem shall be assessed after a score system to estimate roughly the market potential. In case of a positive outcome, a normal or, preferably, a compressed product development can start. The compressed development lasts from the idea up to production in the half time.

Implementation of Developments

The development of an innovative or improved product requires a highly motivated team and engaged team leaders. In large-scale projects, the core team should consist of a chemist, a chemical engineer, and a member of the marketing team. The successful realization demands close cooperation internally and with the customer. In case of major innovations, the measures of product design lead to a significantly faster market entry and an increase in the market success. This can be shown by the gain of market shares and higher profits.

In many small and large companies, the developers cannot or only partially work according to the rules of product design. At first, product design necessitates a change of the management culture. At the beginning, it seems strange and difficult that marketing people, chemists, and engineers should work closely together because there are many prejudices and ignorances on all sides. But many employees are interested in the other field and ready to learn from each other. That helps accelerate the project and increase the chances of success. Second, the developers need an idea about the market opportunities now and in the future.

Therefore, marketing must be involved from the beginning. The marketing people are in contact with the customer to find out their needs, for example, through direct surveys, questionnaires, and discussions in focus groups. They try to identify the market sizes in different regions and the local market opportunities. The sooner there is clarity, the more secure the sales. Another important point is the value for money of a new product, wherein the customer specifies the value. If the product allows advantages in the application compared to products of the competition, then the value rises.

Procedures of the Product Design

Product design describes in detail the methods to adjust the specified properties in the final product, elaborated on by the responsible employee of the application laboratories in cooperation with a customer. The parameters include the performance, handling, and esthetics, apart from, sometimes, the durability. The procedure starts from raw materials that are solids (mostly particulate) or liquids (in continuous phase). Typical processes for solids are the agglomeration, mixing, and coating, and crushing as well as chemical reaction. For liquids, there exist primarily the spray drying and spray agglomeration or prilling and pastillation, besides dropletization and crystallization. The ways to create final products from particles differ according to the resulting sizes, which can be increased and decreased or remain constant. Out of a continuum (solution, suspension, and melt) arises, usually, the particles. The surface of the particles can be modified in various ways by different coating methods. Some possibilities are coloring, scenting, setting the rate of dissolution, increasing the stability and abrasion, and so on.

As starting material, the methods of product design use not only particles but also liquids, solutions, suspensions, pastes, masses, melts, and emulsions. In many cases, the complex final product is a particulate. The relevant operations as well as the diversity of processes are shown in detail for detergents, enzymes, plastics, perfume oils, cosmetics, cellulose ethers, and surfactants. In addition, there are many practical examples from the chemistry, consumer goods, food, pharma, and biotech industries, and many others. Some further examples demonstrate that raw materials and their chemical conversion, including the downstream processes, affect the design, but are partially controllable.

This book tries to depict the field of product design clearly and understandably, but especially to be usable in practice. In many parts, it waives textbook knowledge and formulas. The described experiences as well as the discussed unit operations and products come from the own industrial practice, which limits the selection of the detailed examples.

Acknowledgments

The images from the industry help understand the machinery and processes (companies listed in Chapter 18). Without the many measured values and prepared product samples of my staff, such a book would not have been possible. For about more than 10 years, Ovid Dicoi has accomplished interesting measurements in large spray towers during operation, supplemented by laboratory investigations and theoretical explanations. Furthermore, he performed the described measurements in Chapters 12 and 15, besides the penetration of precious oils into the skin as mentioned in Chapter 13.

Several scientists and friends helped by proofreading individual chapters or by providing equipment:

> Professor Dr. Dr. h.c. Bernhard Blümich, RWTH Aachen (Chapter 13)
> Professor em. Dr. Dr. habil. Gerhard Fink, Max-Planck-Institut, Mühlheim (Chapter 9)
> Professor Dr. Karl-Heinz Maurer, AB Enzymes GmbH, Darmstadt (Chapter 10)
> Dr. Bernd Larson, Henkel AG & Co KGaA, Düsseldorf (Chapters 3–16)
> Professor em. Dr.-Ing. Dr. h.c. mult. Manfred Pahl, Uni-Paderborn (Chapter 7)
> Professor em. Dr. techn. Peter Walzel, TU-Dortmund (Chapter 8).

My wife has tolerated with great patience the constant work at the computer for a long time and my frequent mental absence.

Thanks to all.

Düsseldorf, June 2014

Wilfried Rähse

Prior Publications

The content of this book is based on several articles, which were mostly published in German.

Rähse, W. (2009) Produktdesign disperser Stoffe: Industrielles Partikelcoating. *Chem. Ing. Tech.*, **81** (3), 225–240.

Rähse, W. (2009) Produktdesign disperser Stoffe: Industrielle Granulation. *Chem. Ing. Tech.*, **81** (3), 241–253.

Rähse, W. and Dicoi, O. (2009) Produktdesign disperser Stoffe: Industrielle Sprühtrocknung. *Chem. Ing. Tech.*, **81** (6), 699–716.

Rähse, W. and Dicoi, O. (2009) Produktdesign disperser Stoffe: Emulsionen für die kosmetische Industrie. *Chem. Ing. Tech.*, **81** (9), 1369–1383.

Rähse, W. and Dicoi, O. (2010) Produktdesign von Flüssigkeiten: Parfümöle in der Konsumgüterindustrie. *Chem. Ing. Tech.*, **82** (5), 583–599.

Rähse, W. and Dicoi, O. (2010) Produktdesign disperser Rohstoffe durch chemische Reaktionen. *Chem. Ing. Tech.*, **82** (10), 1655–1670.

Rähse, W. (2010) Produktdesign von Kunststoffen für die Waschmittelindustrie. *Chem. Ing. Tech.*, **82** (12), 2073–2088.

Rähse, W. (2011) Produktdesign von dispersen Feststoffen über Zerteilprozesse. *Chem. Ing. Tech.*, **83** (5), 598–611.

Rähse, W. (2011) Produktdesign von Cosmeceuticals am Beispiel der Hautcreme. *Chem. Ing. Tech.*, **83** (10), 1651–1662.

Rähse, W. (2012) Komprimiertes Entwickeln und Umsetzen von Innovationen. *Chem. Ing. Tech.*, **84** (5), 588–596.

Rähse, W. (2012) Enzyme für Waschmittel. *Chem. Ing. Tech.*, **84** (12), 2152–2163.

Rähse, W. (2013) Design von pulverförmigen und granularen Waschmitteln. *Chem. Ing. Tech.*, **85** (6), 886–900.

Rähse, W. (2013) Design of skin care products, in *Product Design and Engineering*, vol. 3, Chapter 10 (eds U. Bröckel, W. Meier, and G. Wagner), Wiley-VCH Verlag GmbH, Weinheim, pp. 273–313.

Rähse, W. (2014) Production of tailor-made enzymes for detergents. *ChemBioEng* (reviews), to be published.

Rähse, W. (2014) Werkstoffe für Maschinen und Apparate. *Chem. Ing. Tech.*, to be published.

1
Chemical Product Design – a New Approach in Product and Process Development

Summary

Chemical product design is a novel and comprehensive approach in product development. Chemical-based products are tailored to the application requirements by involving the customers via marketing. The development process is managed by a core team (product, process development and marketing), controlled by a steering committee. This interdisciplinary development represents a win-win situation. In case of major innovations, the principles of product design lead to a significantly faster market entry and an increase in market success.

1.1
Definitions

This elaboration about product design relates to raw materials, chemicals, chemical-based or chemically treated products, preparations, and their processing technologies. People of the following industries are mainly affected: chemistry and consumer goods, pharmaceutical, biotech, cosmetic, food and plastics, agriculture, textiles, and ceramics. Because the fundamentals of product design are similar in all areas, a transfer of many principles is possible, or even innovations arising from diverse viewpoints. Learning from neighboring areas (how others handle product design) requires a correspondingly broader teaching range for students to get a wider view of topics based on similar theoretical foundations.

> Product design includes learning from adjacent areas, where already a chemical or technological problem solution exists.

The core issue in "product design" is the development, in an optimal way, of the product desired by the customer. The clear definition of Cussler and Moggridge [1] is:

> Product design is the procedure by which customer needs are translated into commercial products.

Another definition includes the elements of product design: "Product design describes the development of customized products, which satisfy all requirements

Industrial Product Design of Solids and Liquids: A Practical Guide, First Edition. Wilfried Rähse.

of customer regarding to performance, handling, and design" [2]. Solid and multiphase products are predominantly of interest in product design, as also liquid mixtures and emulsions. The focus of all activities is the new or improved product and the further utilization or processing by the customers. The question for development of the product is this, "What do the customers do with the product?"

On the one hand, there are industrial customers. Employees of companies in the supply chain department buy raw materials or other substances and formulations for further processing in the production or in pilot processes. These and the people from the application laboratories are referred to as *industrial customers* in this chapter. On the other hand, every individual is also involved in buying something or other, and is therefore also a customer. *Consumers* are people who buy such daily necessities for themselves and their families.

A design of molecules (molecular modeling) toward a particular substance [3, 4] is not what is meant here and is not covered by the term *product design*, also because computer-aided molecular designs are quite removed from the usual customer products.

New products generally require new or changed processes. This needs a close cooperation between product and process development. Some people, mostly chemists with postgraduate degrees in chemical engineering, are involved both functions.

> Product design brings together product and process development, because they belong together.

In the development of chemical-based products, the customer's needs are in the main focus [5]. These include the critical product performance such as the decision to buy, as well as handling and appearance of the product (Figure 1.1). Customers decide not only on these bases but they are also interested in value for money. Therefore, customers are willing to pay more if their (special) requirements are met.

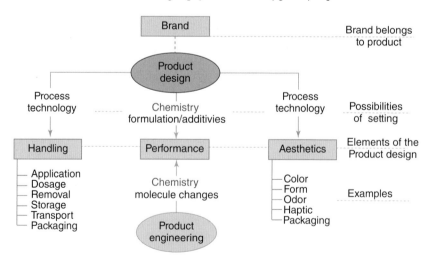

Figure 1.1 General structure of product design.

Chemistry realizes product performance either through changes in the molecule, commonly referred as *product engineering*, or by formulation of various substances. Appropriate technologies (especially design technologies) enable the adjustment of handling and aesthetics. Accordingly, the success of product development is ensured by the *chemistry* and *chemical engineering*. However, *marketing* should be involved from the start, because the responsible person in this area knows the market and is linked with the customers. Marketing must answer the question whether the new product makes sense economically. A strong product or manufacturer brand allows for successful marketing. A part of the brand and product design is in the packaging and packaging design.

- In large-scale development projects, product design requires the cooperation of product and process developers with the marketing people, including the customer.
- The members of the core team should gather the knowledge required for the development and for launch of the product in question, in order to be able to discuss properly any possible problems that may arise (this can be done by learning from others and from literature).

In case of major innovations, the principles of product design lead to a significantly faster market entry and an increase in market success. Chemical-based products are tailored to the application requirements of customers.

The elements of product design, namely, "performance, convenience (of handling), and aesthetics," and for water-containing products, the additional requirement of microbial stability, refer primarily to parameters that the customers perceives. They buy products, when good product performances are guaranteed for long periods. Customers experience the product performance in terms of quality. There are two other elements that contribute to quality. Performance is controlled by chemicals, either directly, by chemical reactions, or by varying the formulation of some of the chemicals. The handling and aesthetics aspects are influenced by the technologies in the preparation processes. In disperse products (particles, suspensions, and emulsions), chemical and dispersion properties (Figure 1.2) determine product design. Concrete possibilities for setting quality depend on product type, hence this is discussed in the appropriate place.

A few successful companies work with the principles of product design for some time when they take on major projects. Although many managers are aware of the elements of advantageous product design, they do not follow them. Switching to consistent product development strategies requires a series of far-reaching measures in business organization, in thinking, and in the organization of work.

1.2
Customer Involvement

In the development of new or modified products, whether the product is based on the developer's ideas or on those initiated by customers, the needs of customers

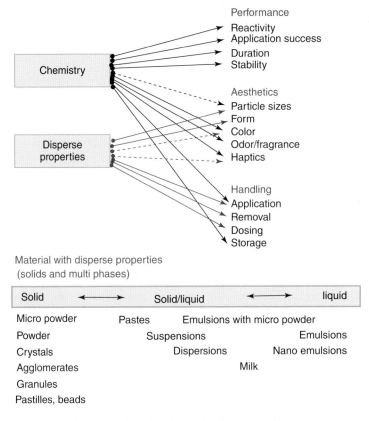

Figure 1.2 Adjustment of product design for disperse products.

are in of utmost importance. Before starting work in the laboratory, the following questions should be answered through a discussion with the customers:

- Do the customers really need this product?
- In what shape do they want the product?
- Do the customers gain their objective in the form of (additional) benefits?
- Do the customers like the product?
- Will the customers accept the value for money?
- How have the customers' experiences been with manufacturers and brands?
- Will customers buy the product?

Success in market presupposes positive response to these questions. Customers will buy a product in order to realize an important improvement for themselves or for their own product. The more accurately that a customers' needs are met, and at acceptable price levels, the better the products sell. For customers, production processes are normally of no interest.

Cooperation with industrial customers in the development stages, usually with the involvement of one to four employees from the applications technology, and

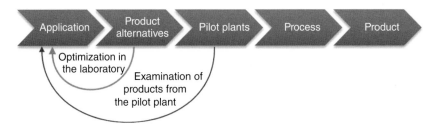

Figure 1.3 Product development in-house or in cooperation with a manufacturer of raw materials, based on the needs of customers.

with several consumers, is important for the success of a novel product. This includes product testing by customers. Formerly, selected *industrial customers* checked out the new products developed by the raw material suppliers for usability, often without a concrete testing proposal. Today, the trend is just the opposite. In fact – as in biotechnology – developments start from the applications point of view (see Figure 1.3). Industrial customers, mostly product developers, discuss their problems in application with different raw material suppliers, in order to obtain information or solutions. These in turn generate attention to new products at fairs, at trade shows in the customer's company, and in conferences. Furthermore, substances are presented in publications and brochures or to business contacts, more frequently delivered to customers as product samples for targeted experiments. In addition to product samples, the customers receive a technical description, a certificate of analysis, and safety data sheets.

Industrial customers want a comprehensive solution to problems, also known as a *system solution*. Therefore, modern developments deal with the applications of customers. If no suitable solution is available that meets completely the demands of the users, then development department starts on a new project. This project is either internal or in cooperation with a raw material supplier. Project partners will be chosen on the basis of the range of their existing products and on their expertise. With the inclusion of raw material manufacturers, executed development projects bring profit to both parties (a win-win-situation). Application chemists of the raw material supplier, design qualified alternatives in consultation with their customers. Industrial customers check the developed products in relation to their needs and then optimize formulations with the new substances.

Because industrial customers order large volumes of products for a considerable sums of money, they examine new product samples intensively. Tests are performed in several applications in comparison to competition and to previously used products. In case of a positive evaluation of the product, they complete longer term contracts with fixed prices for minimum purchase quantities or with a price scale for the expected quantities. In large companies, especially well-trained buyers negotiate purchase agreements for new products. Products are extensively investigated in advance. For understanding of the extent, we consider an example from the detergent industry. The raw material "sodium percarbonate," a hydrogen peroxide-containing bleaching agent in detergents, must pass many tests (see

Test area	Product properties

Analytical chemistry	• Active(oxygen) content
	• Coating layer
	  – Quality
	  – Composition
	• Heavy metal content

Physical chemistry	• Particle size distribution • Dissolution and release rate • Particle stability (mechanical and chemical) • Safety checks

Application technology	• Storage(3 month,30° C/80% relative humidity)
	  – Pure substance
	  – Matrix incorporated(detergent)
	• Effects on soils in different detergents
	• Interactions with enzymes

Process engineering	• Pneumatic conveying • Storage in silos, silo design • Dosing • Separation tendency • Dust collection and processing, safety

Figure 1.4 Testing of raw materials through industrial customers using the example of sodium percarbonate: (a) crystallization – small crystals and (b) spray agglomerization – spherical granules of any size.

Figure 1.4). Extensive testing in various departments indicate the magnitude of the time needed and costs involved. Investigations include a safety assessment of the production facilities.

As team members, industrial customers bring in their product requirements and, if necessary, they demand modifications in each stage of development. In joint developments of a solution, both sides take responsibility for the outcome. Both parties may use the solution. How and when marketing starts is regulated by previously signed agreements. In this context, industrial customers differ significantly from consumers, who have only a very limited impact on product development (see Figure 1.5).

Because *consumers* do not transmit their thoughts, ideas, and desires directly to the company, brand manufacturers consult selected people. This allows to tune products better on the tastes of consumers and reduces risks of a flop. Under instruction from the manufacturing company, agencies invite 12–15 clients to discuss new or modified products. This panel is called the *focus group*. In the group, customers express their wishes and suggestions to improve product and packaging, as also brand name and color design of the logo and lettering. The final questions are, "Would you buy the product?"; "Would you buy at the price XY?" If less than 7 of 12 agree, this product idea is not accepted. Under certain circumstances, the developer adapts the product according to the results of a discussion, and places it

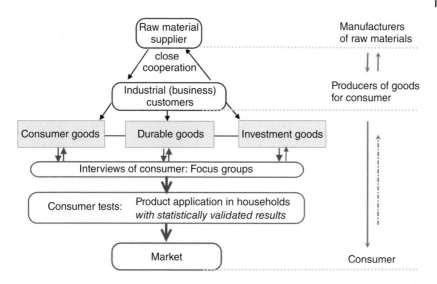

Figure 1.5 Path of a new product from the idea into market.

again in another group discussion. When there are eight or more positive votes, the next step follows.

The next step is to review the product in a concept test (CT). This test runs monadic, that means without reference product, with 100–250 customers (m/f) on the basis of a written and oral product description. The product is only viewed, but not tried. Participants evaluate the featured product ideas on the basis of the following criteria: buying intention, uniqueness, personal relevance, and credibility in school grades 1–5. Customers can be divided into groups, for example, identified as users of a predecessor or competing product, and classified into different age and income groups. A separate evaluation of the different groups is possible. Customers specify, up to what price the offer is interesting to them ("price meter").

In some cases, it is important to clarify haptic feeling in a "sensory assessment" and/or the product odor (smell or perfume) in an "odor evaluation board" (OEB). Twenty to thirty participants in a discussion board judge the products, often in the presence of other products or fragrances as benchmark. In the case of worldwide sales, these tests should be repeated in some distant regions of the world.

With positive evaluation of concept, odor, and haptic test, an extended CT, called the *concept-to-use test (CTU)*, starts. This test includes product applications by the consumers, usually taking place at the customer's home without reference products, as single tests. After product usage, participants fill out a questionnaire and report their experiences and impressions. If use of the product convinces skeptical consumers of its benefits, the CTU is successful. The results bring more clarity regarding market acceptance.

Another consumer test involving about 200 customers (m/f) represents the home-use test (HUT). Participants receive two products for testing, sequentially or simultaneously (successive or simultaneous pair comparison). The first product is the new development; the second may be a competitor's product or predecessor. After application, the participants fill out questionnaires. On the basis of these results, it is possible to determine and quantify the strengths and weaknesses of the test product.

For a full test, on the one hand, customers evaluate product quality, packaging, and design, and on the other hand, there are the advertising claims of the brand and company. Filled up in a white packaging without any information, the product runs through a blind test. The term *partial test* covers the review and evaluation of specific aspects, for example, the need to characterize products with true, credible, and convincing advertising claims (selling proposition). Unfortunately, test results often depend on local customs, so that a test market is only of limited value. Therefore, performance of HUT's in different regions would be more promising. Ultimately, it is the customers who decide whether they will buy the product.

Customers for consumer goods are much less involved in development, compared to customers for industrial goods. Only a very small fraction of the clientele is questioned regarding the product. Furthermore, surveys are usually unknot representative of the population. However, everyone who participates contributes to design. The developers take all opinions seriously and consider all ideas suggested. Accordingly, they modify the product, packaging, or advertising claims. An assessment of market success results from the inclusion of essential customer information after a HUT or CTU. Hence, these methods will apply despite the substantially high costs.

1.3
Specifications

The developer translates all identified customer' needs into the language of science and elaborates a specific product list (Table 1.1). Required chemistry and adjusted measurements, derived from former application testing, form the basis of development work. After extensive laboratory tests, an appraisal takes place with individual product characteristics in order of their importance and volatility. The description of the product with fixing of measured values including their allowable fluctuation are the features of a specification. Using the specification, an identical product is producible anywhere in the world.

1.4
Tasks of Development Team

For steering and supervision of significantly large development projects, two or three employees form a core team. The team leader is usually a chemist, who controls the product development. In addition, a chemical engineer for process development as well as a staff from marketing must be integrated into project management, to cover all areas. This core team appoints suitable employees to participate in the group or to collaborate in subprojects.

Table 1.1 General framework for specification lists of disperse products.

Particle characterization (disperse parameter)	Aesthetics	Physicochemical characteristics
Particle size and distribution	Size/form	Water content
Flowability and silo storability	Color	Dissolution rate
Pneumatic conveying	Whiteness	Porosity
Abrasion resistance	Odor	Specific surface
Explosive dust	Taste	Melting point/glass temperature
Strength, hardness, and elasticity	Haptic	Flocculation point
Crystallinity	Freshness	Density
Analytical parameters		*Application technology*
Composition/purity		Bulk density
Grade of polymerization or substitution		Dosability
		Dust content
Active matter		Sterility
Toxicity		Composition/concentration
		Insoluble residue
		Incorporability
		Storage stability
		Wettability

Employees of marketing are mainly economists, but businessmen or chemists are also included. The focus of their duties differs depending on the product type and customer. These are either industrial partners or consumers. For consumer goods, there are four main tasks of marketing , which are partially distributed among several people:

- Management of brand (brand core, design, advertising claims, target groups, extensions, and internationalization).
- Sales control, especially at events and product launches.
- Hiring of agencies, which execute customer surveys, market analysis, market research, and HUTs and, furthermore design the advertising appearance. The marketing manager evaluates the results and implements the marketing strategies.
- Participation in major development projects.

The key components to success of a project are the contacts with customers and consideration of their desires and experiences. Marketing initiates product tests, weights results, and market analysis. Knowledge of the market with own products, as also from activities of competition, enables to determine the direction of development work.

Furthermore, marketing people bring in market and brand-specific elements from the beginning. Examples include elaboration of key statements, determination of target groups, price–cost relation, design of the packaging, and product shape or color (see Figure 1.6). In the advanced project stage, marketing plans and manages the market entry.

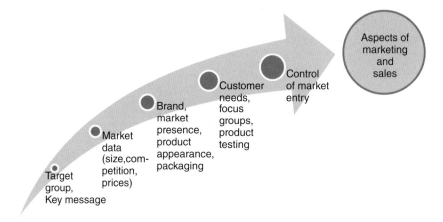

Figure 1.6 Contributions of marketing staff to project management in relevant projects.

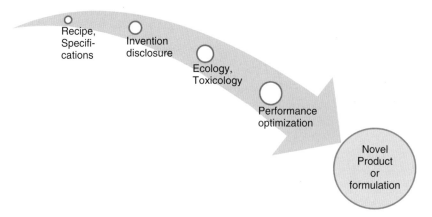

Figure 1.7 Contributions of the product development.

The responsibility for *product development* is in the hands of experienced chemists, pharmacists, or chemical engineers. In the time period from laboratory to pilot plant, several laboratory assistants and technicians are involved, directed by one or two chemists. According to information provided by the project management, they work out formulations, preparation methods for the laboratory, and specifications (including allowable fluctuation of individual components). The responsible chemist takes care of ecology and toxicology, passes formulations to the patent department, and arranges performance tests, in particular, storage tests (Figure 1.7). He reports results weekly to the project core team.

A technical chemist or a chemical engineer with his team (several technicians) executes the *development of process*. The focus is the development of a procedure that allows production of specified products in the most effective way. The so-called design technologies that combine basic operations with shapings are preferred. Following this, a coating procedure is possible for controlling some application

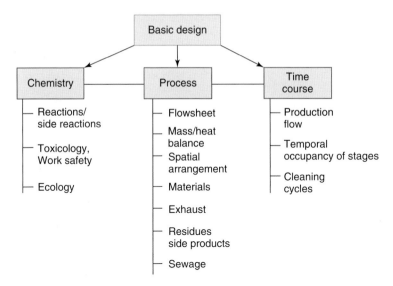

Figure 1.8 Elements of basic design.

parameters. The product design is limited not only to solid, mostly dispersed products, but operates also to pastes, suspensions, emulsions, and liquid mixtures (see Chapter 5).

The work of the process developer focuses on basic design (Figure 1.8). This includes discussions of the chemistry with toxicology and ecology, especially descriptions of procedures with balance sheets. The central part represents the process flow diagram, design of individual stages, and proposals for the arrangement of machinery. For batch processes, it is important to coordinate the time required for product and cleaning in each stage.

After determination of all process stages, production of samples (total of about 20–200 kg) starts in a pilot plant for testing by customers. Thereafter the scale-up from pilot plant to production scale follows (Figure 1.9). Production facilities, if not available, will be designed in detail and built. All phases of design and construction are coordinated by the process developer as a responsible member of the core team, monitored by a steering committee. Further, the process developer controls common subprojects with the packaging department, and with plant construction and production. If the process is new and interesting for the company, it is advisable to apply for a patent. After completion of the production plant, the process developer directs startup. Once the projected capacity is reached, the production people take over the management of the plant.

1.5
Steering of Projects

Depending on the size and importance of project, different levels of management and control ensue. Figure 1.10 shows the steering of two alternative project

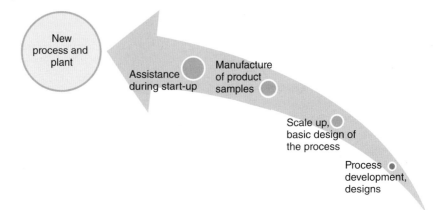

Figure 1.9 Contributions of process development.

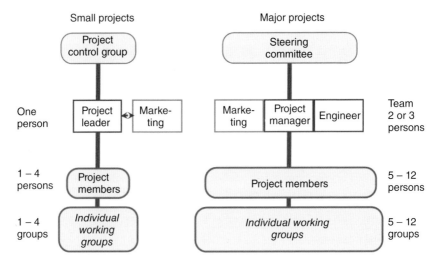

Figure 1.10 Steering of projects.

situations. Small projects are controlled from time to time by several superiors in meetings that take place routinely. For large projects, a high-level steering committee will convene two to four times in a year. The project manager presents the results as well as particulars of expenditure incurred.

Toward the end of a project, a presentation takes place in a large circle. The *management board* of big companies shows great interest in new, improved, and innovative products, furthermore for brands, markets and competitors, product benefits, as well as for required investment and production costs. Usually at the presentation of a new product, the production process is not discussed with the management board (also not with customers). This does not include very new technologies or processes that allow significant cost advantages. Therefore, the development process focuses on aesthetic, easy-to-handle products, meeting the requirements of customers. Only professionals demonstrate interest in process alternatives or novel solutions. All processes that lead directly to the shaped final product, are preferred. Intermediate steps should work fine, but find no attention.

For the implementation of product design, there are a number of alternative variants described in the literature [1, 5–7]. The theoretical background with several examples for the product design of solids and liquids are shown in three volumes [8]. Besides chemical-based products, there are many other applications for "product development according to consumer needs" and the theoretical background [9, 10].

1.6
Learnings

√ Product design stands for development of new products, including marketing and customers.

√ Predominantly solid and multiphase products are of interest for product design.

√ Product design covers the product performance, handling, and aesthetics.

√ Product design realizes comprehensive solutions for the customer's problems.

√ The development process is managed by a team (product, process development, and marketing), controlled by a steering committee.

√ For industrial products, customers are members of the development team; for consumer products, an indirect participation of few selected customers takes place.

√ In all cases, customers test the new products and verify the progress in comparison to former solutions.

√ The HUT is the best known test panel involving 200–250 households.

√ Product design shortens the time to market and reduces the risk of flops.

References

1. Cussler, E.L. and Moggridge, G.D. (2011) *Chemical Product Design*, 2nd edn, Cambridge University Press, Cambridge.

2. Rähse, W. (2007) *Produktdesign in der chemischen Industrie, Schnelle Umsetzung kundenspezifischer Lösungen (Product Design in the Chemical Industry, fast Implementation of Customized Solutions)*, Springer-Verlag, Berlin.

3. Ng, K.M., Gani, R., and Dam-Johansen, K. (eds) (2007) *Chemical Product Design: Toward a Perspective Through Case Studies*, Elsevier, Amsterdam.

4. Kontogeorgis, G.M. and Gani, R. (eds) (2004) *Computer Aided Property Estimation for Process and Product Design*, Elsevier, Amsterdam.

5. Wesselingh, A.J., Kiil, S., and Vigild, M.E. (2007) *Design & Development of Biological, Chemical, Food and Pharmaceutical Products*, John Wiley & Sons, Ltd, Chichester.

6. Ulrich, K.T. and Eppinger, S.D. (2012) *Product Design and Development*, 5th edn, Irwin McGraw-Hill, New York.

7. Seider, W.D., Seader, J.D., and Lewin, D.R. (2008) *Product and Process Design Principles: Synthesis, Analysis and Design*, 3rd edn, Wiley Global Education.

8. Bröckel, U., Meier, W., and Wagner, G. (eds) (2007) *Product Design and Engineering*, Vols. 1 and 2, Vol. 3 (2013), Wiley-VCH Verlag GmbH, Weinheim.

9. Tooley, M. (ed.) (2010) *Design Engineering Manual*, Elsevier, Amsterdam.

10. Pahl, G., Beitz, W., Feldhusen, J., and Grote, K.-H. (eds) (2007) *Engineering Design: A Systematic Approach*, 3rd edn, London, Springer.

2
Diversity of Product Design

Summary

The design of products, based on natural or chemical substances, covers almost all areas of life. In many fields, the product design is already diverse, but it does not always meet the customer needs. Owing to increasing specialization, the need to improve quality, and for offering simplified forms of supply, the number of products is increasing. Examples are the various forms of supply for food and plastics materials as well as for consumer goods. For consumer goods, packaging and labeling represent an integral part of product design. Increasingly, the chemical industry uses the findings of product design and develops new products in consultation with customers. Now, chemicals are provided with brand names, and bags and containers are marked with large labels for publicity.

2.1
General Remarks

Product design depends not only on the requirements but also on the preferences of customers. Product design is already a diverse field, but specialization therein will continue to grow. In the field of drugs (Table 2.1), established forms of supply are comprehensible. These forms have to deliver the drug at the right place, in the right amount during a fixed period. Product design allows this control to be effected. However, in chemistry, innumerable designs are possible. Some of these products, namely, disperse substances and liquids, are listed in Table 2.2. Additionally, many plastics can be made in any shape into a product or part of a product, manufactured with adjustable properties. This is so in the food industry as well. For almost every product, different, interesting types of design exist, quite often more than a 100 varieties. For example, for sugar, there are more than 40 application shapes. For butter, probably considerably more, depending on the taste of the consumer and quantity being packed – from a few grams to several kilograms, in various forms and packagings.

   Example: Butter with emulsified water (max. 16%) has the following forms: without or with addition of canola or sunflower oil; without salt or with sea or rock

Industrial Product Design of Solids and Liquids: A Practical Guide, First Edition. Wilfried Rähse.
© 2014 Wiley-VCH Verlag GmbH & Co. KGaA. Published 2014 by Wiley-VCH Verlag GmbH & Co. KGaA.

Table 2.1 Product shapes for drugs.

Solid shapes	Solid/liquid	Thickened liquids	Liquids
Tablets (in various shapes and sizes)	Shaking mixtures (such as talc in aqueous glycerol)	Ointments	Injection solution
Granules and powder	Active substance plasters	Lotions	Infusion solution
Dragees	Capsules	Gels	Drops (eye)
Capsules	Pastes	Creams	MDIs[a] (liquid propellant)
Suppositories	—	Juices	—
Powder inhalers (promotion on air)	—	—	—
Chewable tablet	—	—	—

[a] Metered-dose inhaler (aerosolized medicine)

salt. As sour cream, sweet cream butter, or mildly soured cream, it is available in full, three-quarter, or half-fat. Of course, butter may be made from cow, sheep, or goat milk, and the more exotic buffalo or yak milk. The following substances may be used depending on the flavor: lactose, minerals, proteins, fat-soluble vitamins, lactic acid, and flavorings. For most foods, there are enough designs to bring out a special book on them.

Before discussing product development based on innovation in Chapter 3, individual examples demonstrate how diverse the demands on product design can be.

2.2
Customizable Developments

There are many positive examples in the field of "product design" of customer-desired products, especially for foodstuffs and plastics. Selected products are briefly addressed here. For the food industry, product design has a long tradition and is of high priority in the expansion of the product range and for differentiation from competing products. In this industry especially, the customer's wishes and the ideas of many employees influence product design. The diversity of pasta, pizza, chocolate, and mustard demonstrate what is meant by product design – customers can choose the shapes and flavors agreeable to them. This multiplicity in product design is possible, because the variants usually can be easily produced and marketed. Even more possibilities for tailoring and extending offers exist in plastics, where material properties, shape, and color are adjustable. For many other products, this state has not yet been reached, or is difficult to realize. The following chapters in this book describe, for example, how such product and process innovations or improvements are implemented.

Table 2.2 Shapes for disperse products and liquids in chemistry.

Solid particles	Solid shape body	Solid/liquid	Thickened liquids	Liquids
Powder	Tablets (single or multiphase)	High-filled solids	Thickened macro- and microemulsions	Pure liquids
Micropowder	Capsules	Suspensions	Lotions	Mixtures
Nanopowders	Fibers and fabrics	Pastes with macro- and nanoparticles	Creams	Nonmiscible liquids
Powdered particles	Pastilles	Melting with particles	Gels	Aqueous solutions
Coated particles	Pearls	Emulsions with particles	Milk	Organic solutions
Instant powder	Flakes and chip	Dispersions	Liquid concentrates	Melting
Granules, pellets, and extrudates	Briquettes	Capsules	—	Liquefied gases
Crystals	Moldings (compact and hollow)	Liquid-filled shaped body	—	Macroemulsions
Prilled melt (micro, macro)	Foamed shaped bodies	Pouches	—	Mini- (nano) emulsions
Cut foils	Spheres, rods, and needles	Supersaturated solutions	—	Microemulsions

2.3
Foodstuffs

Noodles (see Figure 2.1) consist of (hard) wheat flour, eggs, oil, and salt. Industrial manufacturing takes place with the addition of water in temperature-controlled extruders using the tools for kneading, shearing, and homogenizing. Coloring takes place with the addition of red pepper or spinach powder. Besides, noodles are also handmade, quite often (bio pasta). In most cases, various forms of the pasta are formed through the extruder by a means of a headplate with drilled holes. Different shapes are obtained by changing the headplate. In the production of spaghetti, an extruder with 90° deflection is used, so that strands exit downward. A coupled blade, which runs on the headplate, cuts the noodles into the correct lengths. This is followed by drying takes to a level of below 13%. Owing to the diversity in manufacture with the use of integrated shaping, which is implemented in other processes as well (e.g., in the plastics and detergent industries), the extrusion used is known as *design technology*.

The headplate of the extruder has multiple functions. It is not only the shape that is generated, but the material of the plate affects the surface structure and thus the smoothness of the pasta, as also the taste. Manufactured in teflon-coated nozzles, the resulting noodles are so smooth that they cannot stay on the fork and they do not mix with sauce. The opposite effect arises with the use of nonferrous metal nozzles, for which the word "bronze" on the pasta packaging represents a quality feature. This provides an example of how material (Chapter 16) and technology influence the quality (product performance).

Chocolate is available in more than 20 flavors; in addition, bio and diet versions as well as "summer" chocolate are also on offer. Separately packaged dices and minis serve for a one time consumption, whereas the other 100 and 250 g bars are for storage and later consumption (see Figure 2.2).

Figure 2.1 Shaped pasta (macaroni, gnocchi, green tagliatelle, wide noodles, trulli, penne, farfalle, spiral pasta, vermicelli, fork spaghetti, elbow macaroni, funnel, and spaghetti – from top left to bottom right).

Figure 2.2 Chocolate in different sizes and flavors (Courtesy of Ritter Sport).

Other foods need to be cut or ground before they are eaten (see Figure 2.3). Powdered mustard is obtained by dry milling the white, brown, and black seeds or by coarse grinding. The origin and ratio of white to brown or black mustard seeds, the freeness of flow, and the cider or vinegar that is used determine the different flavors and textures of the pasty mustard. Other flavors and typical colors are realizable via addition of sugar, caramel or honey, horseradish and cayenne pepper, herbs and spices, cinnamon, lemon juice or wine, garlic, tomatoes, or

Figure 2.3 Dry grinding to lumpy products and wet grinding or pulverizing to pastes in the food industry (Courtesy of FrymaKoruma).

Butter	Condensed milk
Malt extract	Tomato soup powder
Meat extract	Orange powder

Figure 2.4 Dried and powdered products in food industry (Courtesy of GEA/Niro).

peppers, and the tastes range from sweetly mild to spicy and very spicy. Mustard is normally offered in tubes and jars of different sizes.

In other cases, difficulties arise because of some special features of product do not allow the implementation of customer requests. Global food manufacturers (milk, spices, meat extracts, and soups) concentrate their production to a few locations, so they are at a relatively long distances from the consumers. For water-containing products the risk of contamination arises during transport over long distances. Therefore, many products are dried, so that become free-flowing powders or granules, which then retain their quality when they reach the consumer. Sometimes, usual drying in spray towers is not possible. Some solutions and suspensions cannot tolerate high temperatures, show hygroscopic behavior, go through highly viscous phases, or agglomerate and form large lumps. With a new process development, these liquids can be transformed into dry powders (see Figure 2.4) with properly adjusted operational conditions. The method consists of a combination of a co-current spray dryer with a belt dryer, the so-called hybrid dryer (Filtermat®).

2.4
Chemicals

In the basic chemicals industry, several alternative forms of supply usually exist for each product. The customer might want to use certain solid forms or solutions at

Figure 2.5 Typical forms of supply of sulfur, powder – flakes – pastilles (Courtesy of Sandvik).

different concentrations. For example, caustic soda (NaOH) with a technical purity up to > 99% is available as prills, microprills, flakes, pearls, and as mini or regular pastilles; furthermore, it is also available as standard solutions in concentrations of 50%, 32%, (8%), and 4%. Formerly, in addition to pastilles, powdered variants were also customarily available. As customers did not value these often clumped products, they disappeared from the market. Overall, today the customer has a much wider range to select from.

Another basic chemical, sulfur, is mainly offered in the form of ground or prilled powder, spheres, flakes, pellets, and pastilles (see Figure 2.5).

Fine chemicals, which are meltable without decomposition, were formerly prilled by atomizing of melt in a tower. Alternatively, they were processed on a flaker, which is a cooled drum on which the melt solidifies in flakes that are scraped off. Customers have not been accepting these qualities since the 1990s, because of the poor dosing properties of the flakes and also because they produce large amounts of dust, resulting in the hazard of dust explosion. As a result, new solidification processes were devised. Three methods have established themselves: micropastillation, production of melt drops via spraying and solidifying in a the cooling channel, and targeted crystallization (see Figure 2.6).

Figure 2.6 Nondusting fine chemicals (a) micro pastilles (lauric), (b) crystalline product (polyvinyl alcohol), and (c,d) solidified melt drops (hydrogenated castor oil with resin and wax) (Courtesy of COGNIS).

2.5
Cosmetics and Pharmaceuticals

A toothpaste is characterized by the brand name and packaging. The ingredients, formulated in different ways, determine the performance. Overwhelming importance of product design is demonstrated by the variety of toothpastes available in the market. One part of toothpaste is in the form of a relatively thin, transparent, liquid gel (the solids remain invisible owing to the refractive index). Another part represents a thick, opaque, colored gel with and without stripes or beads. The third part is a foam-forming powder. Consumers can choose the toothpaste variant that they prefer and fulfill their requirements (this is part of product design; see Figure 2.7).

The pharmacists increasingly combine a therapeutic substance with other useful active ingredients. A successful example of diversification is the case of "Aspirin," with acetylsalicylic acid as the active ingredient. Normal tablets as well as coated pills with and without additives are offered; further, effervescent tablets that may also contain vitamin C and chewable tablets are also available. In addition, there are various types of granules that dissolve directly in the mouth or are first dissolved in a glass of water(see Figure 2.8).

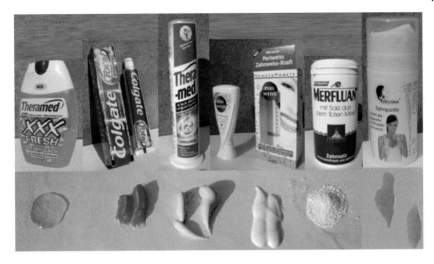

Figure 2.7 Dentifrices as gel, thin, or thick, and in powdered form; far right: toothpaste in an airless dispensers with an additive for quitting smoking.

Figure 2.8 Eight forms of supply for the active ingredient acetylsalicylic acid (Courtesy of Bayer AG).

The tablets exist in different shapes and sizes with or without coating, depending on the drug delivery technology. Further diversity is related to the amount of active ingredient, pharmaceutical formulation, and rate of intake. Effervescent tablets are large and flat. Enteric-coated pills dissolve in the intestine and are characterized by a thick and glossy coating (see Figure 2.9). Some tablets must be large, because of their high drug content.

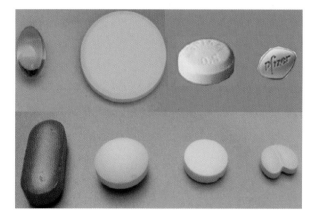

Figure 2.9 Various shapes and sizes of tablets and capsules.

2.6
Polymers and Plastics

Latex paints (wall paints) consist of solvents and binders, as well as fillers, additives, and pigments. The solvent is water, while acrylic resins (for latex paints: also polyvinyl acetate) mainly serve as binder. White color is generated by finely ground titanium dioxide. Possible fillers may be calcium carbonate, quartz powder, and fine silicates. Further additives are present for adjusting the viscosity, stabilizing, preserving, and foam suppression. The amount and quality of the resin controls the water resistance. Titania effects the opacity depending on amount and particle size distribution. Besides the usual, low-drip dispersions, there are the so-called solid colors. Set thixotropic, they liquefy under pressure when painting, but these could not prevail in market. The application of normal dispersions is easy (in terms of handling) and provide better optical results (product performance).

Ironing is an unpopular activity in households. In particular, the smoothing of shirts requires time and great skill. Therefore, cotton fibers are highly refined with chemical agents after dyeing, to achieve the easy-care properties. In cotton fibers, crystalline with amorphous regions alternate. They shift against each other, causing wrinkling. By cross-linking reactions, the fibers in cotton are stiffened (durable press/permanent press). For this purpose, a reaction with oligomers (UF (urea-formaldehyde) resins, formaldehyde-urea condensates) is of help. The substances penetrate the fiber and react by polycondensation. Although the methods are known, only a few manufacturers manage a convincing result for the non-iron form of cotton.

Surface structures of textile fibers change with a permanent finish. The haptic (handle) alters, as also physical properties. Haptic includes stiffness or softness, luster, density, smoothness, and flexibility of fibers. In addition, water-resistant, antistatic and flame retardant, as well as antimicrobial finishes (silver thread) are common. Textile fibers provide a good example for product design. The customer wants special properties that are realizable by a corresponding chemistry.

Formulated adhesives are used not only by individuals in household and for home improvement (DIY = do-it-yourself). During the last 20 years, in industry, tailor-made adhesives have conquered many markets. Examples of these can be found in the metal, automotive, and electronics industries, as well as in aerospace and transportation. In this context, it is exciting to see, for example, how bonding of the wing on an aircraft with optimized adhesives achieves very high mechanical stability in the temperature range from +50 to −80 °C. Apart from construction materials, all adhesives have one factor in common: adjusted polymers that provide for the two-dimensional connection.

Chemically modified renewable resources form the basis for wallpaper adhesives that find household use. These are dissolved cellulose ethers in water, optionally mixed with polyvinyl acetate for heavy wallpaper. Other physically setting adhesives are hot melts and contact adhesives. Super glue (cyanoacrylate) and two-component epoxy glue belong to the category of chemically curing adhesives. Figure 2.10 shows a small selection of 50 variations encountered in household adhesives.

In recent years, there has been much progress in the development of car tires with regard to reliability on the road, aquaplaning resistance, durability, tire noise, high-speed suitability, handling, comfort, and rolling resistance. Chemicals guarantee a high degree of cross-linking of rubber during vulcanization, a protection against aging and gas diffusion. Improved wet skid resistance as also reduced rolling

Figure 2.10 Adhesives in household use: Pritt: paper glue; Loctite: instant adhesives; Metylan: wallpaper adhesive; Sista: jointing compound; and Pattex: two-component glue, hot glue, and power tape (Courtesy of Henkel).

Figure 2.11 Design of tire profiles by G. Giugiaro (Courtesy of Vredestein).

resistance and abrasion can be achieved by addition of nanoparticles of carbon and silica in combination with organosilanes. The desired combination of function and aesthetics satisfies the customer (Figure 2.11).

Such results are possible because the desired quality is adjustable, not only by polymerization (using copolymers and molecular weight distributions) but on numerous additives and polymer mixtures (known as *blends*). In adddition, the molding process allows adjustment of all parameters for product design and can be tailored to the required application (see Figure 2.12). In later chapters, especially in Chapter 9, the settings of product design are described in greater detail.

More and more polymers possessing a variety of colors, shapes, and qualities are used in cars, depending on the manufacturers. The following are some examples of the use of a variety of polymers:

- *Polyurethane foam*: instrument panel, seats, and coverings.
- *Thermoplastics*: levers, buttons, engine cover, air and oil filters, and shock absorbers.
- *Duroplastics*: hood, boot lid, cable trays, covers for lamps and turn signals, reflectors, windows, mudguards, and pump housing.
- *Elastomers*: cable sheathing, components, gaskets, and tires.

Figure 2.12 Application-optimized products in polyurethane or with a polyurethane coating: designer chair by Gaetano Pesce, Alcantara microfiber based on polyester and polyurethane (Courtesy of Bayer AG).

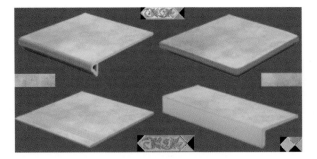

Figure 2.13 Tiles, steps, socket and corners, and decorative elements (Courtesy of Korzilius).

2.7
Ceramic Industry

Kitchens, bathrooms, and terraces are areas that have to look perfect. Therefore, the ceramic industry has come up with not only simple tiles of high quality and in numerous colors and patterns for for indoor and outdoor use but has also profiled plates for levels as straight elements or as a corner pieces, stair tiles, base parts, and decorative elements (see Figure 2.13). Furthermore, the tiles exist in different stress groups (attrition) and slip strength (R9–11), so that customers can find the right products to suit their needs.

2.8
Packaging

Packaging is also a part of product design of consumer goods. It is often made of glass or plastics (bottles, foils, and blister), paper, cardboard, laminates, or aluminum and is decorated with overprints or labels. Packaging fulfills many tasks. First, packaging represents the brand in shape, size, colors, and labels (see Figure 2.14). Second, packaging enables protected transport as well as storage, and allows simple, metered removal. For example, adapted containers or bottles with holes, nozzles, pumps and sprayers, sponges, brushes, needles, measuring

Figure 2.14 Typical packaging for consumer goods representing a brand.

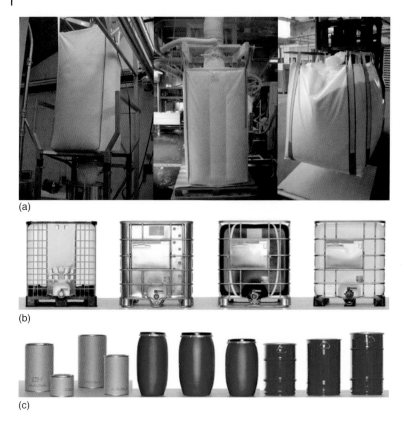

(a)

(b)

(c)

Figure 2.15 Transport containers for industrial products. (a) Solids in big bags (Courtesy of Flexxolutions). (b) Liquid container.(c) Drums for solids and liquids (Courtesy of Menke).

cups, or other removal systems are in use. For some products such as cosmetics, wrapping protects the content from UV radiation, which causes chemical reactions. Furthermore, it reduces or prevents the penetration (diffusion through the wall) of oxygen and water vapor. The entire packaging should match the appearance of the brand and product type, be unique, and generate attention. In addition, there are packages only for transport, such as cardboard packaging for the six-pack of beer or six-in-foil welded polyethylene terephthalate (PET) mineral water bottles.

Figure 2.15 shows examples of how industrial products, solids and liquids, reach the industrial customer. Large quantities (> 20 t) are transported in tank and silo vehicles, whereas big bags, containers, or drums are used for smaller quantities.

2.9
Brand

The brand is another element of product design. This distinguishes goods and services from different companies (original function) by a word/character/symbol.

Officially registered trademarks (registration mark, Section 4 No. 1 trademark act) are marked with an ®, while unregistered trademarks frequently in business carry the ™ sign. In the United States the additional designation ^SM^ for service mark is also used. Each brand manufacturer tries to design the products through an external appearance that is unique and easily identifiable. Brand name, graphic design, and typical colors characterize the brand appearance. An unusual but attractive design promotes awareness of the brand. For wrapped products, packaging design plays a significant role including the shutter opening. In a brand family, the products are similar to each other, typified by the same lettering, colors, and packaging.

Brands are widespread especially in consumer goods as manufacturer and/or product brands, supported by a trademark symbol (logo). In advertising, brand-typical sayings are often heard. The following discussion of a well-known example from the automotive industry explains the various types of brands. The company Toyota (manufacturer brand) assigns a name to all models, such as Yaris, Prius, and Verso (product brands). The logo consists of three linked ellipses; in the middle, a T is visible. It symbolizes the company's philosophy: customer satisfaction, innovation, and quality. The brand saying goes as follows: "nothing is impossible ... "

Large companies for consumer products use either an umbrella brand strategy with a corporate name (Hewlett Packard, BMW, McDonald's, and Intel), or with both company and product name (Toyota Corolla and VW Golf), or only with a product name (Coca Cola, Cartier, Nivea, and Whiskas). Individual brands are

Figure 2.16 Importance of product and manufacturer brands, depending on regional distribution.

available, particularly in pharmaceutical companies (Novartis: Lamisil, Otrivin, and Voltaren) but also in the chemical industry (Styropor, Nylon, and Persil). The value of a brand depends on its commonness (see Figure 2.16) and recognition (Apple, Google, and Microsoft). All mentioned product and company names are registered trademarks.

2.10
Learnings

√ Product design, based on natural or chemical substances, covers almost all areas of life.

√ Brand and brand design are essential elements of the product design for consumer products, more and more also for industrial products, especially in case of worldwide marketing.

√ The principles of product design can be used anywhere. Previously applied in the food, consumer, and plastics industry, the method extends now to all areas.

√ Product design leads to a broadening of product offer, but also to higher revenues.

√ For consumer goods, packaging and labeling represent a part of product design.

3
Generation and Assessment of Ideas for Novel Products

Summary

An innovation represents the response to a market problem. The company analyzes the situation correctly and starts the search for a suitable product. A novel product with high customer benefit helps the company to improve its market share. In most cases, a creative chemist/chemical engineer or another qualified employee suggests a corresponding product that may meet the customer's needs. This is followed up with an assessment of the market opportunities for this product idea, taking 15 valuation points into consideration. If the idea obtains a positive score, its development is taken up by a highly motivated project team. Some considerations for ensuring the project success are discussed in this chapter.

3.1
Innovation

The term *innovation* is from the Latin verb "innovare," which means "to renew" or "to innovate." The chemical industry understands the term "innovation" to mean a new product that has market relevance but differs from its predecessor in some essential feature. This difference may refer to performance or to handling properties, or affect the aesthetics. Upgraded or new products enable the company to grow further and to generate additional revenue [1, 2]. Therefore, each company endeavors to launch a certain number of innovations into the market every year. An innovative product should provide price and/or performance advantages, simplify the application, and improve the appeal through aesthetics.

In addition to the product innovations discussed here, some process innovations play an important role, if they spell out economic and ecological benefits. It is only when the innovative process provides a visibly different product that customers possibly take into account the production procedure. Innovative, ecological processes are thus very important for the company's image.

This chapter describes how a modern company manages innovative ideas; how new products arise; and how they are marketed. Practical action is in the foreground, and not how it could be theoretically improved upon.

Industrial Product Design of Solids and Liquids: A Practical Guide, First Edition. Wilfried Rähse.
© 2014 Wiley-VCH Verlag GmbH & Co. KGaA. Published 2014 by Wiley-VCH Verlag GmbH & Co. KGaA.

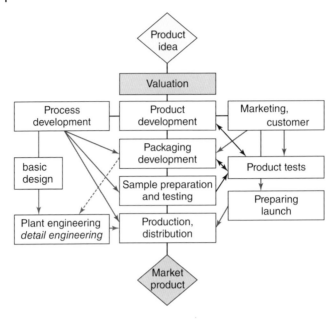

Figure 3.1 General description of the development of an innovative product with the participation of the customer.

3.2
Implementation of a Product Idea

The path from idea to launch in the market is shown in Figure 3.1. First, a responsible person analyzes the proposal, produces samples, and calculates the production and investment costs. Possibly, it may be necessary to conduct a feasibility study and risk analysis, as also an assessment of market potential. A suitable method to assess an idea is presented in Section 3.5. Following a positive evaluation, the installation of project team with core team is taken up. The core team, with a project leader, takes over project coordination and a plant engineer, the management of the construction of the new facility. In large projects, the work of the team and the progress of the project are monitored by a steering committee, consisting of high level managers. The team leaders take charge of development, as described in Chapter 4, until the prototype is tested and high-quality products become available.

3.3
Project Success (Some Personal Reflections on the People Involved)

Employees, who want to be successful in product and process development, should follow and adopt the suggestions and guidelines given in Table 3.1. The table illustrates the social skills, thinking, working, and implementation processes that are

Table 3.1 Requirements and guidelines for employees in innovation processes.

Will, drive, and social behavior	Thinking	Working	Implementation
Professional and positive attitude	Multidimensional and analytical thinking based on knowledge	Only activity and endurance (persistence) lead to complex targets	Open communication both internally and externally
Motivation	Consideration of own and others experiences	Plan and implement carefully	Inclusion of others in the success (team spirit)
Activity	Various aspects considered (physics, chemistry, and engineering)	Self-criticism	Use of all connections
Courage, happiness, and fun	Implement conclusions	Learning from each other	Danger: success creates envy
Communication and team spirit	Creativity and determination		Avoid arrogance
Contacts	Basically think in alternatives		
Acceptance of others	Simple solutions are often the best for clear and feasible objective analysis of possible ways to choose the right path		
Conflict resolution skills			
Self-confidence			

required for an employee with the desire to succeed. The last step, the implementation of a new product or process into practice, requires tact because innovation affects some people directly or indirectly, and they fear impact of these changes. The table was created after 20 years of practice in the process and product development of a large company, representing their experiences. The suggestions that have been mentioned increase the likelihood of a successful outcome and enhance the self-motivation (with the knowledge of ways to achieve success) of the employee.

A simplified stage model (Figure 3.2) displays the project work of team members, specially controlled by a motivated leader. In a larger team, the steps can be executed one at a time or in parallel. Each step should be followed as it represents a substantial aspect of project success. [Multidimensional thinking is encouraged, namely, the search for a solution is not confined to a particular field but should also look into neighboring areas. A simple example of this is the use of disintegrants or

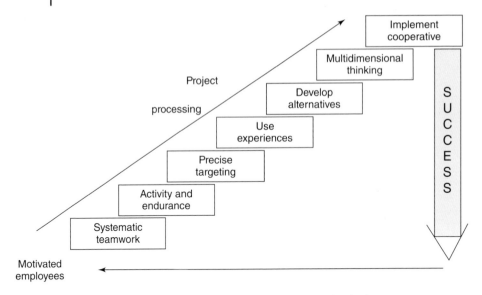

Figure 3.2 A simplified stage model of the way to success of a development team.

effervescent chemicals (chemistry) greatly accelerates the dissolution large tablets (physics), which is otherwise a very slow process.]

3.4
Generation of Innovations

The first step in an innovation process is to suggest an interesting product idea. The idea must also include information about how product design is realizable. In addition, the estimated market potential for the product idea should be presented. The generation of an idea for a new product or process is a creative one. In the chemical/technical field, only few people are capable of generating novel solutions. Suggestions come from among the three "C's," namely, customers, company employees, and competitors. Good product ideas originate often from development departments of the company. There are numerous systematic approaches for finding innovations [3, 4], which will not be discussed here, because they are too theoretical (otherwise, the innovation manager would have to be the best innovator).

The following describes some systematic approaches of top management. The exemption of young, just hired employees from work for six months or a year in order to generate new product ideas, often meets with very little success. These employees overlook neither the business nor do they have the necessary product knowledge. Brainstorming, a systematic approach for new ideas, usually brings an interesting overview, but no breakthrough. The ideas that are thrown up are useless in the absence of concrete proposals for their implementation. Brainstorming is always an effective tool when it comes to simple problems. With difficult tasks, knowledge and expertise are needed, which only a few scientists have. The collection of proposals from the entire staff usually brings about only minor improvements; real innovations very rarely take place. Systematic training of academic staff in innovation management also does not result in any great success.

Every big company identifies its innovators. Furthermore, the identification of potential innovators succeeds with the use of relevant psychological tests, which are executed in several companies. Innovators are people that always stand out with good ideas. Their creative proposals come from chemistry, chemical engineering, or from neighboring areas. These creative persons have profound technical knowledge, usually in several fields. They do not give up during the solving of problems, but seek alternatives. Thinking up a solution affects the entire life of innovators, their work and leisure, travel, and even their sleep. Good ideas often provide answers to market or technical problems. Innovative employees need free spaces to create new things, to check their usefulness, and to first manufacture laboratory samples. In addition, a great number of people express an interesting idea at least once in their lifetime.

As examples, some personal experiences in the development activity at Henkel are explained in the following. These involve several innovations that have been implemented both for processes and products. Most of the examples are described greater detail elsewhere in the book. The innovations were developed mainly in response to major problems and market pressure.

- 1. Cellulose ethers (circa 1984)
  Problem: During the methylation of cellulose fibers, two by-products that were formed were methanol and dimethyl ether, which escaped via exhaust air and wastewater.
  Solution: Separation of these substances by distillation processes (internally implemented).
  Innovation: Reaction of the by-products with gaseous hydrogen chloride to produce methyl chloride, using a novel highly active catalyst. This realization enables a complete recycling of the by-products (externally implemented).
- 2a. Enzymes (circa 1989)
  Problem: Using genetically modified microorganism, it must be guaranteed that the production strain cannot enter the drains. A modified strain should be ready within one year.
  Solution: Sterile filtration of very dilute fermentation broths.
  Innovation: Cross-flow microfiltration with inorganic membranes in combination with sterile filtration for simultaneous separation of the production strain.
- 2b. Enzymes (circa 1993)
  Problem: The enzyme pellets smell unpleasant.
  Solution: Removal of odors from the fermentation broth.
  Innovation: Development of a novel deodorizing process (with CFD), to remove the odors with superheated steam in vacuum under minimization of entrainment.

- 3. Powdered detergents (1994)

  Problem: The residential area in Vienna moved closer and closer to the factory. Location of the spray drying tower was problematic and a ban threatened.

  Solution: Removal of exhaust gas plume.

  Innovation: Use of superheated steam instead of hot air in the exhaust-free spray tower, with energy recovery by condensing the excess steam.

- 4a. Detergent concentrates (Megaperls®, 1992)

  Problem: Detergent powder with higher bulk densities were introduced in Europe.

  Solution: Development of a new generation of detergents: rounded pellets instead of powder.

  Innovation: A new extrusion process, using an aqueous ingredient of the formulation as lubricant, and the extruder headplate with a subsequent spheronizer for shaping.

- 4b. Detergent concentrates (Megaperls, 1996)

  Problem: Owing to the average sphere size of 1.4–1.6 mm compared to 0.4 mm of the powder, the solubility was significantly lower; in addition, residues remained in some cases.

  Solution: Improving the solubility of the spheres.

  Innovation: The extruder compacted largely water-free powders at elevated temperatures with the help of a polymer melt. After shaping and cooling down, the spheres harden. During the dissolving process, these spheres disintegrate into the initial powder.

- 5. Detergents (2003)

  Problem: For normal powders from the spray tower and the concentrates from extruder, there are different production plants with varying full utilization.

  Solution: Production of all types of detergents in a large-scale plant.

  Innovation: Development of novel granulation (core and shell), which can cover relevant areas of size and bulk density for detergents (not yet implemented).

- 6. Medical detergents and softeners (2004)

  Problem: Chafing and skin irritation on laundering.

  Solution: Skin care via washed laundry.

  Innovation: Add nanoparticulate skin care oils to liquid detergents, and transfer them from the laundry to skin. The very small, transferred amounts were proved analytically (not implemented).

- 7. Skin care products (2003)

  Problem: Novel creams containing nanodroplets are not accepted in the company ("does not work, because the emulsions disintegrate directly on the skin").

  Solution: Comparative measurements of the penetration of mini and macro oil droplets into the skin.

  Innovation: Worldwide novel NMR method, developed by an employee in cooperation with university, measures the penetration depending on skin depth (every 15 µm). There are significant differences between macro and mini emulsions.

- 8. Toothpaste (2012; ATS License GmbH)

  Problem: Free spaces for smokers are increasingly restricted. Many want to quit smoking, but they need support.

Solution: Toothpaste with a special ingredient and the desire for cigarettes is reduced.

Innovation: Toothpaste with a natural substance (essential root oil) that reduces the desire to smoke.

3.5
Evaluation of Product Ideas

Before a development project is commenced, some employees evaluate the product idea. Exceptions are variants of already manufactured products. They can be produced simply, and inexpensively introduced into the current product range. In case of a flop, a removal is possible without big money losses. An assessment of all

Table 3.2 Assessment of market opportunities for innovative products by estimations (market standard = 0; negative ratings lead to point reductions, positive ones to an increase of score).

Negative feature	Score	Positive feature
Product variant	0 ... +10	Completely new product (market innovation)
Product design below market standard	−10 ... 0 ... +10	Attractive product design
Handling below market standard	−10 ... 0 ... +10	Consumer-friendly and easy to use
Product performance below standard	−10 ... 0 ... +10	Clearly superior product performance
High development costs	0 ... +10	Low development costs
No spec. know-how in the company	0 ... +10	High product know-how in the company
Cost uninteresting (too high)	−10 ... 0 ... +10	Cost of interest (low)
Several strong competitors against the new product	0 ... +10	No competitor against the new product
No patent barriers against competitors	0 ... +10	High patent barriers against competitors
Tight regional market	0 ... +10	Wider global market
Expectation of low demand	−10 ... 0 ... +10	Expectation of high demand
Distribution channels not available	−10 ... 0 ... +10	Global distribution channels available
No brand available	0 ... +10	Strong brand or umbrella brand
Customer benefits below competition	−10 ... 0 ... +10	High customer benefits
Value for money below competition	−10 ... 0 ... +10	Attractive price/performance ratio
	Total points P (sum)	
Score $S < 0$; high risk and uninteresting	Total score S: $S = (P/15) \times f$	Score $S > 5$; low risk and high potential

other product ideas are provided in the specified schema in Table 3.2, designed for large organizations. An additional factor (f) serves for consideration of the level of innovation. At the top of the table are ideas that represent a world first ($f = 2$). Then come proposals that are new in regions where the company operates, and have not yet been introduced ($f = 1.3$). All other products get a factor of 1, especially if comparable products exist in-house already.

Approximately 6–10 competent employees from different departments, who are familiar with the product idea, should give their assessment according to Table 3.2 and then discuss the results with each other and other participants (team members). An average value of a total score greater than 5 indicates a large market opportunity that should be exercised. Values less than zero lead to termination of the activities, unless a positive decision is important for strategic reasons. If values of total score between 0 and 5 arise, the discussion needs to be deepened, because business risk increases with decreasing score. Decision makers should take this relationship into consideration. However, the incorporation of a new product in the existing product range makes sense in some cases.

3.6
Learnings

√ An innovation represents the response on an urgent market problem, devised by a creative chemist/chemical engineer.

√ Innovations with high customer benefit help the company to gain market shares.

√ Project success requires highly motivated team members.

√ Before starting development, team members assess the product idea.

√ The score helps to estimate the value of idea, and to decide whether to start development.

√ High value for money improves the chances of market success.

References

1. Kanter, R.M., Kao, J., and Wiersema, F. (eds) (1997) *Innovation: Breakthrough Thinking at 3 M, DuPont, GE, Pfizer, and Rubbermaid*, Harper Collins, New York.

2. Meffert, H., Burmann, C., and Koers, M. (eds) (2013) *Markenmanagement: Identitätsorientierte Markenführung und praktische Umsetzung*, 2nd edn, Verlag Gabler, Wiebaden.

3. Zobel, D. (2007) *TRIZ für alle: der systematische Weg zur Problemlösung*, Expert verlag, Renningen.

4. Gundlach, C., Lindemann, U., and Ried, H. (2007) *Current Scientific and Industrial Reality: Proceedings of the TRIZ-Future Conference 2007, Frankfurt, Germany, November 6–8, 2007*, Kassel University Press GmbH.

4
Compressed Development and Implementation of Innovations

Summary

To improve competitiveness in the global market, innovative product ideas must rapidly be placed in the market. The innovator gets higher prices, gains market shares, and benefits from a positive image. Developing in half the time requires not only support of top management but also the immediate involvement of customers. A core team, consisting of a product and process developer, as also a staff member from marketing, should lead the project. Time can be gained by competent, parallel processing of subprojects and by reducing the number of attempts for designing the machines and equipment. Communication within the team and with the leadership and the absolute determination to bring a new product in market are crucial to the success of the product.

4.1
Preliminary Remarks

The author's first chief at Henkel[1] initiated the idea of compressed process development in the 1980s, which resulted in short development times. This method, has been supported by the author, taken over, and later also transferred in some cases to product development. In accelerated development, experimental work in the pilot plant and planning of a process plant including the orders overlap. Therefore, the project starts immediately after acceptance of an innovative product proposal. The goal of compression is to halve the time from idea to market.

If a good idea is present, the promotion and steering of innovations represents a task for the top management. The work of an independent department that understands innovation management as systematic planning, management, and control is ineffective in many cases because they are far away from market problems. Normally, this central department is not involved in the company's product

---

1) Henkel AG & Co KGaA, Düsseldorf, Germany.

Industrial Product Design of Solids and Liquids: A Practical Guide, First Edition. Wilfried Rähse.
© 2014 Wiley-VCH Verlag GmbH & Co. KGaA. Published 2014 by Wiley-VCH Verlag GmbH & Co. KGaA.

development, which is performed in operational units. Therefore, innovation management [1, 2] is not discussed at this point.

However, in practice, the development presupposes an existing and feasible product or process idea at the start. The compilation of the author's own observations and practical experience as a member and team leader of projects contributes to this discussion. This is a personal view and not a scientific approach to problem solving with statistical relevance. In the following sections, the reasons and opportunities for, as also the risks and barriers to accelerated implementation are explained. This is followed by description of formerly common "central projects" at Henkel,[1] after which compressed development is explained in general, with examples, outlining the targets.

4.2
Reasons for an Accelerated Product Development

In order to succeed and survive in the chemical industry, owing to global competition, an increasingly rapid deployment of innovative products is required. Concrete ideas and implementable suggestions for novel products with significant market potential are rare even in big companies. Therefore, rapid implementation of such innovations is of particular importance. An established matrix (see Table 3.2 and [3]) allows estimation of market opportunities and risks before a company starts on a development project.

Following acceptance of the idea by responsible managers, compressed development must be initiated without any delay. It is then that there is a big chance to be first in the market. In the case of a consumer product, many tests should support the assessment results. Market analysis [4, 5], customer surveys, and product tests at households of customers accompany large development projects. Marketing hires some agencies to execute the tests in different countries. The whole process involves not only rapid implementation in production but also preparation for early market entry. This preparation begins with information and training of sales staff, placement in different outlets (markets) in addition to press releases and advertising.

The time from product concept to implementation in production is particularly important for financial reasons. The accumulated development costs rise with increase in the duration. Nevertheless, it is crucial that early market entry brings in money by way of product sales, at additionally higher prices. Furthermore, the innovator gains market shares, significantly more than imitators do. Ultimately, the market determines the success of the development. Innovative products should be patented. This can enable a privileged position in the market or even a long-term monopoly. A second important reason for a rapid launch is that based on good communication, it brings an improved image for the company, resulting in a positive attitude to the whole product range.

4.3
Risks

Generally, the financial risks of process development and of implementation depend on the company's know-how and on diversification. Several important factors are illustrated in Figure 4.1. If the innovative product is a possible part of the core business, then chances of success considerably outweigh those of failure. In cases of backward integration or buildup of a new business, the risks increase. If a project is stopped in the development phase, the company makes losses [6]. These are the cost of market research, further cost for laboratory and pilot tests, and for the fees paid to consultants. Investment in special equipment in laboratory and pilot plant could possibly add to the losses incurred. Stopping development at a later phase would involve, in addition, the cancellation charges for buildings as well as for machinery and equipment.

Henkel successfully carried out all types of diversification (product examples: enzymes and tetraacetyl ethylendiamine (TAED); detergents: spheres, granules, tabs, and water-free liquid detergents). In the early 1980s, enzyme development and manufacturing was new for Henkel and, therefore, associated with very high risks. This diversification is called *lateral* diversification. Another high risk case demonstrated is that of TAED, which was also new but was offered as raw material. Other supply forms of laundry detergents (pearls and tabs) are higher value products of the core business and depict low to medium risks as horizontal to vertical diversification. Whether executed in normal or compressed development, the risks were always financial. In some projects, small problems

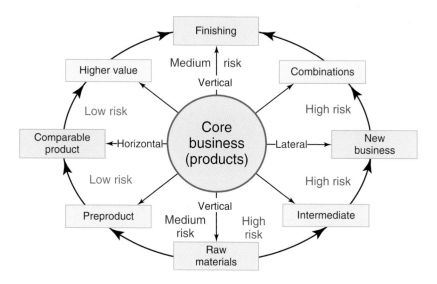

Figure 4.1 Risks associated with implementation of innovations, depending on the type of diversification.

arose that resulted in higher costs than had been planned during the development or investment stages. There were also some significant problems with the actual sales falling short of projected sales.

This raises the question regarding the need for compressed development. There are higher risks, when possibly the quality of the product is not as originally intended or when manufacturing is not optimally tuned in all basic operations. These arguments do not really speak against rapid development. Customers see the product in the developed form, without perceiving process deficiencies or improvements. Costs for improvements are covered by the ongoing product sales. Using compressed implementation, it is acceptable to have a few minor changes in line with the expectations. Moreover, early market entry results in a faster route to lower production costs over the learning curve.

4.4
Barriers in Development Projects

In innovation projects, it should be accepted that new and unknown situations triggers uncertainty in the implementation phase for most employees. In the course of development, the manager must explain the impacts of the new product to the affected employees. Such projects require strong commitment on the part of all parties involved as the execution process creates stress that can sometimes be extreme. For example, employees who already headed the start-up of new facilities or were involved in developing the know-how would already be faced with time demands of 14–16 h a day, including weekends. Additionally, the home office can ask from time to time, why implementation has not speeded up. This background makes it clear that even influential people in leadership positions argue against a project or project extension, for example, as follows:

1) This product will not meet the demands (i.e., specifications).
2) There is not enough time available for the necessary time-consuming tests, inspections, and storage arrangements.
3) Scale-up and implementation in production do not happen overnight. Delivery times for individual machines/devices alone take a year or more.
4) Change in production processes cannot be establish in such a short time.
5) Hasty introduction involves high risks in the market.

There are always people who oppose any innovation, either openly or covertly, because they fear and hate changes. Therefore, at all levels, the persons concerned discuss the uncertainties in combination with the high costs. Such discussions lead to nothing, except when they deal with suggestions for improvement. The constant nagging brings uncertainty among the project participants and decision makers and also delays the project. Rapid development presents many advantages, but in prudent manner, it is better to schedule an extended period for the start-up plant to reach the designed capacity.

Basic principles of product development that have been published [7, 8] are in most cases not tailored to chemistry and do not treat the compressed working. The following description shows the author's personal experiences with accelerated developments and introduction of innovative products (detergents, biotechnology, adhesives, cosmetics, and chemicals).

4.5
History

Until about the mid-1980s, the development of new products used to be executed in separate phases. First, ideas were collected. Sifting and evaluation yielded favored ideas, which were released for working in the research department. After some time (about 2–3 years), the researcher, together with process developer, discussed the progress. They determined a pioneering product for the future. For such an innovation, the process department examined implementation already present in existing productions or started a new process development. Thereafter, the production people, product and process developer, researcher and plant engineers went deeper into the details for implementation. These discussions took place at the level of the head of department (±1). Here, they decided on possible locations, capacities, and raw materials; furthermore they also calculated required energies, materials, wastes, and emissions, as well as product costs. Plant engineering took over the construction of the production facility or the specific changes that were needed in the existing factory. After clarification of open issues, the project team informed the marketing staff, which then started surveys and household tests. This approach led to project times of ~6 years and more.

For major innovations, the company (Henkel) constituted the so-called central projects with accelerated completion. Editing was done with support and supervision of the corporate management. Once a year, the project leader reported and defended the development work/costs to the entire board, including the chair. In these projects, there was a close link between R&D, process technology, and production, besides partly involving the marketing staff. In the hot project phases (construction and start-up), rapid processing was ensured by convening monthly meetings with participants from all levels.

4.6
Compressed Project Processing

While previously the plant engineering held a strong position and influenced the development, it is now simply a service provider who is not involved in development and decision making. The crucial position of production people in former times no longer exists. In the final phase, some employees of the selected production site work with the development team in order to incorporate their experience and to take into account regional specificities. The influence of central research department on decisions is unlikely, as most researchers now work in operational units.

The integrated way of working, which was introduced for larger innovation projects, has evolved with changing conditions over the years. Today, product and process developers work together with marketing in a core team and arrive at the decisions on product launches. Reactions are faster and easier because crucial activities are within one operational unit. Problems at interfaces to other areas have been eliminated. The main changes compared to the past are as follows:

1) Major projects, supported by top management, take place with the participation of marketing from the beginning.
2) A core team, consisting of product and process developer, further of a marketing employee, controls all processes and mandated internal and external work. This team prepares all facts for decision.
3) Customers are involved through interviews.
4) All operations run parallel.
5) A short, flexible project plan is recommended.

The possible timing illustrated in Figure 4.2 shows that project launch to market can be achieved in less than half the time in a compressed project.

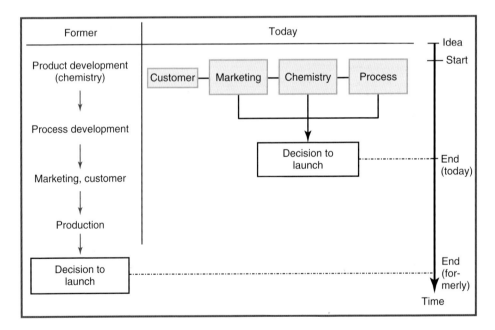

Figure 4.2 Collaboration of marketing with product and process people in compressed innovation projects reduces the development time.

4.7
Project Leadership

In compressed projects, the project manager (usually the idea finder with leadership skills) designs a useful, clear project plan with milestones. The plan contains all essential activities, but without going into finer details. Work takes place after progress as reported and not after plan execution. In this learning development, planned research activities shall not be continued, if after few trials the wrong direction is evident. A strong commitment to achieve success motivates all team members.

In parallel processing, no employee awaits results from another, before starting. Laboratory experiments start and go on, assuming that at least one alternative formulation meets the requirements of the three-month storage test. Thus, the process engineer can order elements of the pilot plant using a flow sheet diagram, without knowing the exact composition of the product. Marketing examines possible distributions and markets. Customers are able to express their ideas immediately. Packaging department designs possible packaging. A suitable production site is available after some discussions. Figure 4.3 illustrates a schematic project flow, controlled by the core team.

Figure 4.4 displays the main tasks of members in the core team. The core team is responsible for project success and, therefore, meets every week. It invites those scientists who are working on project parts. A member of the core team should be self-confident and well networked within the company, and should be able to increase and accelerate cooperation in the development of the new product. Technical and interpersonal skills of members enable rapid progress. In meetings, the core team confirms or changes the development directions, and also assigns new tasks or has discussions with external consultants. The successful acceleration of a project requires high decisiveness of the core team in the cancellation or

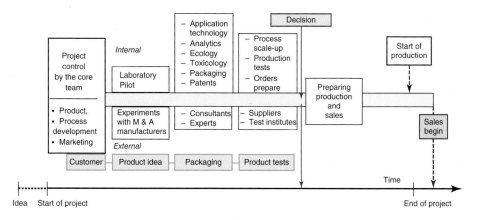

Figure 4.3 Timing of compressed product development for washing and cleaning agents under control of a core team.

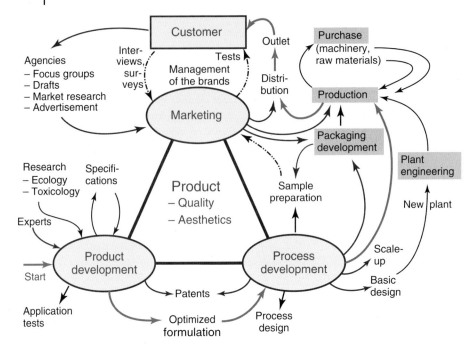

Figure 4.4 Linking up of individual steps in the development of a novel product; the short-est way from product development (start) via an existing plant to the customer is indicated by the thicker lines.

assignment of additional development work. The core team, in particular, the project leader, must inform top management regularly and candidly about the status and costs.

Product developers have to create the new product. They are responsible for for-mulations and specification, the optimization as well as for ecology and toxicology. Their task is to check the products in application, immediately after preparation and after storage (3 months, 30 °C, and 80% RH). The process developers establish an economical method for production. They take into account existing production plants and try to integrate the new product. In some cases, the developer, in cooperation with the plant engineering, must plan a new production plant on a greenfield site. In addition, the developer calculates the production costs of the adopted procedure and of alternatives. Scale-up and production of samples in association with the packaging department are also among the tasks of process developers. Marketing is especially active in product and packaging aesthetics; it leads the brand, speaks with customers [9], and arranges household tests [10–14]. It also coordinates the launch (purchase, production, and sales), in some aspects in cooperation with agencies (advertisement).

4.8
Teamwork in Projects

In almost all cases, the innovative idea comes from an individual, often also the basic analysis, even if ideas originated in a group discussion. However, in big companies, many employees participate in the implementation of the idea into a valuable market product. The transfer of the idea into a product is the object of a great team. The members come from various departments and work together in the project team for a limited time (see Figure 4.5). These employees in the project team must be exempt from other work wholly or partly. For rapid development, technical and social skills are crucial, especially of the core team and of the project manager, but also of technical staff. Stress between individual academic staff arises when tasks are not clearly defined and delimited. It is not always possible to ensure that the selection of team members should not lead to widely overlapping skills of two scientists at the same level. In cooperation, the least problems arise with people of different, specific training. For example, when chemists work together with physicians, economists, engineers, or microbiologists, then fruitful collaboration is the rule.

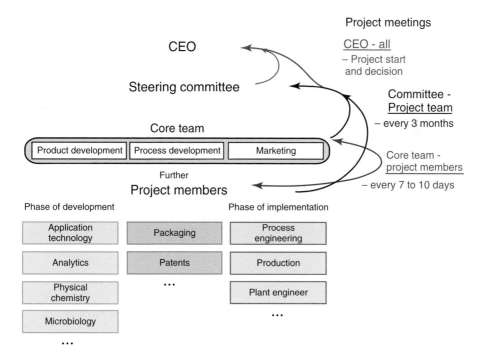

Figure 4.5 Changes in project organization with project progress and periods of project meetings/reporting; information flows in large projects on the basis of past experiences.

Members of the core team operate with the full support of their coworkers in subprojects. Information management may be prescribed or carried out as required. Intervals between meetings should not be too long as this would increase pressure.

4.9
Conditions for Success of Compressed Project Work

In large companies, success depends on the support of leadership and key employees at all levels. Right at the start of the project, responsible managers provide the necessary funds and appoint the project leader and other members of the core team. A dedicated supporter of the top management helps in particular in the initial phase with organizational problems (personnel, facilities, and priorities). Speed requires immediate and targeted work in all affected units, which means parallel work of all involved departments.

Commitment and ideas to solve problems, which always occur, are usually available from project participants. Bureaucracy should not restrict the development work, such as raising money for staff/equipment/facilities/premises, or extensive reports. Chemists and chemical engineers who are involved in experimental work must be able to access their employees directly. It is only then possible to discuss the test results on a daily basis. This includes taking into account new insights in order to adapt and correct the test program directly for the next day. For motivation, it makes sense to entrust the (academic) staff with sole responsibility for a project or task.

Project success needs precisely specified, clear, and consistent product specifications. Critical values for the product performance, handling, and aesthetics must be set at the beginning of development work. Subsequent requests/demands lead to extension of the project period up to termination. This error represents the most common cause for the failure of a project. Successful implementation also requires constant communication with parties involved, especially within the project team. As additional motivation, top management should communicate that fast market entry affects the total costs positively. This is valid even when improvements in the manufacturing plant reduce the profit from premature sale of products. This aspect is often not seen or ignored. (The ghost of failure lurks in the background.) Usually, the flop of a product does not depend on early market entry, but on its acceptance by the customer in regional markets.

From the very beginning of the development, marketing deals with issues of product positioning in the market. How does the product differ from other market products? What advantages or features can be turned out and how can they be described? Which brand should be assigned? What is the design of product and of packaging? At the request of marketing, a responsible agency invites some typical consumers. Results of discussions in consumer groups (focus groups [10–14]) provide answers to questions of product and packaging design. It is only after this that the finished product is presented and tests in the private households follow

[3, 15]. Marketing coordinates all activities related to the launch of the product. In the case of industrial customers, products originate in cooperation with selected customers already involved in the development phase. Their experiences and desires are considered. Industrial customers, on the other hand, test the samples repeatedly and intensively.

4.10
Design of Production Plant

After the start of the project, the responsible process developer works out a feasible flow sheet diagram [16] for the complete plant, possibly with an alternative, in total or in parts. In most cases, this work is performed by one of company's experts efficiently and without problems, using special software tools. For processing of solids, "SolidSim" is a suitable tool that also delivers mass and energy balances.

First, the expert knows the technologies needed for core business. Eventually, the company utilizes unit operations in a similar manner or has used them somewhere already. Secondly, in targeted experiments, the design of not yet proven operations is set. The company's engineers prepare the necessary documentation for technical implementation. The planning staff recognizes critical steps and unit operations of the whole process. Steps that can lead to problems (circulations, sewage, exhaust gas, by-products, wastes, and residues) need to be tested. Experiments take place in the company's own pilot plants or are conducted by manufacturers of machines and apparatuses (M&A). For manufacturing the product, a competent engineer designs the complete process flow diagram according to EN ISO 10628 (European Standard, released by CEN, European Committee for Standardization, 2000). In the case of production expansion, only a possible new part is detected. For very new plants, the engineer considers all elements from raw material to finished product delivery in the flow sheet. On this basis, taking the capacity into account, he calculates mass and heat balances as well as individual dimensions of the machinery, similar to the procedure for preliminary cost calculations.

For reasonable assessment of equipment sizes and energy demands, the engineer must know sizing and scale-up rules as well as the procedures characterizing dimensionless groups. Equally important is the specification of used materials. Chemical plants consist predominantly of the well-known CrNi-stainless steel qualities (Mat. 1.4301 (V2A) AISI 304 and 1.4571 (V4A) AISI 316Ti); other grades are also processed (Mat. 1.4435-316L, 1.4404-316L, and 1.4539 AISI 904L) or plastics that are resistant to aggressive liquids (see Chapter 16). The M&A manufacturers who are approached disclose the prices of machinery on the basis of design and required material. Responses to price inquiries allow a rough estimation of expected production costs and of possible alternatives, regardless of location.

In all cases, the process phases (solid, liquid, and gas) play an important role for design. First, the process takes place in homogeneous phases (gas or liquid). Here, the chemical engineers calculate the exact design of the machines and other equipment required. Examples are pumps, fans, containers, tanks, agitators, distillation plants, and mixing equipment. Second, unit operations with heterogeneous phases usually require experimentation in the pilot scale. The design of a chemical or biotechnological reactor necessitates experimental determination of basic parameters. The same applies to targeted separation and concentration of a dissolved substance (protein) from fermentation broth or for creating a stable emulsion in cosmetics.

Furthermore, difficulties in sizing increase in the presence of solids (suspensions and dispersed substances), particularly due to the recycling of off-spec materials. Design of grinding/sieving, coatings, precipitation for producing nanoparticles, extruders with a headplates for shaping and agglomeration processes requires many years of experience and meaningful experiments. These tests may be run in pilot plants, which reflect the conditions of the planned production facility in part or entirely. The inclusion of recycled materials is vital because it negatively affects product quality. However, large recycle streams increase the size of several steps within the plant.

In contrast to fluids, the behavior of solids and multiphase systems cannot be characterized by material properties. Owing to deviating product properties of solids, pastes, or suspensions, there is more frequently no exact calculation possible (granulation, coating, grinding, and sieving). Therefore, these operations always require tests to clarify the design. Then, an experienced developer estimates the sizing on the basis of some pilot tests, scale-up rules, and application of dimensionless groups with 20% surcharge on the estimated capacity. Most of this surcharge is needed after a short period, mostly for reasons of capacity, product modifications, or because another product is added to the process plant. Otherwise, it helps to prove the required capacity during start-up. Easy-to-clean equipment is made of stainless steel and characterized by smooth, polished surfaces (Chapter 16), without any dead space in the piping and machinery. There should be preferably automatic activation of cleaning cycles.

Right from the early stages, the process developer works together with a plant manufacturer, who heads the project to "build up a new process plant." He designs the detailed process flow diagram with measuring and control equipment as well as with an installation plan in existing or new buildings, called the *piping and instrument diagram (P&ID)* [16, 17]. All experimental results are directly incorporated into planning. The plant engineers contact suppliers for selected machinery, taking into account the infrastructure of the proposed production site. He estimates the capital costs more closely and creates concrete installation plans. On this basis, production and product costs are estimated more precisely. Management decision has to be obtained for two factors, namely, investment and product costs (see Figure 4.6).

Consistent reduction of the development phase from time to time requires ordering of machinery based on some orientating experiments in the pilot scale. For adaptation and optimization to be planned in a large plant, there is enough time available up to the delivery of M&A. In extreme cases, very long delivery times imply the compressed development of a purchase order before deciding to enter

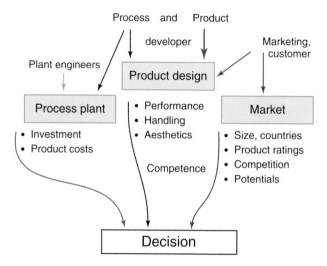

Figure 4.6 Ten key decision-making points for top masnagement.

the market. All this should be doable but one of the risks is loss incurred in case of cancellation.

4.11
Biotechnology

Rules for the fastest possible implementation in practice can be learned from biotechnology: if an interesting product is fermented in the laboratory, management decides to transfer it to production. All decision-makers know that the fermentation broth from the laboratory does not correspond to quality out of production. The suspension of the optimized production fermenter will have higher concentration of the product and there will also be a change in its composition and viscosity. Top management has obviously a strong confidence in the abilities of their employees. They let build a building and order the crucial machinery. In most cases, the basis is only the first pilot test and experience from similar projects.

On the question to biotechnology developers, whether this procedure does not bear high risks, the answer was, "It always succeeds to realize an optimal process." The high value of the product allows/demands unusual decisions. For a quick start, possible and costly rework is accepted. This is not really a problem because all people involved assume process improvements. In addition, the product is already making money. Order changes have been frowned upon so far in chemistry.

In addition, the chemistry should consider some business aspects: early sales of the product bring in money much sooner. This rationalizes the achievement of precise but lengthy, hence expensive process design. All should also be aware that over time the demands on the production line may alter for many reasons. Experience shows that compressed development does not require extensive reworks. Nevertheless, the benefits of quick market entry can often be realized.

4.12
Maximum Speed-to-Market (Examples)

One positive example for a faster implementation up to market are the Megaperls ® [18], a novel detergent generation of Henkel introduced in the early 1990s. For Megaperls, the product developers formulated new concentrated recipes, which together with increased bulk density of about $680\,g\,l^{-1}$ allowed visible reduction of volumetric amount of detergent per wash load. The innovative idea was the spherical shape of the particles, small beads for concentrated washing power. The process developers designed a completely new technology for this purpose. A mass of raw material was mixed, extruded, and thereafter shaped to beads using a spheronizer and then finally cured. The goal was to get consumers to immediately recognize the product and positively evaluate it with respect to product performance, handling, and aesthetics (see Chapter 11).

Initially, a technical chemist and a marketing colleague (core team) directed the project. Both presented the project results to the board of the company including the chair. Shortly thereafter, the shareholders' committee and the supervisory board agreed on the project. A strong commitment of all project participants and support of top management ensured very fast processing, although there were many critics and opponents at all levels. The process was new and unknown in the detergent sector. Extrusion of detergent spaghetti with diameters of 1.2 and 1.4 mm and with the target of cutting into small cylinders demanded modifications in the existing extruders. Especially with a planned capacity of $5\,t\,h^{-1}$ per extruder (later the extruder output was significantly increased), there was no technical solution available. The new technology in combination with concentrated formulations led to a series of challenges. These were enormous tasks for a small team, which had to solve all problems in a short time. The total investment amounted to about 100 million € (excluding the investment in filling lines) with a scheduled total capacity of 160.000 t/a, spread over three locations.

The timing can be summarized briefly. The idea of the innovation with a practical solution originated 1989. As test market, Switzerland was supplied (1991) with the products from a small process plant identified for further scale-up. In 1992, the first large-scale plant with four extruders in Belgium was ready and was housed in an existing building. A new facility in Italy started in 1993, and one year later, the second major plant in a new building in Düsseldorf. According to schedule, the plants in Belgium and Germany were to go into operation in the same year. However, the German authorities demanded many additional documents for an approval of the building and production plant. This preparation took some time. The completed documents were transported in a rented VW van to the authority (in Belgium the authority was with one person). The time required was 3 years, from the idea to the new production plants in Belgium. (In Germany, it took 5 years, thus there were two additional years without revenue). The test market had already taken place within just 2 years. In retrospect, besides the company being successful in the market with this innovation, the positive image of entire business was also boosted.

The second example is the implementation of a new spray-drying technology. The idea was to dry detergent slurries with superheated steam instead of hot air. This innovative process offers a number of advantages. After installation, there exists no exhaust plume, because steam is recirculated and excess steam is condensed. Plant safety and energy recovery are the other features that represent the innovation. After a decision to convert a production spray tower in Vienna, the pilot plant was designed and ordered. The pilot plant corresponded totally to a production plant. The following questions had to be clarified, mainly in pilot scale. How is the temperature set in the circulation loop and how is the excess steam removed? How does the recovery of heat energy work? The planning of a large steam-drying tower started long before first pilot trials ran. After some very positive pilot trials, the plant engineer ordered all the machinery for the production plant. The period from the decision to implement the idea up to the start of production was two years.

Start-ups are much faster than large companies are, as could be demonstrated by a small enterprise for development and marketing of cosmeceuticals (ATS). In February 2011, a published reference [19] described the antifungal effect of lavender oil. The elaboration led to the idea to develop skin cream for care for the feet, which at the same time could fight against athlete's foot. On the same day, a discussion of the idea took place with the partners in ATS [20]. The next day, the needed, additional ingredients were purchased. Ten days later, the first laboratory approach run took place in accordance with the formulation that had been developed in the meanwhile. At the same time the wording for the label and a compilation of the INCI (International Nomenclature of Cosmetic Ingredients) list were taken up. Further, a qualified security assessment was also commenced. The assessment took into account the toxicological profile of the individual ingredients, the chemical composition, as well as possible interactions and exposure by its intended use.

For risk assessment and determination of effectiveness, 12 probands were found, 4 with athlete's foot, the others, without the problem. The athlete's foot disappeared completely after 3–5 days due to the prolonged exposure to the active ingredient in the formulation. The probands used the cream for another 5 days to get totally smooth feet. A few days later, labels were printed and the performance results were published on the Internet. The product allocation number was applied for sale in pharmacies. The local authorities received the composition for registration. Following the successful small-scale production, regular sales commenced. The duration – from idea to market (investment in equipment was not necessary) – was 6 weeks. Large corporations need an estimated 18–24 months.

4.13
Relationship between Compressed Development and Product Design

The doctrines of product design [21], such as design of chemical products according to the wishes of clients, cause not only progress in generation of new product ideas (not discussed here in detail) but also provide accelerated development of innovations:

1) The product idea offers the realization of customer needs or even the solution of an important customer problem. Development concentrates on the correlation between performance, function, and form.
2) Marketing must contribute from the start in order to have the idea verified by customers, to explore the buying interest, and to consider the suggestions of customers from beginning. This accelerates the development and reduces the risk of a flop.
3) Product design utilizes all opportunities to improve the performance of the product (such as optimal recipe, coating, dust removal, set of particle sizes, and bioavailability) and, if desired, to transfer it into another "physical state" (solidification of liquids, emulsions, gels, and pastes).
4) Processes, which include a shaping step, are preferably used ("design technologies") because they are easier and less expensive in general (such as drying in a mill, spray-drying, fluidized bed spray agglomeration, pelletizing, and extrusion).
5) In addition, handling and aesthetics contribute to product performance. In particular, product aesthetics allows a differentiation from the competitor and thus supports market success.
6) Packaging is part of the product for transporting, protecting, storing, and for easy application.

A study of the product design helps in developing optimal products and processes. But market success brings to fruit the unconditional shared desire to create something new.

4.14
Outlook

For increased competitiveness, compressed developments are gaining more and more in importance. Development and implementation as a team is the future, and not the months of thinking by one employee, who alone solves parts of the entire plant. Constant discussions of experimental results among team members generate new suggestions and accelerate the momentum of work. Development is not only faster but overall also less expensive. Involvement of the customer reduces the risks of a flop. The advantages of this approach are so convincing that they should affect a rethink by employees and top management.

4.15
Learnings

√ Global competition enforces a rapid launch of innovations. For being first on the market, an accelerated implementation is particularly economical and image promoting.

√ Additional risks due to the acceleration are manageable and rework is possible. The benefits outweigh the costs.

√ Positive and negative impacts of innovation (market opportunities, work processes, and jobs) must be discussed openly at the start, in order to reduce uncertainty.

√ Major keys to a successful, rapid product launch lie in the organization of the operational unit and support of the top management, project organization, and parallel processing, in addition to involvement of customers from the beginning.

√ A core team (product and/or process development, marketing staff) leads with expertise and communicates the project status honestly.

√ Professional and socially competent employees with responsibilities for sub-projects receive clear lines and enough time for work, including the required funds and resources.

√ Future changes in the specifications must be avoided, because they represent the most important causes for failure of a project.

√ Compressed development restricts design experiments/simulations for lack of time. Often, exploratory experiments suffice; in heterogeneous phases, further pilot testing becomes necessary. Scheduling a surcharge of 20% on the design value minimizes the risk of a too scarce design.

√ Fast projects require more courage and commitment of the leadership and confidence in staff. Rapid implementation brings faster market success and revenues.

References

1. Masson, P.L., Weil, B., and Hatchuel, A. (2010) *Strategic Management of Innovation and Design*, Cambridge University Press.

2. Salomo, S. and Hauschildt, J. (2011) *Innovations Management*, Franz Vahlen, München.

3. Rähse, W. (2004) Produktdesign – Beschleunigte Entwicklung von Produkten durch Einbeziehung des Kunden. *Chem. Ing. Tech.*, **76** (3), 220–231.

4. Meffert, H. (2000) *Marketing: Grundlagen marktorientierter Unternehmensführung*, Meffert Marketing Edition, 9th Aufl., Gabler.

5. Scheck, H. and Scheck, B. (2007) *Wirtschaftliches Grundwissen für Naturwissenschaftler und Ingenieure*, 2nd edn, Wiley-VCH Verlag GmbH.

6. Disselkamp, M. (2012) *Instrumente und Methoden zur Umsetzung im Unternehmen*, Springer Gabler, Wiesbaden.

7. Schäppi, B., Andreasen, M.M., Kirchgeorg, M., and Radermacher, F.-J. (2005) *Handbuch Produktentwicklung*, Carl Hanser Verlag, München.

8. Barclay, I., Dann, Z., and Holroyd, P. (2000) *New Product Development*, Butterworth-Heinemann, Oxford.

9. Rähse, W. and Hoffmann, S. (2002) Produkt- design - Zusammenwirken von Chemie, Technik und Marketing im Dienste des Kunden. *Chem. Ing. Tech.*, **74** (9), 1220–1229; in English: Product design – interaction between chemistry, technology and marketing to meet consumer needs. *Chem. Eng. Technol.*, (2003), **26** (9), 1–10.

10. Wehner, C. (2006) *Konzepttests in der Produktentwicklung*, Grin-Verlag (Studienarbeit).

11. Kolb, B. (2008) *Market Research*, SAGE Publications Ltd., London.

12. Stewart, D.W., Shamdasani, P.N., and Rook, D.W. (2007) *Focus Groups: Theory and Practice*, SAGE Publications, London.

13. Krueger, R.A. (1998) *Analyzing and Reporting Focus Group Results*, SAGE Publications, London.

14. Krueger, R.A. and Casey, M.A. (2009) *Focus Groups: A Practical Guide for Applied Research*, SAGE Publications, London.

15. Chitale, A.K. and Gupta, R. (2011) *Product Policy and Brand Management: Text and Cases*, PHI Learning Private Limited, New Delhi.

16. Sinnott, R.K. (2005) *Chemical Engineering Design*, 4th edn, vol. **6**, Elevier, Oxford.

17. Silla, H. (2003) *Chemical Process Engineering: Design and Economics*, Marcel Dekker, New York.

18. Morwind, K., Koppenhöfer, J.P., and Nüßler, P. (2005) Markenführung im Spannungsfeld zwischen Tradition und Innovation: Persil – Da weiß man, was man hat, in *Markenmanagement: Identitätsorientierte Markenführung und praktische Umsetzung*, 2nd, Kapitel 5 ed (eds H. Meffert, C. Burmann, and M. Koers), Gabler, Wiesbaden, S. 621.

19. Zuzarte, M., Gonçalves, M.J., Cavaleiro, C., Canhoto, J., Vale-Silva, L., Silva, M.J., Pinto, E., and Salgueiro, L. (2011) Chemical composition and antifungal activity of the essential oils of Lavandula viridis L'Hér. *J. Med. Microbiol.*, **60**, 612–618. doi: 10.1099/jmm0.027748-0

20. ATS License GmbH (2013) Marleen Foot Cream with Lavender Oil Against Fungi, *www.ats-license.com* (accessed 26 November 2013).

21. Rähse, W. (2004) Produktdesign – Möglichkeiten der Produktgestaltung. *Chem. Ing. Tech.*, **76** (8), 1051–1064.

5
Product Design of Particles

Summary

The production of granules from powdered solids takes place in dry conditions under pressure or with the addition of liquids with subsequent drying/cooling. These processes are well established in industry worldwide. Fertilizers, detergents and cleaning agents, iron ore pellets, and expanded clay (building auxiliaries), as well as granules for tableting of therapeutics are examples of typical products that are available as granules. Wet granulation offers several possibilities for improving powders in product design. In comparison to primary particles, the granules produced show increased size and improved shape (round), stability, and flowability. From differences in particle size distributions of starting particles (very wide/mono- and bimodal or very close) result in three types of granulation, which differ significantly in regard to the duration of the process. Differences in particle diameters determine the agglomeration. In contrast to large particles, fine particles adhere immediately to the coarse core material (CM). In agglomerates of equal-sized particles, liquid bridges are repeatedly dissolved to form new ones, until stable agglomerates are formed. Scaling up of the granulation process in horizontal mixers is based on similarity of type, geometry, and identical Froude numbers. All operating conditions set are comparable. The representative product samples require three successive granulations. The second and third batches need the addition of over- and undersize particles from the milling/sieving step and filter dust. The targeted product design influences the process management in the desired manner.

5.1
Dry Agglomeration Processes: Pelleting and Tableting

Production of particles from solids takes place in three ways (Figure 5.1). Fine powders as starting material agglomerate in mechanical processes, without (Section 5.1) or with the addition of liquids/auxiliaries (Section 5.2). Coarse solids must first be crushed for further processing (Sections 7.4 and 7.5), mostly in a cutting mill or in single or in multistage grinding/sieving processes. Coating helps to set the properties of the solid particles, such as powders, granules, or tablets (Chapter 6). Coatings also allow the adjustment of dissolution times.

Industrial Product Design of Solids and Liquids: A Practical Guide, First Edition. Wilfried Rähse.
© 2014 Wiley-VCH Verlag GmbH & Co. KGaA. Published 2014 by Wiley-VCH Verlag GmbH & Co. KGaA.

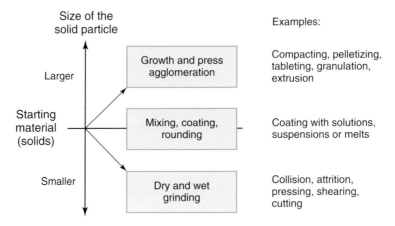

Figure 5.1 Procedures for product design of solid particles from solids.

The agglomeration of dry powders is known as *compaction, dry granulation, agglomeration, press agglomeration, press granulation,* or *pelletization*. For the compaction of dry powder, a little water is often used. Water is added by spraying or as steam. Sometimes, a supplement of special substances is also necessary to ensure that the particles adhere together and form flakes or pellets. These substances may be various binders and lubricants as well as fillers and auxiliaries (Table 5.1). Pellets, which are used as starting material for tablets, also require these additives.

Table 5.1 Auxiliary substances for dry granulation and tableting.

Binders	Lubricants	Fillers	Excipients
Starch	Magnesium stearate	Starch	Starch
Cellulose ethers (hydroxypropyl methylcellulose, hydroxypropyl-, methyl-, and ethyl-cellulose)	Calcium stearate	—	Mannite
Microcrystalline cellulose	Stearic	Microcrystalline cellulose	Lactoses
Powdered cellulose	Talc	Powdered cellulose	Powdered cellulose
Sugar alcohols (mannitol and sorbitol)	—	Monosaccharides and disaccharides	Disaccharides
Lactoses	—	—	—
Polyvinylpyrrolidone	—	—	—
Vinylpyrrolidone vinylacetate copolymer	—	—	—
Polyethylene glycols	—	—	—
Magnesium carbonate	—	—	—
Calcium carbonate	—	—	—

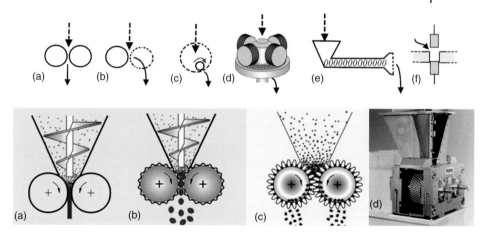

Figure 5.2 (A) Principles for compaction and pelleting of powders: (a) double roll press; (b) double roller press, pelleting on one or two perforated rolls; (c) annular die press with one, two, or three rollers inside; (d) annular flat die press; (e) extruder with a die plate at the head, a basket or a ring die; and (f) tablet press. (B) Principles for compaction and pelleting of powders: Left side: in a compaction between smooth or fluted rolls, large flakes arise and granules with increased density after flake crushing, and right side: briquetting between two rolls with the appropriate shapes on their surfaces and a briquetting machine (Courtesy of Hosokawa Alpine).

Starting from powders, dry agglomeration represents a method for shaping and increasing the bulk density and for dedusting. To increase the bulk density, a compression between two smooth rollers is necessary with subsequent grinding. Two different methods exist for the pelletization of conditioned powders (Figure 5.2A). Under the first, compression takes place between two rollers. The rollers can have a smooth surface, or one or both can be perforated. Alternatively, the compaction takes place between rollers and a ring die or an annular flat die. In the second method, the conditioned powder is compacted with screws and pelletized on the headplate with cutting blades in a high pressure process. In the low pressure extrusion, the dry powders are granulated inside out through a perforated ring die (not shown here).

Organic powders obtained from spray drying often have sticky surfaces, low bulk densities, and poor flow properties. Existing dust levels complicate the handling, increase the risk of a dust explosion, and require precautions for occupational safety. For the production of coarse, heavy powder from fine powders, the material is compacted and then again carefully ground (Figure 5.2B). Another method is the formation of pellets in briquetting machines or between rollers with perforated dies in the boreholes (Figure 5.3).

For compaction and grinding of organic compounds, particularly, encapsulated systems are suitable (Figure 5.4). The encapsulation enables an inerting with nitrogen and the operation in combination with an inertized co-current spray tower, which can be directly coupled. Other supply forms (strands, cushions, and cylinders, see Figure 5.5) originate directly in fluted or perforated dies. For pellets,

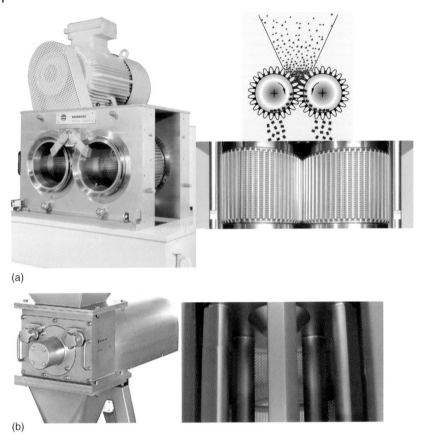

(a)

(b)

Figure 5.3 Machinery for the formation of pellets: (a) Pellets are obtained from the powder by pressing the material through boreholes in two opposing rollers. The product leaves the machine as cylindrical pellets through the interior of the rollers. (b) Dry granules are obtained by gentle grinding with an internal sieving; right: view from above of the grinding tools and the sieve (Courtesy of Hosokawa Alpine).

cylinders, and heavy powder, an annular flat die is also used for pelleting, as shown in Figure 5.6.

Furthermore, in extruder processes, the addition of powdery substances takes place with packing screws. Some products show lubricating properties (paraffin, soaps, and surfactants) alone or in mixtures and can be easily extruded. Otherwise, for compaction of hard powders, added lubricants help in the extrusion process (see Table 5.1). Many powders, containing large amounts of inorganic materials, require high pressures, depending on the type and amount of lubricants. Owing to the high energy input, the temperature rises despite cooling of the walls. Therefore, some extruders have an additional cooling by holly screws.

For solids, twin-screw extruders (see Figure 5.7) with special headplates generate the best results. The headplate contains many small plates with borings for

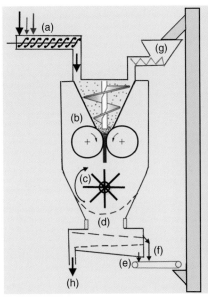

Figure 5.4 Increasing the bulk density of powders, obtained by roller compaction and grinding in air or in an inert atmosphere (encapsulated system): On the left side is an illustration of the principle (a) addition of powders and excipients; (b) packing screw; (c) mill; (d) sieves; (e, f) mechanical recycling of over- and undersize particles; (g) dosing of recycled powders; and (h) product outlet (similar to Fitzpatrick). On the left is the implementation of the principle into practice: roller compaction with two downstream screen mills in an encapsulated system for the pharmaceutical industry (made of polished stainless steel) (Courtesy of Hosokawa Alpine).

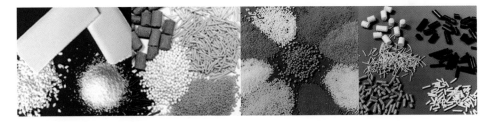

Figure 5.5 Some product shapes from process plants for dry compaction (Courtesy of Hosokawa Alpine).

the spaghetti extrusion. These small plates are easily exchangeable in case of blockages. A special feature of extrusion is the flexibility of the extrusion screw. Single-screw elements can be put together according to the task in hand. As a result, many basic operations are possible in one operation: mixing, crushing, dyeing, filling with a liquid, compression, removal of air, and shaping. The design of nozzles in the small plates determines the shape of product, whether it yields spaghetti, macaroni, or stars.

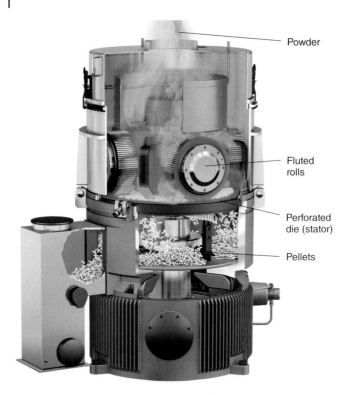

Figure 5.6 Pelleting machine with moving rollers on an annular flat die (Courtesy of Amandus Kahl).

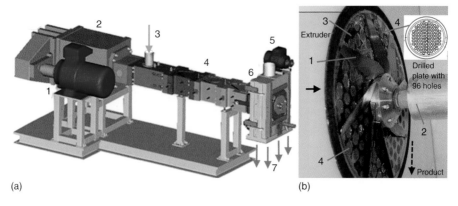

(a)   (b)

Figure 5.7 Twin-screw extruder with head-plate and cutter: (a) 1: motor for the extruder; 2: gears; 3: product feed, usually by packing screw; 4: heated or cooled segments; 5: motor for cutting blade; 6: headplate with cutter; and 7: product outlet, pellets. (b) 1: eight-bladed knife, 2: drive-shaft; 3: headplate for receiving the small plates; and 4: small plate with 96 boreholes (Courtesy of Lihotzky and Henkel).

Granulation   Tableting

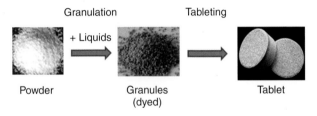

Powder   Granules   Tablet
                (dyed)

Figure 5.8 Tableting of agglomerated powders.

A process for producing the shaped bodies from dry particles represents tableting. Some tablets should be dissolved in water within a relatively short time. Extremely fine powders as starting material compress significantly during pressing. This results in hard, water-insoluble tablets. Therefore, powders are first transformed into granules (wet granulation, extrusion, and pelletization), with the fine (< 0.2 mm) and coarse fractions (> 2 mm) being sieved out. To optimize the tablet, a much narrower particle size distribution can be taken (0.4–1.2 mm), before the compression takes place in the tablet press (Figure 5.8).

In addition to particle size, many other particle properties affect the resulting tablet. Operating conditions and machinery designs represent additional parameters for optimization (see Figure 5.9). Disintegrants are added in some cases, to ensure immediate disintegration of the tablet on contact with water. Starches and celluloses, and polyvinylpyrrolidone (PVP) in pure form or chemically modified (as Kollidone-types from BASF) are in use as disintegrants. Another common agent is microcrystalline cellulose.

Predominantly, in production, tablet rotary presses are operated; and in some cases, eccentric presses are also used for a large molded body. Rotary presses allow the manufacture of two-phase tablets by separately operating the upper and lower punches. Figure 5.10 shows a separately pressed tablet, which is inserted into the upper phase of a two-phase tablet. This type of tablets cleans dishes in machines.

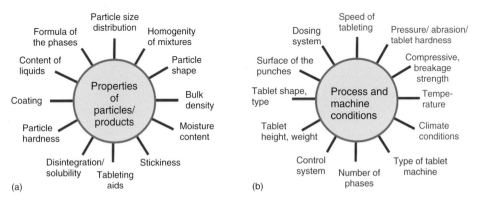

Figure 5.9 (a,b) Influencing the results of a tableting through particle, product, process, and machine parameters.

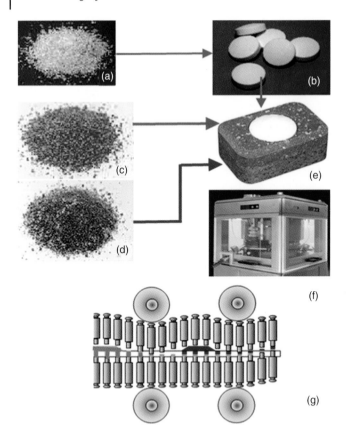

Figure 5.10 Two-phase tablet, suitable for dishwashers, manufactured in a rotary press, with an inlaid small tablet. (a) White granules for the inlaid tablet (b), (c) red, enzyme-containing phase, (d) blue, bleach-containing phase, (e) finished two-phase tablet, (f) rotary press (closed), and (g) pressing process for two phases with the upper and lower punches (Courtesy of Henkel).

5.2
Wet Agglomeration Process: Granulation

Wet granulation is a very common process worldwide. The industries discussed here use this type of granulation for product improvement.

5.2.1
Definition of Granulation

In the strict sense, granulation is an agglomeration with addition of liquid. The adhesive forces of the liquid, make the powder particles (primary particles) stick together. They form coarser agglomerates (granules), which show improved characteristic product properties after drying. This process is called *granulation*, or *growth/buildup agglomeration*, or *wet granulation*. It features solids that are valuable substances in the form of particles, and a limited amount of granulation liquid (usually < 15%). Material bridges that are formed between particles characterize the granules. These bridges arise during drying/cooling from the liquid droplets in combination with solid particles. Owing to the formation of granules during the process, the mean diameter grows, compared to that of the primary particles. This particle growth and drying/cooling takes place in separate equipment, such as in the combination of a mixer and a fluidized bed [1]. The growth (buildup) agglomeration method used worldwide for a variety of product designs is discussed here.

Rumpf [2] understood by the term *granules* a grainy material, whereby the granules ("pellets, granules, and granulates") show relatively uniform particle size distribution. Therefore, in the broader sense, the term *granulation* means all procedures leading to granular materials. These are agglomeration processes, such as fluidized-bed spray granulation [3, 4] and press agglomeration [5, 6], or a reduction process such as the underwater granulation of extruded plastic strands. The definitions and principles of granulation come mainly by Rumpf [2, 7, 8].

In this chapter, the term *granulation* means the adhesion of small particles by liquid additives [4–6, 9]. This type of wet granulation is used in many industries, for example, in chemical (basic chemicals, fertilizers) and food industries (instant coffee), for pharmaceuticals (drugs), in biotechnology (enzymes, animal feed), for washing and cleaning agents, for minerals (iron ore) and building materials (concrete and mortar additives), and for pigments and recycling of blast furnace dusts. Further, there are a large number of other, interesting applications.

5.2.2
Tasks of Granulation

The main task of granulation is an optimization of product design [1, 10], consisting of product performance, handling, and aesthetics. Product design here refers to design of individual particles and of larger particle quantities. In comparison to the starting material, the granules show better qualities at least in one dimension. The modifiable properties, listed in Table 5.2, relate primarily to particle size, particle shape, and their densities.

Growth granulation causes a shift of particle size distribution [11, 12] in the direction of larger diameters, wherein through this process the fines partially disappear. Well-executed granulation improves the roundness of particles, and flow behavior and stability of granules. Sticky products need an additional step of powdering of granules into very fine material. Sieving adjusts the mean diameter and width of particle size distribution. A very narrow distribution requires large amounts of coarse and fine particles to be recycled, which leads to significant larger equipment.

As an essential task, granulation reduces or prevents segregation effects of the starting material. To circumvent segregation in the powder, admixed powders must

Table 5.2 Objectives of a (wet) granulation process.

Product design by granulation for improving the quality
1) Enlarge the particle size
2) Increase the shape factor
3) Reduce the segregation
• Adaptation of particle distributions (size and width)
• Combining of different powders in one granule
• Adjusting of bulk densities in case of mixtures
4) Dedusting
5) Modification of bulk density
6) Loading with liquids
7) Changing the aesthetics (color and smell)
8) Influence of dissolution (instant and retard release)
9) Improving the flow and storage properties
10) Improving the particle stability

fit not only in particle size distribution and in form of particles to each other, but must also have similar density and corresponding bulk density. Granulation of powder represents a simple method to solve the problem, if each granule contains all or nearly all the components of the mixture. Very finely powdered particles satisfy this precondition better, because then the granule contains considerably more primary particles.

Despite the large surface, the fine particles are not allowed to react with each other. A reactive powder, or a chemically incompatible ingredient, may be formulated in a separate granulate. The formed pellets can additionally carry an inert layer, if necessary. The coating [13] prevents the absorption of moisture and the direct contact of the agglomerates among themselves. The remaining powder mixture granulates in such a way, that the mixture of the two granulates do not segregate.

Compared to starting powders, the bulk density of resulting granules can be adjusted to be higher or lower, depending on the process parameters and liquid properties. To obtain low bulk densities, agglomeration preferably takes place in fluidized beds. The progress of granulation is measurable by particle size distributions over time, on the amounts of over- and undersize particles after sieving. With relatively small amounts of liquid, there remains a part that is ungranulated, with the ratio of over- to undersize particles staying below 1. This setting allows the production of low density products with high solubility (instant character).

High bulk densities require greater amounts of granulation liquid and another type of granulator. At the start, the powders are moistened under high mechanical stress in a mixer (crushing conditions). Subsequently, granulation follows without the use of choppers. The ratio of over- to undersize particles then significantly exceeds 1. The granules obtained are larger and denser than the starting materials. Therefore, they show comparable or lower solubility. For the optimal bulk density setting, a variation in concentration, type, and quantity of the granulation liquids helps. A surfactant may improve the wetting, accelerate the granulation, and in some cases further increase the bulk density.

For influencing the solubility, for increasing the bulk density or for improving the aesthetics, the resulting granules receive a liquid load at the end of granulation and before drying or while passing the fluidized bed. This may be small amounts of nonionic surfactants to accelerate the solubility, of silicone oil for retarding the dissolution, or of a dye solution for surface coloring. Further modifications take place by a subsequent coating process.

5.2.3
Theoretical Basics

In the agglomeration by particle buildup, depending on particle size distributions of starting materials, three different types (Figure 5.11) exist. Type 1 starts with the normal powder or mixture of various powders or sometimes powders with granules. This type of granulation occurs very frequently, and is characterized by a broad, monomodal size distribution of the starting material. In the course of granulation, larger particles grow at the cost of small particles. First, the finest fractions [14] disappears, then the slightly larger, until finally the larger particles also agglomerate. The granules produced consist of two to eight starting particles (without consideration of dust). Average diameter d_{50} increases characteristically for type 1 by a factor of about 1.3–2.

The characteristic of type 2 is a very broad, bimodal distribution. The powdery mixture consists of coarse CM and of fine particles for growth-oriented agglomeration, wherein the fine particles stick in large amounts on the core particles. This situation allows the designing of the size and shape of the particles [15–18]. For type 2a, the CM ($\sim$ 5–30 wt%) arises from sieving of the starting material or from a component. The sieve covering determines the mean particle size of the granules. All other components of the formulation and the remaining residual of the starting material (overall 70–95%) undergo fine grinding before granulation. They leave the milling in a size range of $d_{50} < 20\,\mu$m. Finely ground powder adheres to the CM in layers at a high speed. Agglomeration of 10-μm particles is 10-fold faster than with 100-μm particles [19].

Owing to the large mass of ground material and the rolling motion of particles in granulation, there occurs an additional rounding of the agglomerates. The shape factor improves considerably in the direction of a spherical shape. The mean diameter increases in comparison to d_{50} of the CM. Depending on the mass of material that is sieved, the growth of the diameter amounts to 20–50%, with the shape factor in the range 10–25%. Each individual granule of type 2 granulation consists of over 100, mostly more than 10 000 primary particles. This means each granule represents the complete recipe. Segregation is no longer possible.

In contrast to type 2a, the process for 2b often takes place continuously. As an essential characteristic, the CM is identical in chemical composition to fine

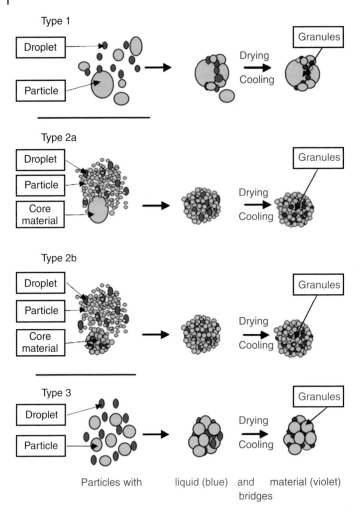

Figure 5.11 Mechanisms of different granulation processes: Starting materials for type 1: powder with monomodal particle size distribution; type 2a: coarse core material and finely ground particles; type 2b: granular core material from previous batch or undersized sieved granules and finely ground particles; and type 3: similarly sized particles.

particles. It originates from undersized granules out of the process or consists of smaller agglomerates from the previous batch. Anyway, the mean diameter increases to the desired sizes and much more, because fine particles are available in unlimited quantities. The granules are mostly elliptical or spherical in shape.

Type 3 particles display a narrow particle size distribution. The granulation occurs between approximately equal-sized particles and leads to the well-known "blackberry structure." Granules increase significantly, that is, around a factor of 2 or more in diameter.

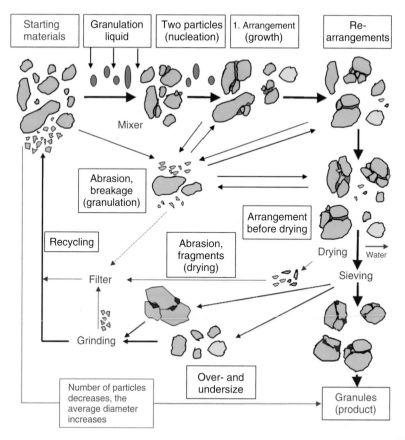

Figure 5.12 Mechanism of granulation type 1 with formation of fragments and abrasions, further recycling of ground under- and oversize particles (not participating particle is yellow, liquid bridges are blue, and material bridges are purple.).

5.2.4
Mechanisms of Granulation

From a sprayed granulation liquid, bridges between arbitrarily shaped particles of a monomodal size distribution originate (type 1). Firstly, they form two cores (nucleation), consisting of a small particle that adheres on a larger one. The twin cores start the subsequent growth phase [20] and form unstable arrangements with other particles. Influenced by particle movements in the mixer, rearrangements take place, that is, on one hand liquid bridges solve, and on the other, they form new, stable agglomerates (Figure 5.12). In parallel, unwanted processes expire. Primary particles and agglomerates constitute abrasion and break [21], particularly with the high energy demand of the mixer.

Abrasions as well as large and small fragments predominantly participate in the granulation process, depending on the quantity and quality of the granulation liquid. In case of low liquid amounts, some particles cannot agglomerate. Using too much liquid, big granules are formed. The "right" quantity of granulation liquid depends on sieves for adjusting the distribution and on the desired mean diameter. As a guideline, the ratio of the weights between under- and oversize particles is at the most 1, usually slightly lower.

Particles that are not granulated flow downstream via a sieve as undersize particles into the mill. There grinding takes place together with the oversize particles. The regrind is fed back to the starting material, as well as the filter dust from the drying and milling.

For type 2, large quantities of very finely ground particles (GPs) agglomerate on the CM. These fine particles immediately adhere firmly on the CM as a result of high adhesive forces. For this reason, during granulation with very fine particles, no rearrangements take place and process is very fast. Differences in chemical composition between the CM and fine particles characterize type 2a, depicted in

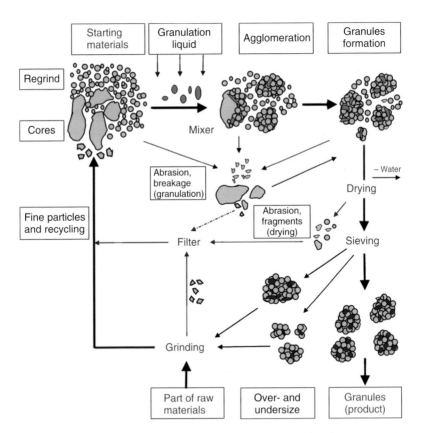

Figure 5.13 Mechanism of granulation according to type 2a; formation of product, including that of breakage (green) and abrasion; recycling of filter dust and of milled over- and undersize material; (liquid bridges in blue and material bridges in violet).

Figure 5.13. During this granulation, extensive adhesion of fine particles on the CM and on some larger fragments occurs, predominantly in a plurality of shells (layers). Abrasion arises from particle movements in the granulator and dryer from particle-to-particle contactor as a result of the contact of particles with the steel walls or mixer tools. The majority of the abrasion sticks immediately on the granules with the help of the granulation liquid. A small part emerges in the dryer and flows with drying air into the filter.

During the drying process, the water of granulation liquid evaporates. A sieve separates the product from the over- and undersize particles. Oversized agglomerates originate from some large particles of the CM, which further grow in size by the granulation of the fine particles. Also, two or more granules may stick together, leading to the formation of larger grains. Undersize particles arise from granulated breakage. Fragments develop in the granulator because of the mixing tools and shear forces.

The rolling movement of the granules and adhesion of the regrind from the raw materials result in an improvement of the shape toward that of a sphere. The reasons for this are that fine particles fill the caves and additionally unloaded positions ("shadows") behind hills.

Particle size distributions, displayed in Figure 5.14, explain a little better the course of this granulation. The CM (15% ± 10%) is screened from the source material. Most of the starting material flows with the other formulation ingredients into a mill for extensive grinding. Regrind with particle sizes between 1 and 50 μm agglomerates in several layers on the CM. The subsequent sieving allows the production of dust-free products, which are desirable in many cases. Choice of mesh sizes on the sieves enables a setting of any particle sizes, as also very narrow distributions.

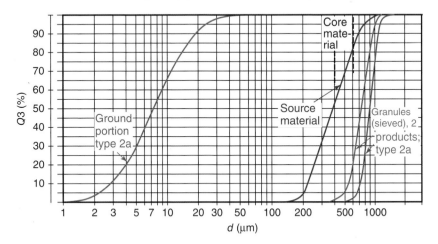

Figure 5.14 Particle size sum distributions (cumulative frequency) of powdered starting material (in the middle), ground portions (left), and finished granules (right), formed by growth agglomerations with differentially screened core materials; shape factor of starting material about 0.7 and of the granules > 0.85.

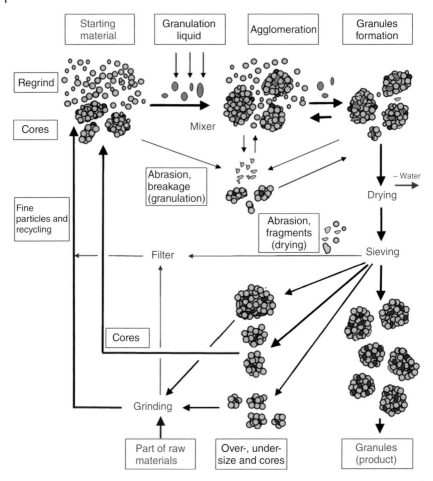

Figure 5.15 Mechanism of granulation according to type 2b with slightly undersized product as core material and milled particles from starting material and recycle as main component.

CM of type 2b granulation consists of finished granules (see Figure 5.15), but with slightly smaller sizes. This means all cores exhibit the same composition as the product. Granulation runs fast and without rearrangement, forming spherical pellets under rolling motion. In case of customary continuous operations, some undried agglomerates always remain in the granulator as further CM. This granulation method is suitable to produce unusually large agglomerates (20 mm or more), by controlling the residence time and the added amounts of fine particles. In this way, new shells are continually applied on existing pellets. When using plenty of granulation liquid, dense particles with high bulk densities arise in the high shear mixers. However, granules with low bulk densities develop with the use of small quantities of granulation liquid and utilizing a low shear granulator, such as a fluidized bed.

Drying of the granules is followed by sieving, preferably by separating the coarse and fine fractions from the desired product, but also a middle fraction for gaining cores. This middle fraction lies slightly below product sizes and is fed back for further granulation. The over- and undersize particles as well as the predominant part of the starting material are ground and recycled. Filter dusts flow via recirculation lines metered directly into the granulator. The product yields depend on the materials, liquid, and especially sieving limits (approximate value: 50–70%).

Connection of nearly equal size primary particles [22] according to type 3 is difficult and slow, because liquid bridges constantly dissolve and form new ones, until stable agglomerates arise. Rearrangements occur between starting materials, recycled particles, and fragments; furthermore with abrasion of the starting material and of granulates, rearrangements occur with fragments of agglomerates as well. In the drying stage, they flow with warm air into the filter. Stable granules may consist of several particles, forming a blackberry shape. The diameter d_{50} increases in comparison to that of the starting material in many cases by a factor of 1.5–2.5. When sieved particles are ground, the granulation is faster. Here, the possible objective of producing uniform grains with "blackberry structure," loses its importance. The mechanism of this procedure is described in Figure 5.16.

5.2.5
Industrial Granulation

Granulation is preferably carried out at atmospheric pressure in the temperature range 20-70 °C in drums, mixers, or in fluidized beds. Horizontal granulation mixers (high shear mixers) provide good results at Froude numbers from 1 to 6 (better than 1.5–3.5 according to past experience). These Froude numbers, depending on mixer design, are needed for intensive mixing without destroying the particles. The working temperature is decided primarily by the viscosity of the granulation liquid. Highly viscose solutions, suspensions, or melts require elevated temperatures in some cases. A granulation liquid should be as highly concentrated as possible and should also exhibit adhesive properties for facilitating the drying. Depending on the quantity to be produced, batch or continuous processes are preferred.

For carrying out batchwise granulation, starting materials are presented and mixed with the recirculated powder within 15–30 s. Addition of granulation liquid lasts for about 1 min. With large differences in sizes, the particles instantly form liquid bridges during wetting with the granulation liquid. Granulation starts and is not completed after addition of the liquid. In most cases, reorientations of the particles follow, forming more stable granules (type 1). Nearly same size particles (type 3) gradually establish agglomerates first. The granulation time depends strongly on the ratio of the diameters of the particles involved, and is in the range of 1.5 to more than 10 min. Subsequently drying occurs, which usually takes place in a downstream step. The desired particle size distribution is constituted by sieving with the right mesh sizes.

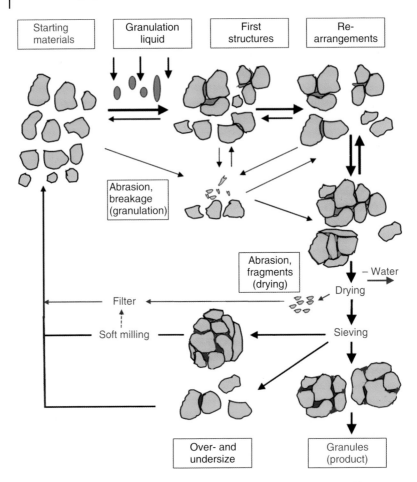

| Starting materials | Granulation liquid | First structures | Re-arrangements |

Abrasion, breakage (granulation)

Abrasion, fragments (drying)

– Water

Drying

Filter

Soft milling

Sieving

Over- and undersize

Granules (product)

Figure 5.16 Mechanism of granulation according to type 3, forming a blackberry structure; the illustration depicts the formation of fragments (green) and abrasion as well as recycling of undersize (oversized material goes through a mill).

The diameter ratio between coarse and small particles determines the type of granulation. In the conventional type 1, the entire material is derived from one particle size distribution of starting material (example 1: $d_{10} = 200\,\mu m$ and $d_{90} = 600\,\mu m$). These values result in a diameter ratio β of (200/600) 0.33. In accordance with the curve in Figure 5.17, based on experience in granulation of detergent powders in 30–600-l scale and with particle sizes of $5\,\mu m < d < 1\,mm$, the batch granulation lasts an average of 4 min. This value can be only conditionally generalized depending on the material and size dependency. At significantly shorter times, the granules present should be checked to confirm the efficiency of granulation. It is possible that only dust sticks on larger particles, because fine particles agglomerate appreciably faster (factor of 2). By considering the whole duration of granulation inclusive of the time for mixing

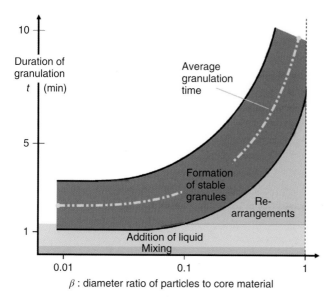

Figure 5.17 Dependence of the granulation time from size ratio of the particles present in a batch process (differences caused by the starting materials and the operating conditions; $\beta = d_{10}/d_{90}$ or for two distributions d_{50GP}/d_{50CM}).

and liquid addition, a 60-μm dust with a β of 0.1 is bound within 2 min on the CM.

In granulations of the types 2a,b, GPs adhere on the CM (example 2: $d_{50GP} = 10\,\mu m$, $d_{50CM} = 500\,\mu m$, and $\beta = 0.02$). The actual granulation occurs within a few seconds. However, in a batch process, it may be advisable to add both the granulation liquid as well as the fine particles over a period between 30 and 90 s. The granulation liquid can be poured with a momentum onto the moving bed of solids. A spraying is not required. Type 2 granulation usually results in a period of about 2 min. From the filling till the discharging of the batch, the procedure lasts at most 3 min for an automated production at room temperature; the total time taken is about half the time compared to type 1 and one fourth the time compared to type 3 granulation.

Probably owing to the insufficient stability of granules originating from particles of similar size (type 3; β about 0.75–1), the equilibrium state arises with constant changing between formation and disintegration of agglomerates. This type of granulation works better, only if many small fragments emerge in the presence of relatively large primary particles (~0.2–1 mm). In other cases, with extremely fine powders as CM, the granulation starts after first forming large granules. In both examples, β decreases, so granulation speed increases significantly.

Selecting the type of granulation liquid and its concentration, as also the determination of the optimum amount of liquid takes place in some pilot experiments. The parameters depend on starting materials (powder mixture and liquid) and on operating conditions. In many cases, the amount of liquid constitutes 5-15% of the

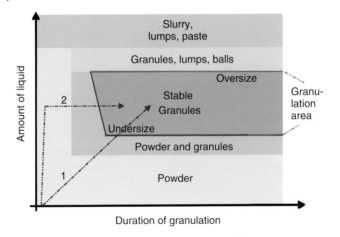

Figure 5.18 Formation of stable granules in dependency of liquid (quantity and quality) and time; 1: continuously dosing liquid and 2: complete liquid addition at the start.

solid mixture. With a too small an amount of liquid, no reasonable granulation is possible. Only a few unstable granules arise, according to Figure 5.18. Excessive amount of fluid leads to wet materials forming coarse agglomerates, lumps, and big balls, significantly larger than the usual oversized granulates.

The granulation liquid is distributed evenly over a minute. In another method, the entire amount of liquid is added initially and it directly enters the granulation region. In this way, the fastest granulation is achieved. This may be an interesting variant, especially for type 2. The amount of granulation liquid depends on particle sizes [23]. Fine particles usually need more fluid, because of the higher specific surface area and poorer wetting.

In production, particularly for type 2b, the continuous procedure is preferred. This mode, without further consideration here, runs in large plants. Control of the fluid ensures that the agglomerates in spaces behind the particle supply always come into the granulation area (see Figure 5.18). The short duration of the agglomeration allows high throughputs.

For addition of liquid, five possibilities exist. A method consists in spraying via (single-fluid) nozzles, another in pumping with a free outlet, the third in the pressing of granulation foam with gas, next in dosing of water vapor, and the last in the addition of a meltable substance. The third method is little known. This method allows a very effective distribution due to the large volume of foam. The viscosity of the solution should be set high, so that the liquid does not penetrate into particles. A sticky, dissolved substance remains almost completely on surface for forming the liquid bridges. Very high viscosities (cream) arise especially when foaming with gas (air), supported by a suitable surfactant. For this reason, it is also very favorable to apply the granulation liquid in the form of foam [24].

Customary excipients for granulation are displayed in Table 5.3 [1, 25]. Popular polymers represent the well-adhesive cellulose ethers. They cover all areas in solubility and viscosity. A number of cellulose derivatives are in use, some preferred

Table 5.3   Selection of granulation liquids.

Phases	Excipients for the granulation
Single phase and aqueous	Water glass
	Surfactants (alkylpolyglucosides)
	Carbohydrates (saccharides)
	Natural polymers (cellulose ethers and starch)
	Synthetic polymers (PEG, PVOH, and polyacryl acids)
	Biopolymers (xanthan)
	Salts
Single phase and low in water	Water-containing organic solvent with swollen polymer
Single phase and water-free	Melt (fatty alcohols and -acids)
	Di- and trivalent alcohols
Two phases and aqueous	Emulsions
	Surfactant mixture/air
	Suspensions (dissolved and suspended)
Two phases and water-free	Solid-containing melt

PEG = polyethylene glycol and PVOH = polyvinyl alcohol.

ones being methyl cellulose, methyl hydroxyethyl cellulose, and hydroxypropyl cellulose, as also hydroxyethyl cellulose and sodium carboxymethyl cellulose. Polymers such as hydroxypropyl cellulose and polyvinyl pyrrolidone have proved to be strong binders in the pharmaceutical industry [26]. With the exception of the pharmaceutical industry, large-scale production is performed usually with aqueous systems. The excipients are fully dissolved or dissolved and suspended, and then emulsified or rarely melted.

5.2.6
Scale-Up

In the buildup agglomeration, performed in a discontinuous mixer (ploughshare mixer), the experimental procedure in pilot scale is pilot scale is oriented to the desired production mixer. Both mixers have to be of the same type and similar in geometry (see Table 5.4). A suitable scale-up factor amounts to 10–30, so that a 600 l mixer in production requires a 30 l mixer for experimentation. To choose an even smaller pilot machine is not helpful and does not provide reasonable results. A possibly existing cutter is turned off because the cutter effect is not replicable in scale-up. However, it can be used in manufacturing during the addition of liquid for about 30 s to accelerate the granulation, for a short time and to protect the arising granules.

The intensity (speed and strength) of agglomeration must be determined through experiments, owing to the properties of the solids (chemistry and dispersity) and liquids (type, amount, and concentration) used. The calculated batch cycle time for the production depends on the type of granulation. This value provides the basis for initial pilot tests. A test program may start with a Froude number of 3 and duration of

Table 5.4 Implementation of scale-up experiments in pilot scale and implication for production (discontinuous mixer).

Parameters	Comparison of mixers/granulators in pilot-/production plants	Actions
Starting materials and recycled powders	Comparable amounts according to the scale-up factor	Check
	Same particle distributions of the product	Measure
	Same moisture contents	Measure
Granulation liquid (-foam)	Identical	Check
Excipients (stickiness)		
Concentrations		
Temperature		
Viscosity		
Similarities		
Type of mixer	The same	Procure, if necessary
Geometry	Comparable ratios of the dimensions	Pay attention exactly
Energy input (driveshaft)	Comparable, based on the mass (specific values)	Measure and calculate
Cutter	Only for some seconds, when starting the production (see the text)	Granulation without cutter
Operating conditions at the mixer		
Temperature	Identical	Scribe
Mixing time	The same	Control
Inflow of the liquid (time and strategy)	Identical	Accordingly dosing
Granulation/batch time	Identical and coordinated with the production	Measure
Rotational speed of the mixer shaft	Same Froude number	Calculate and measure
Filling degree	Identical	Measure and determine over weight
Process		
Moisture before and after drying	Comparable	Control
Dust	Comparable	Control
Over- and undersize	Relative amounts comparable	Measure
Mass balances	Comparable	Calculate
Quality of the onsize[a]	Comparable amounts, distributions and shape factors	Measure and control
Mill	Moisture and particle size distributions	Measure and control

[a]Without the application properties, such as solubility, bulk density, color, odor, storage stability, flowability, stability, dosing, and product performance.

1 min for the addition of liquids. The search for a suitable granulation liquid follows next in the course of experimental work. The determination of the amount of liquid takes into account the product moisture, which can play a major role. All relevant times are the same in both the pilot and production scales. Optimization of amounts and times are required for feasibility in production. The targets are represented in terms of the yield of normal size granules as well as their quality and appearance.

In scale-up experiments, all relevant operating and process parameters (Table 5.4) are taken into account and adapted to production conditions. For the experiments in production, the ratios of solids (source and recycled material) and fluids should be equal. The process requires constant monitoring of the key parameters. An identical specific energy input (zero power subtracted), calculated from the measured values results for mixers of the same type that are geometrically similar. This implies that the degree of filling (about 65–70%) should be the same, with relatively the same amounts of liquid.

The manufacture of a representative product sample in the pilot plant needs at least three granulations. This depends on the selected procedure. The second and third tests consider recirculated material, ground material, and filter dusts. Particle size distributions and moisture contents of starting and recirculated material should be measured to compensate for deviations. The first granulation without recycled material leads to granules that perform better in the application tests (e.g., in terms of color and dissolution rate) but represent the later process inadequately. Granules from the third trial may characterize the achievable product quality.

Intensive research has led to advances in the simulation of processes for solids, especially for spray agglomeration in fluidized beds. Despite population balances [27–30] and computer simulations [31], the industry has developed largely empirically, because theoretical approaches have not yet considered the material properties. Only trials allow statements about nature, amount, and concentration of the granulation liquid. Further the quality, depending on the agglomeration of dust and recycled material (particle size distribution after grinding), can be determined only in targeted experiments. Population balances may be helpful for optimization of quasi-continuous and fully continuous processes.

5.2.7
Applications

Three main technical methodologies exist for the process of wet granulation, namely, the buildup agglomeration in balling discs or drums, in mixers, and in fluidized beds. In the chemical industry, the choice of procedure depends on the granulation behavior, capacity, and throughput, required bulk density of the product, and the need for drying. The machines and equipment used by the industry for discontinuous or continuous granulation are shown in Figure 5.19. For products with high bulk densities, there are several vertical (Eirich) and horizontal high shear mixers (Loedige) in use, preferably plowshare mixers, in continuous operations often combined with a ring-layer (CM) or Schugi mixer. The production of low bulk densities necessitates fluidized beds, if the particles are not too fine.

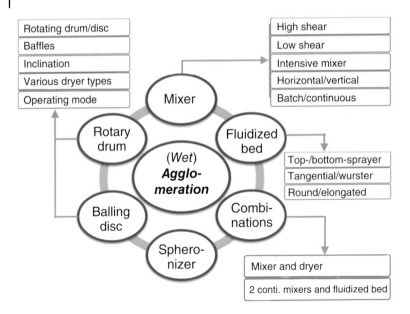

Figure 5.19 Machinery for the wet granulation.

Inclined balling discs or cones and rotary drums operate for mass products. The drying/cooling takes place under strong product movements. In batch processes the treatment takes place partially in the same machinery, but preferably downstream in separate fluidized beds or rotary dryers, especially in continuous processes. Sintering/cooling may be followed in the minerals industry.

Construction materials, fertilizers, and feed usually agglomerate on inclined balling discs or cones and in rotating drums, because these materials often stick together during the rolling movement. So ground clay with low calcium is mixed with organic components, granulated under addition of water on balling discs, selectable in the size range of 4–40 mm. Ball-agglomeration follows the type 2b-mechanism. Finished granules are formed on decomposition of organic matter, in which gases puff up the clay in a rotary kiln at 1200 °C. In the furnace, the granules sinter only at the surface. The internal structure of the granules displays closed pores. The construction industry uses this expanded clay for mortar and concrete (lightweight concrete). Because of its open structure, expanded clay is lightweight (floats on water) and shows good heat insulation. The gardener needs these granules for hydroponics for water storage.

For preparation of calcium ammonium nitrate – a *fertilizer*, ground dolomite (calcium/magnesium carbonate) is agglomerated after addition of ammonium nitrate melt [32] in a rotating drum. After drying, the sieving and cooling of the balls follow. In this case, one molten component of the product represents the granulation liquid. In other cases, the buildup agglomeration of powdery fertilizer mixtures ensues after addition of water or fertilizer-water mixture in a rotary drum or on a balling disc with subsequent drying in the rotary kiln [9]. In addition, fertilizers are also producible by pressure agglomerations in pellet presses or on rollers as well as extruders. Therefore, in many cases, dry agglomeration needs no drying step.

One of the major applications of buildup agglomeration is granulation of *iron ore* dust ($< 50\,\mu m$) for use in blast furnaces. In 2001, more than 200 million tons of iron ore pellets were manufactured [9]. After addition of the binder (2% bentonite) and flow agent to the water-containing ore, granulation is carried out either on huge balling discs or in large rotating drums, optionally with the addition of water. The green grain dries and then sinters at 1350 °C.

An economic granulation of *washing powders* is relatively difficult, because a given bulk density must be achieved (see Chapter 11). The detergent industry works in large continuous plants, in which granulation consists of a combination of a fast and slow running mixer with a fluidized bed dryer downstream [33]. First, in the fast running ring-layer mixer, the supplied powder is homogenized, milled a little, and sprayed with partial amounts of the granulation liquid. Actual granulation occurs during addition of the remaining fluid in the second mixer, which is operated at relatively low speeds [1]. Finally, air streams through the bottom of a fluidized bed for drying and cooling the pellets. In some recipes, the powder can be continuously granulated, dried, and cooled in a simple process, on a long fluidized bed, with all the steps taking place in one apparatus.

In the manufacture of enzyme pellets, powdery mixtures of excipients are agglomerated by the addition of an aqueous enzyme concentrate in a mixer or extruder. Drying of the granules takes place in a fluidized bed. Subsequent coating with a protective layer follows downstream discontinuously, preferably in a round fluidized bed coater [13] (see Chapter 6).

The food industry uses the agglomeration processes for instantization of beverages, particularly coffee, after the introduction of steam as a granulating aid. For dragees, the pharmaceutical industry utilizes the rotating discs and the drum coater for buildup agglomerations, in addition to mixers [27], extruders, and pellet presses. In the batch fluidized bed with an integrated filter, coating, drying, and cooling are also possible. The agglomeration and drying takes place preferably in drum coaters or in fluidized beds (closed system and for organic liquids). Granulation of homogenized active ingredients and excipients is necessary so that the pressed tablets dissolve in water. For this reason, granulation occupies a significant role in many pharmaceutical processes.

Some product examples for different granulating processes can be seen in Figure 5.20A,B.

5.2.8
Design of Particles by Granulation

The choice of granulation type, quantities of recirculation as well as material properties influence the product design. The more sophisticated the chosen particle design is, the more complex is the process and the greater are the mass flows. A rather simple process, with typical mass balance, followed in the detergent industry, is represented by type 1 granulation (Figure 5.21). Two powders agglomerate by

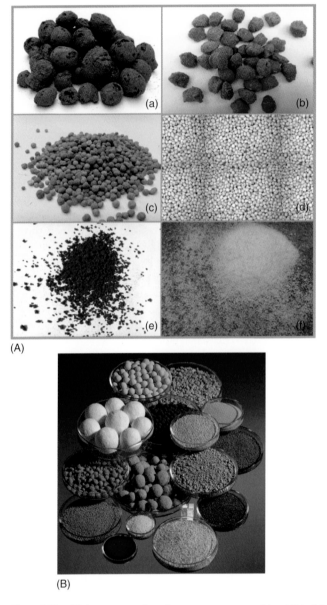

Figure 5.20 (A) Large, normal, and very small granules, emerged from the wet granulations: (a) Building materials: expanded clay as excipient (or for hydroponics 8–16 mm), (b) animal feed: pellets of bird feed (about 8 mm), (c) fertilizers: NPK fertilizer (2–6 mm), (d) biotechnology: detergent enzymes (about 0.4–0.6 mm), (e) food: instant coffee, and (f) therapeutics: painkiller in combination with a cold remedy in the form of mini granules. (B) Several small and large granules out of an intensive mixer (Courtesy of Eirich).

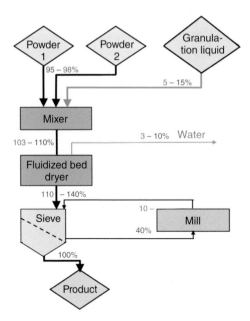

Figure 5.21 Simple granulation process (A) including the mass balance; dust recycling not shown.

addition of the liquid in a mixer. Then water from the granulation liquid evaporates in the fluidized bed dryer. After screening, the dried oversize flows into a mill. There, the large particles are crushed gently, for instance, by two rollers. The regrind and the filter dust recycle directly to the sieve. The product consists not only of granules but also of nongranulated parts, as also of dust as well of oversize GPs. Overall, a rather unsatisfactory particle design with poor aesthetics results.

The process described in Figure 5.22 is very similar. In contrast, after drying, a screening takes place on two coverings for separation of the coarse and fine fractions from the onsize one. For reasons of product design, the outsize particles should always be separated, sharply ground, and recycled into the granulation step.

Ground material agglomerates quickly and easily on cores in the granulation mixer. This approach ensures that the product consists completely of granules and provides a uniform appearance. Depending on the material, fine particles and dust may be recirculated directly back into the mixer without sacrificing quality. Nevertheless, in most cases, a grinding of undersize particles leads to better granule quality. Between these two processes, the first one very easy and the other a little complex, but the best for product design. There are a number of less advisable variants, such as the addition of the dust into the product or direct recirculation of undersize particles into the dryer.

Figure 5.23 shows a continuous granulation process, used in many cases (minerals and fertilizer) for high production levels. Balling discs/cones or rotary

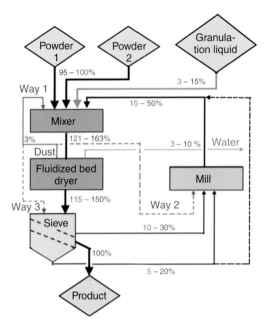

Figure 5.22 Granulation process (B) with recirculation of dust from the filter and of over- and undersize particles from the screen, as ground particles; typical values for the mass balance.

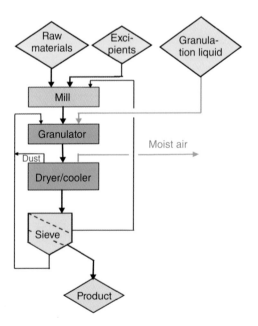

Figure 5.23 Block flow diagram for large-scale continuous granulations (C).

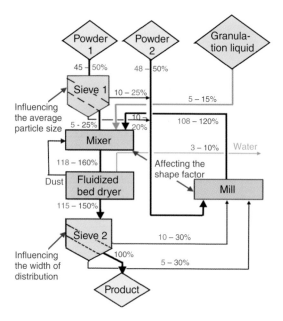

Figure 5.24 Optimized granulation (D) according to type 2a with several possibilities to adapt the particle design.

drums are the usual granulators. Dehumidification takes place mainly in rotary dryers. Some minerals agglomerate on turntables (rotating inclined discs) so well that further optimization to improve the particle design is not necessary.

Depending on the granulator and operating conditions, the desired spherical shape normally is not achievable in type 1 granulation. A successful approach to optimum shape ensures the type 2. Here, not only are the oversize particles ground but also the undersize particles and major part of the starting material. In type 2a, the cores consist of sieved starting materials, as shown in Figure 5.24 (method D). Type 2b uses the slightly undersized material as cores. This process E, displayed in Figure 5.25, needs two sieves or one sieve with three different coverings.

The mass balances in the block flow diagrams demonstrate typical ratios based on several detergent granulations. Extremely finely GPs agglomerate on rolling CM at free sites. Therefore, a round shape is formed. Type 2 processes allow controlling of particle size, the width of distribution and shape factors, and adjusting them within wide ranges. The fixing of an average granule size ensues by proper sieving of the CM. In several cases, the mean size of CM is set at $\sim -20\%$ of the desired granule size, calculated from mass fractions. The calculated value is arrived at by choosing the mesh sizes of the screen 1. Mesh sizes of the sieve 2 determine the distribution width. Further, the shape factor is influenced by the milling and mixing processes. Table 5.5 recommends process variants for each type of granulation.

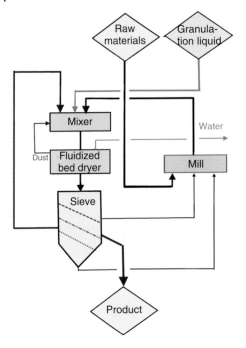

Figure 5.25 Optimized granulation (E) according to type 2b with several possibilities to adjust the particle design.

5.3
Learnings

√ The enlargement of particles sizes takes place by agglomerations. These processes are used worldwide to a great extent.

√ Agglomerations improve the product design (size, shape, bulk density, dissolution, flow properties, and less dust) and prevent particle segregations.

√ Dry agglomerations (roller, pellet, and tablet press) work almost liquid-free under high pressure, while in the wet agglomerations (mixers, rotary drum, and fluidized bed), a sticky liquid connects the particles to the granules.

√ Owing to the broad range of sticky liquids, wet granulation is universally applicable.

√ The granulation liquid forms bridges between particles, which convert during drying to material bridges.

√ The mechanisms are very complex, owing to abrasion and breakage as well as changes in arrangements and rearrangements of the involved particles.

√ The time required depends on the particle size distributions. Very small particles adhere fast and strong on coarse materials. Equal sized particle stick together hardly.

√ A rolling motion of the particles supports the buildup granulation.

Table 5.5 Choice of suitable processes.

Type of granulation	Features	Particle size distribution	Process
1	Monomodal particle size distribution		**B**, Block flow diagram, Figure 5.22 mechanism, Figure 5.12
2a	Bimodal particle size distribution; coarse cores from the raw material and finely ground particles		**D**, Bock flow diagram, Figure 5.24; possibly also an adapted process C, Figure 5.23; mechanism, Figure 5.13
2b	As 2a; but the composition of the cores is identical to that of the finely ground particles		**E**, Block flow diagram, Figure 5.25; possibly also an adapted process B or C, Figures 5.22 and 5.23; mechanism, Figure 5.15
3	Narrow particle size distribution (about the same size)		Adapted to process B, Figure 5.22; mechanism, Figure 5.16

√ Small granules depict a diameter of 200 μm, the biggest ones about 2 cm.
√ Aesthetic granules with good performance characteristics result from the buildup granulation of finely ground powders on a coarse core material.

References

1. Rähse, W. (2007) *Produktdesign in der chemischen Industrie*, Springer-Verlag, Berlin.
2. Rumpf, H. (1958) Grundlagen und Methoden des Granulierens. *Chem. Ing. Tech.*, **30** (3), 144–158.
3. Uhlemann, H. and Mörl, L. (2000) *Wirbelschicht- Sprühgranulation*, Springer-Verlag, Berlin.
4. Salman, A.D., Hounslow, M.J., and Seville, J.P.K. (eds) (2007) *Granulation (Handbook of Powder Technology)*, vol. 11, Elsevier.

5. Pietsch, W. (1991) *Size Enlargement by Agglomeration*, John Wiley & Sons, Inc., New York.

6. Heinze, G. (2000) *Handbuch der Agglomerationstechnik*, Wiley-VCH Verlag GmbH, Weinheim.

7. Pietsch, W. and Rumpf, H. (1967) Haftkraft, Kapillardruck, Flüssigkeitsvolumen und Grenzwinkel einer Flüssigkeitsbrücke zwischen zwei Kugeln. *Chem. Ing. Tech.*, **39** (15), 885–893.

8. Rumpf, H. (1970) Zur Theorie der Zugfestigkeit zwischen Agglomeraten bei Kraftübertragung an Kontaktpunkten. *Chem. Ing. Tech.*, **42** (8), 538–540.

9. Pietsch, W. (2005) *Agglomeration in Industry*, Vols. **1 and 2**, Wiley-VCH Verlag GmbH, Weinheim.

10. Rähse, W. (2004) Produktdesign – Möglichkeiten der Produktgestaltung. *Chem. Ing. Tech.*, **76** (8), 1051–1064.

11. Pietsch, W. (2002) Systematische Entwicklung von Verfahren zur Kornvergrößerung durch Agglomerieren. *Chem. Ing. Tech.*, **74** (11), 1517–1530.

12. Tu, W.-D., Ingram, A., Seville, J., and Hsiau, S.-S. (2009) Exploring the regime map for high-shear mixer granulation. *Chem. Eng. J.*, **145** (3), 505–513.

13. Rähse, W. (2009) Produktdesign disperser Stoffe: Industrielles Partikelcoating. *Chem. Ing. Tech.*, **81** (3), 225–240.

14. Rähse, W., Kühne, N., Jung, D., Fues, J.-F., and Sandkühler, P. (1995) Henkel AG & Co. KGaA, DE 195 47 457 A1, Dec. 19, 1995.

15. Rähse, W. (2006) Henkel AG & Co. KGaA, DE 10 2004 053 385 A1 Nov. 05, 2006.

16. Rähse, W. (2005) Henkel AG & Co. KGaA, EP 1 807 498 A2, 22.10, WO 2006048142.

17. Rähse, W. (2007) Henkel AG & Co. KGaA, DE 10 2005 036 346 A1 Jan. 02, 2007.

18. Rähse, W. and Larson, B. (2007) Henkel AG & Co. KGaA, DE 10 2006 017 312 A1, Oct. 18, 2007.

19. Tu, W.-D., Ingram, A., Seville, J., and Hsiau, S.-S. (2008) The effect of powder size on induction behaviour and binder distribution during high shear melt agglomeration of calcium carbonate. *Powder Technol.*, **184** (3), 298–312.

20. Litster, J.D. (2003) Scaleup of wet granulation processes: science not art. *Powder Technol.*, **130** (1–3), 35–40.

21. Hemati, M., Benali, M., and Diguet, S. (2007) in *Product Design and Engineering*, vol. **1**, Chapter 8 (eds U. Bröckel, W. Meier, and G. Wagner), Wiley-VCH Verlag GmbH, Weinheim, pp. 147–191.

22. Jakob, M. (2002) *Symposium Produktgestaltung in der Partikeltechnologie*, Fraunhofer IRB Verlag, p. 33.

23. Reynolds, G.K., Le, P.K., and Nilpawar, A.M. (2007) in *Granulation (Handbook of Powder Technology)*, vol. **11** (eds A.D. Salman, M.J. Hounslow, and J.P.K. Seville), Elsevier, pp. 1–20.

24. Rähse, W., Larson, B., and Semrau, M. (2001) Henkel AG & Co. KGaA, DE 101 24 430.4.

25. Tardos, G.I., Khan, M.I., and Mort, P.R. (1997) Critical parameters and limiting conditions in binder granulation of fine powders. *Powder Technol.*, **94** (3), 245–258.

26. Bika, D., Tardos, G.I., Panmai, S., Farber, L., and Michaels, J. (2005) Strength and morphology of solid bridges in dry granules of pharmaceutical powders. *Powder Technol.*, **150** (2), 104–116.

27. Sanders, C.F.W., Willemse, A.W., Salman, A.D., and Hounslow, M.J. (2003) Development of a predictive high-shear granulation model. *Powder Technol.*, **138** (1), 18–24.

28. Biggs, C.A., Sanders, C., Scott, A.C., Willemse, A.W., Hoffman, A.C., Instone, T., Salman, A.D., and Hounslow, M.J. (2003) Coupling granule properties and granulation rates in high-shear granulation. *Powder Technol.*, **130** (1–3), 162–168.

29. Wildeboer, W.J., Litster, J.D., and Cameron, I.T. (2005) Modelling nucleation in wet granulation. *Chem. Eng. Sci.*, **60** (14), 3751–3761.

30. Cameron, I.T., Wang, F.Y., Immanuel, C.D., and Stepanek, F. (2005) Process systems modelling and applications in

granulation: A review. *Chem. Eng. Sci.*, **60** (14), 3723–3750.

31. Talu, I., Tardos, G.I., and Khan, M.I. (2000) Computer simulation of wet granulation. *Powder Technol.*, **110** (1–2), 59–75.

32. YARA Fertilizers *www.yara.com* (accessed 26 November 2013).

33. Showell, M.S. (ed) (1998) *Powdered Detergents*, Surfactant Science Series, vol. **71**, Marcel Dekker Inc., New York, p. 28 ff.

6
Product Design of Particles by Coatings

Summary

Surface coating of particles with one or more water-soluble layers improves the mechanical and aesthetic properties, storage stability as well as product safety. Furthermore, the ability arises to influence the timing of dissolution and the pH-controlled release kinetics by an appropriate coating (usually polymers). Applying a coating takes place preferably by simultaneous spraying of the solutions/suspensions on particles and drying in fluidized beds. The method works optimally just between spray-drying and granulation, with accurate adjustment. The industry uses the coating method not only for tablets in pharmaceuticals but also in biotechnology for enzymes, and furthermore in farming as well for dressing seeds, for slow-release fertilizers as well as various chemicals (sodium percarbonate and citric acid). Particle coating provides many possibilities for optimizing product design, which are increasingly utilized in chemistry.

6.1
Processes for Setting the Product Design of Particles

Through an applied superficial layer (particle coating), disperse solids change their properties without any significant change in the particle size. Thus, application tests show improved dissolution in water and mechanical stability. Shape factor, color, smell, and texture (aesthetics) appear more attractive. The core material of a coating process shows particle sizes of about 0.1–3.0 mm, for melt coating when milled down to 5 μm, and for tablets up to 25 mm. Cores originate usually from an agglomeration process. Coating is a downstream procedure, preferably carried out after granulation, extrusion, tableting, pelletizing, fluidized bed granulation, spray-drying, or crystallization. In addition, crushed solids and pieces can be coated with solutions/suspensions or melts. Furthermore, natural particulate materials, such as seeds, are also used as core materials.

Targeted coating of particle surfaces provides a desired modification of the product, wherein the success of coating depends on the material and layer thickness as well as on the uniformity of the film. Chemical technology describes a layer

Industrial Product Design of Solids and Liquids: A Practical Guide, First Edition. Wilfried Rähse.
© 2014 Wiley-VCH Verlag GmbH & Co. KGaA. Published 2014 by Wiley-VCH Verlag GmbH & Co. KGaA.

as a coat, coating, cover, or film, wherein these terms are not limited to particles (tablets, tubes, plates, and stirrers). For applying a stable coating layer, findings from chemistry for the core and coating material as well as knowledge of the mechanical and thermal process engineering (mixing, swirling, spraying, and drying) can help.

Particle coatings are clearly different from microencapsulation, particularly concerning the core material. Particles are solids, such as powders, prills, grains, agglomerates, granulates, pellets, pastilles, spheres, masses, and tablets. The size of the coated particles is usually above $100\,\mu m$. In contrast, the cores of microcapsules predominantly contain liquids and/or powders with the flow behavior of liquids. Such solids are micro- and nanopowders. The content and the layer of a microcapsule may be water insoluble, but swellable or suspendable. Furthermore, microcapsules show a considerably smaller particle size (frequently with $d < 50\,\mu m$). Another difference between coating and microencapsulation is in the manufacturing process [1, 2].

Coated products are generally known in the food industry, especially in confectionery (chocolate or sugar coating) and in the pharmaceutical industry (tablets). The coating of enzymes used in detergents protective layers for product safety has already been known for more than 30 years. A number of products (e.g., fertilizers, seeds, sodium percarbonate, citric acid, detergent base powders, and cosmetic ingredients) receive special coatings for different reasons. The variety of coating materials increases the applications and importance of coating processes significantly. Examples are shown Table 6.1.

6.2
Opportunities for Influencing Particle Design

Superficial coating of particles with one or more layers influences positively some elements of the product design, as shown in Figure 6.1. A number of parameters can be assigned to the product performance, handling, and aesthetics. These also include abrasion resistance, surface properties (roughness and haptic nature), and flowability (stickiness).

In exceptional cases, it may be that the only reason for a coating is an improvement of the aesthetics (coloring, scenting, and smoothing). Normally, the coating layer improves several granule properties. Furthermore, layers provide the possibility to integrate an additional chemical function. The coating must be inert and adhere evenly and well. Its task is the visible improvement of at least one characteristic property. The chemistry of the layer is differentiated from that of the core material.

Four well-established coating types (Figure 6.2) exist according to the material of the layer. On the one hand, there is a powdering of granules with micropowders. The powdering results mostly in reducing stickiness by addition of fine dry particles in a mixer or dryer. A calculation provides the amount of powder adhering on the

Table 6.1 Examples for application of a coating layer.

Industry	Product	Coating by	Coating layer	Improvement of the product design
Biotechnology	Enzyme	Powdering and suspension	Salts, polymers, and titania	Protection, stability (storage, abrasion), and color
Pharmacy	Tablets	Solutions and suspensions	Polymers and sugar	Solubility, controlled release, protection against moisture, stability, masking of odor and taste, and aesthetics
Agriculture	Seed	Suspensions	Polymers, surfactants, pigments, insecticides, and fungicides	Protection against diseases, color, and nutrients
Agriculture	Fertilizer	Suspensions	Insoluble polymer films	Slow release, long-lasting effect (color)
Chemistry	Sodium per-carbonate	Solution	Salts	Protection against abrasion and moisture; Storage stability
Consumer goods	Detergent base powder	Powdering	Salts	Flowability
Food	Sweets	Melt	Chocolate	Taste and aesthetics

surface of slightly sticky granules. On the other hand, a great variety of spraying-on methods are offered, using water-soluble or emulsified and suspendable or swellable substances as well as meltable materials, in most cases, polymers. In the chemistry, the spraying takes place in fluidized beds, operating discontinuously in many cases.

Applying of coating takes place preferably in the form of small droplets with diameters of <5 to a maximum of 30 μm (twin-fluid nozzle, see Chapter 8). The spraying ends when the desired layer thickness of ~10–50 μm is reached, in exceptional cases, also higher or lower. The liquid in the layer must remain flowable for short time and cover the surface completely and evenly before drying takes effect. In some cases, a second coating follows, which is carried out with different liquids or with solids. Core materials as well as layers 1 and 2 are different in composition. Applying the coating as liquid, except as a melt, always requires simultaneous drying with hot air; this is normally carried out during fluidization.

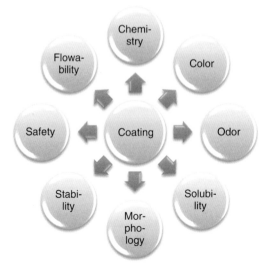

Figure 6.1 Improvement of the particle design by a targeted coating.

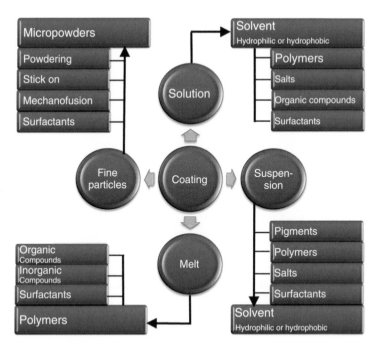

Figure 6.2 Coatings with solutions/suspensions as well as with melts or micropowders in chemistry (excluding the special case of emulsions).

6.3
Tasks of Coatings

The key task of coating is to significantly improve the storage stability for disperse chemical products. The activity/effectiveness of many substances changes over time (within few days, weeks, months, or years) as a result of chemical reactions. These take place because of the contact of particles with each other as well as the penetration of air and humidity, as also the ultraviolet radiation of sunlight. Changes in temperature may also initiate the reactions. In addition, relatively high temperatures accelerate the loss of activity.

Product safety is of high priority in the cases that enzymes (biocatalysts for detergents, animal feed, and textiles) are used (see Chapter 10). The coating layer has to be free of enzymes (proteins), completely enclose the particles, and prevent direct contact with the skin (allergy hazard). Therefore, coated particles may not generate respirable abrasion. Furthermore, the particles should be sufficiently stable and not break up, so that the whole surface remains protected. Ductile or elastic formulations thus have advantages over hard and brittle particles. Worldwide, the manufacturers of enzymes measure the formation of abrasion and breakage according to standardized methods in fluidized beds and in a "shooting apparatus." Measured values must be below severe limits. Under measuring conditions, particles are subject to forces ("worst case") far beyond normal applications. The layer improves mechanical properties significantly, especially the stability of particles. According to Table 6.2, protected particle surfaces increase the storage particle stability as well as the safety of the products. In addition, the layer allows a modification in solubility (hydrophilic/hydrophobic) and in rate of dissolution.

Table 6.2 Tasks of particle coatings in chemistry and pharmacy.

1) Preventing chemical reactions with adjacent particles, with oxygen, water, or UV radiation (storage stability)
2) Exclusion of possible skin contacts with hazard ingredients
3) Prevention/reduction of respirable abrasion
4) Increasing the mechanical stability, in particular of the surface
5) Improving the flow properties
6) Improving the aesthetic (surface) properties: color, gloss, smoothness, and sphericity as well as haptic
7) Change in solubility (instantization, retardation, and hydrophobication) and bioavailability
8) Realization of controlled release (start)
9) Timed release (period of minutes to several hours) via pH-dependent dissolution or diffusion
10) Masking/improvement of smell and taste
11) Additional chemical or biological functions
12) Catalytically active layer on an inert carrier

Tailor-made coating layers allow controlled drug release with a timed start and predetermined duration of release ("controlled release"). For fixing the point of dissolution, several switches are available. The possible switches are mechanical in nature (pressure and friction) or adjusted by suitable chemistry (pH-change, ionic strength, disintegrants, and solvents) as well as by operating conditions (temperature and turbulence).

The pharmaceutical industry has developed numerous coatings and switches for tablets and dragees [3, 4]. Initially some tablets may not dissolve in the mouth (neutral pH), because of which they cause a bitter or unpleasant taste. A coating with cationic copolymers (amino-alkyl methacrylate; >10 µm) prevents even partial dissolution in the mouth and esophagus. The tablet tastes neutral, as it is taste masked. Dissolution takes place only by salt formation in the acidic stomach (pH < 5). A significant delay in release does not occur.

The protection of active ingredients against humidity necessitates a minimal layer thickness of 10 µm, based on methacrylic acid copolymer. With this film thickness, the tablet is not yet protected against stomach acid; therefore, it dissolves in the stomach. This thin coating layer acts as good barrier to water vapor. Owing to the coating, the tablet's efficacy is guaranteed between it is first taken until the expiry date of the medicines. The 10 µm film is also sufficient to color the tablet uniformly. Dragees possess a thick sugar layer (up to a few millimeters), which smoothes roughness and provides a pleasant aesthetics. This layer also represents very good protection against water vapor.

In other cases, active ingredients are unstable in acids. Therefore, anionic acrylic polymers (methacrylic acid) or certain cellulose derivatives (hydroxypropyl methylcellulose phthalate) are used to form a layer that is enteric, that is, insoluble in acidic medium. Such coatings with a thicknesses in the range of about 15–40 µm avoid the incompatibilities of the stomach lining and protect the active agents in the acidic range. The chemistry of a layer is adapted to a several hour passage through the stomach at pH 1–4. In the process, only a slight swelling occurs and the sheath remains substantially impermeable. According to US Pharmacopoeia, not more than 10% of drug may release during this time (P. Kleinebudde, private communication, Universität Düsseldorf, June 2008). When methacrylic acid copolymers are used, the dissolution of polymer film occurs either in the front region of the small intestine (pH range >5.5) or only in the rear part. Adjusted by reducing the number of methacrylic groups, the switch reacts in this case at pH 7. This pH emerges there only briefly. A mixture of both copolymers causes gradual dissolution along the small intestine of pH 5.5 (pH-controlled delay) to 7.0. With optimal settings of layers, uniform drug release with high bioavailability takes place along the entire small intestine [4].

Thirdly, completely water-insoluble coatings are needed. For example, surface films from ammonio methacrylate copolymer or ethyl cellulose become only permeable after swelling in digestive juices. When water enters the capsule, active ingredient slowly dissolves and diffuses through the intact layer. Owing to low permeability, the researchers incorporate hydrophilic pore formers such as sugar, sugar alcohols, or macrogols in the film (P. Kleinebudde, private communication,

Table 6.3 Controlling the release by coating layers of medical tablets.

Tasks	Substance	Layer thickness (μm)	Switch
Dissolves only in the stomach (not in mouth)	Cationic copolymer (aminoalkyl-methacrylate)	> 10	pH < 5
Enteric coating	Anionic copolymer (methacrylic acid)	15–45	pH > 5.5
Enteric coating	Anionic copolymer (methacrylic acid with less acidic groups)	15–45	pH = 7
Retardation	Ammonio-methacrylate copolymer (insoluble)	15–40	Diffusion

Universität Düsseldorf, June 2008), especially in ethyl cellulose. In this way, continuous release is achievable for many hours. The time required also depends on layer thickness. This type of coating is called *retardation* (extended release tablets). Examples for controlling the release are depicted in Table 6.3.

Some polymers based on cellulose are suitable for coatings, such as methyl-, ethyl-, and hydroxypropyl celluloses; acrylic copolymers and any mixtures thereof are also used. Fourthly, the chemistry influences the rate of dissolution, in the case of cellulose ethers by type and degree of substitutions. Processing of the coating materials takes place as a function of its solubility in either aqueous or solvent-containing solution (such as isopropyl alcohol/acetone) or an aqueous dispersion, optionally with the addition of plasticizers. For example, in the case of pellets, the extension of release is possible by applying a lacquer layer of ethyl cellulose and polymethacrylate [5]. The active compound is located thereunder in a depot layer on a sugar sphere (core).

Besides solubility, bioavailability particularly has gained in importance. The therapeutically active agent must be available, absorbed at the site of action, and not immediately excreted. Bioavailability refers to that active proportion in relation to the total incorporated active ingredient and lies accordingly between 0 and 100%. Especially poorly soluble active ingredients may show very low levels of bioavailability. To improve the solubility and bioavailability, a micronization of valuable materials takes place, producing micro- or preferably nanoparticles. These particles show very high surface areas. They are finely distributed in the tablet matrix. A suitable coating ensures the continuous release over a prolonged period to increase the bioavailability.

Medical tablets demand coatings for several applications to control the solubility. However, there are also some examples from the field of chemical products. A delay in solubility of citric acid ensures a coating with fatty acid (FA). Without the film, citric acid dissolves instantly. By hydrophobicization of the surface, the start of dissolution in water is delayed for few seconds. The time required for complete dissolution shifts from instantly to several minutes. This effect ensures a neutral or slightly alkaline pH value after flushing of detergents into the washing machine.

Sodium percarbonate represents a chemical raw material produced from heavy-metal-free soda and hydrogen peroxide. The material is neither transportable in big bags nor storable in silos. In the presence of moisture, decomposition occurs with high exothermicity and some clumping. A coating with pure sodium carbonate or carbonate/sulfate mixtures solves the problem. The coated particles are stable in storage and in abrasion. An advantage is that a possible abrasion contains no decomposable or oxidizable compound. This allows safe handling of the product. For controlling the rate of dissolution (release delay), an already salt-coated pellet gets a second layer, for this specific case, a coating with sodium silicate [6] is preferred.

Other applications in field of *detergents* are the base powders, besides the already discussed enzymes. Owing to the addition of nonionics by spraying, the evolving stickiness of product leads to handling and metering problems. Therefore, a powdering with very fine materials ensues, which adhere on the surface and eliminate the stickiness. Controlling the solubility of some other components is possible [7], but it has not yet been carried out industrially.

Special coatings can realize a slow release with long-lasting effect, which is needed for *fertilizer*. The cover consists of water-insoluble polymers. The release takes place by a diffusion processes (osmosis) after the addition of water. Much later, the empty shells are left behind on the field. Carboxyl-carrying ethylene copolymer may represent a suitable coating for fertilizers [8].

In modern *agriculture*, a multifunctional layer ensures the protection of seeds against pests. The air- and water-permeable layer contains surfactants, dyes (to differentiate), and polymers in addition to the active ingredients. During growth the intake of active substances takes place, in particular, that of insecticides and fungicides. The substances enter via sap flow from the roots to leaves. Thus, the germ is protected from all sides for long time against all diseases and insect attacks. In comparison to spraying of aqueous solutions on plants in agricultural areas, an application to seeds requires only 1% of active ingredient. Furthermore, agricultural land is not additionally loaded by seed treatment. Modern methods enclose the seeds with a drug-containing sheath (shell). Another procedure involves the integration of the active substances in a matrix [9], that is coated subsequently. The pelleting provides uniformly sized spherical particles with improved flowability. Special applications require multiple coatings even for the seeds.

Statements and methods for coating in chemistry often refer to water-soluble, and less often to water suspendable or swellable core materials. The coating layers may be insoluble in water, sparingly soluble, or swellable as well as instantly soluble. Powdering particles are often insoluble in water; they remain invisible owing to small diameters in the micron range and small amounts in the solution. Water-insoluble films swell. The dissolved actives penetrate through this layer. The food industry needs these coatings not for an improvement of solubility, but for melting behavior, taste, smell, and design.

In addition, the coating allows an improvement of *aesthetics*. The superficial roughness disappears, sphericity increases, and color and gloss can be intensified. Other suitable materials [10] bind odors in the layer. Alternatively, the layer contains fragrances. This method separates the perfume from substances in the core. In the food industry, the layer masks bad tastes of the core or the layer adds pleasant taste of a different material.

The decision to use an additional step to optimize the product depends not only on production costs but also on formulation and expected market success by quality improvement. The technical feasibility of treatment must be demonstrated by some pilot tests. Therefore, representative core material with an optimized coating solution must be used.

6.4
Basic Variants of Coating

The selected process of agglomeration influences the particle surfaces. There are many ways exist for subsequent treatment [11] that are mainly based on coating methods. The coating takes place as shown in Table 6.4 in six basic variants.

Powdering means a coating of granules with dry solids without addition of liquids. The chemical industry uses finely ground powders with diameters of about 1–30 μm as the average size of d_{50}. Fine particles adhere on the surface under movement of the particles, as displayed in Figure 6.3. The powdered material, which may not react with the core, is nonsticky, and shows no hygroscopic properties.

Detergent manufacturers often use salts, silicates, or sodium aluminum silicates to improve the flow properties of surfactant-based formulations. Usually, the powdering needs an addition of 0.5–3 wt%. Depending on the fineness of particles, the amount ranges for an interrupted layer or for the microparticles in a "monolayer." The powdering works either due to high adhesive forces of the very fine powder, and by adhesion, particularly on slightly sticky surfaces. Surface powder not only reduces the stickiness but also improves the flowability and storage stability. It protects against humidity. The superficially adhered particles prevent direct contact between the particles and direct contact of the product with human skin.

Table 6.4 Processes for coating of particles.

Possibilities for applying of coating layers
Surface modifications with finely ground materials
by powdering in mixers or spheronizers
by incorporation into the surface (mechanofusion)
by surface granulation with a suitable liquid
Coating with a melt (film coatings) or with melt suspensions
Coating with inorganic and/or organic solutions and suspensions, which become solids on drying
Coating with liquids, solutions, and/or emulsions, which remain adsorbed at the surface
Coating with solutions or melts with subsequent chemical reaction
Combination of two or more methods (double or triple coating layers)

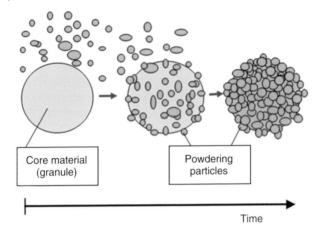

Figure 6.3 Mechanism of powdering.

As shown in Figure 6.4, powdering represents an equilibrium process. Depending on the material and its dispersity, the procedure comprises a constant adsorption and abrasion of fine particles. This occurs as function of the mechanics and operating conditions of the mixer, as well as surface properties of the core material, its specific surface area, and surface coverage. Increasing coverage increases the detachment of powdering and decreases the adhesion until finally all sticky parts are covered. Therefore, often a material input is reached, which only partially covers the surface. Gentle movement of the core material achieves optimal results. The

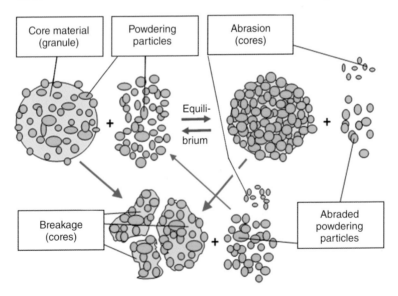

Figure 6.4 Balance between the core material and powdering particles; the breakage and abrasion of cores (dry process) is also displayed.

core material can break from time to time and generate abrasions by mechanics and permanent contacts of particles. These undesirable actions depend on the material properties, particle shape and size, and also the mechanical forces.

The second method, used for nontacky particles, is the sticking of fine powders and core abrasion [12] on the surface with a fluid, which offers adhesive properties. The droplets arise from a liquid by spraying via nozzles. This procedure may be referred to as *surface granulation*. It usually works without drying the small amounts of liquid that are present. Figure 6.5 shows a representation of the principle. Here again equilibrium arises between the coated granules and powdering particles, depending on the type and quantity of the fluid.

Apart from aforementioned methods for modifying the surface with fine particles, there are many coating procedures that operate with sprayed liquids in larger amounts. In a broader sense, all modifications of the particle surface, with or without a liquid, are referred herein as *coating processes*. In the strict sense, the term *coating* means an application of a solid protective film (shell, sheath, and layer) by liquids, which stick firmly and enclose the entire pellet. The film results from the spray and solidifies by drying of the solution/suspension or by cooling of a melt on the surface. Figure 6.6 shows, how the sprayed liquid (twin-fluid nozzle) forms closed layers on granules. A mixer generated mechanically and a fluidized bed hydraulically the moving of all particles.

Sticking droplets, such as those used for surface granulation, may be of hydrophilic as well as hydrophobic fluids. Solutions often include a polymer that forms an adhesive film on the particles after removal of the solvent through drying. With the correct amount, a continuous layer emerges, which compensates some peaks and valleys on the particle surface. This leads not only to an improvement in shape, but also in flowability. The use of suspensions as coating liquid is preferred in many cases: First, due to higher concentration, less fluid needs to be evaporated. Secondly, the addition of pigments enables a coloring or lightening of the particles in the same process step.

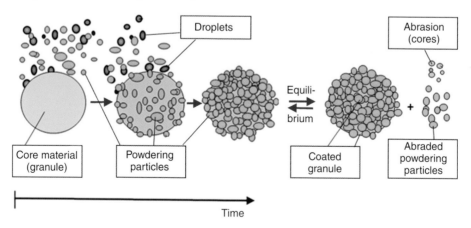

Figure 6.5 "Sticking" (surface granulation) of powdering particles with the aid of a liquid.

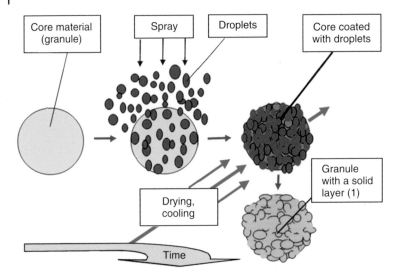

Figure 6.6 Principle of a conventional coating by spraying on liquids (droplets impinge, adhere, spread, run into each other, dry, and cool down).

When the liquid sprayed, drying occurs simultaneously. This results in a target conflict: On the one hand, the hot air allows an immediately drying of the spray in a fluidized bed coater. Then, the dust formed out of the droplets can be already so dry that the fine particles no longer adhere to the core surfaces. The drying air conveys this portion of the spray ("overspray") into the filter, where the dust is separated. On the other hand, in case of a very slow drying, the particle surfaces become wet and dissolution starts. With the superficially dissolved cores, some particles form liquid bridges, which cause particle size enlargements. This unavoidable granulation leads to larger recycling streams and reduces the yield. This indicates to us the following:

> The coating process takes place in the narrow operating range between spray drying and granulation.

Both principles (Table 6.5) are followed during each coating process. The coating procedure is monitored and controlled by the ratio of oversize to undersize particles. The oversize particles in the product consist of big granules, twins, and triplets; the undersize particles are present as dust (size distribution and composition). The optimization of some parameters provides a minimizing of these secondary processes. This includes the choice of spray nozzles (type and number) and of the operating conditions. It is advisable to make the spray as fine as possible. Furthermore, the spray rate, optimum amount of liquid phase per nozzle and unit time, as well as the number of nozzles per unit area, and thus the entire liquid phase (aqueous) per unit area are significant parameters for an optimum coating. At a given inlet air temperature and/or maximum product temperature, the flow rate and desired thickness of the layer determine the spraying duration.

Table 6.5 Typical parameters of the three spray processes in the chemical industry (spray granulation is between coating and granulation).

Parameter	Process		
	Spray-drying	Coating	Granulation
Product	Powder	Superficial layer on powders, granules, extrudates, and tablets	Granules and granulates (agglomerated powders)
Particle size	50–600 μm	0.2–10 mm	0.25–5 mm
Starting material	Liquid (solution and suspension)	Solid core material; coating sprayed on as liquid	Powder; forms with a sticking liquid and larger particles
Spray droplets	Medium sized (20–250 μm)	Small (5–30 μm)	Large; alternative pouring the liquid or addition of foam
Nozzle	Single- or twin-fluid (alternative: rotating disc)	Twin-fluid (or triple-fluid with microclimate for melting)	Single-fluid; (alternative: feed pipe for liquids and foams)
Procedure	High spray rate and fast, simultaneous drying at high temperatures	Low spray rate and slow, simultaneous drying at low temperatures	Addition of liquids and drying successively, medium temperature level
Layer thickness	—	10–50 μm	—

Figure 6.7 contains a summary of the subprocesses that occur during the coating process. As discussed, these subprocesses are granulation and spray-drying, as well as abrasions and breakages of particles. Abrasion takes place starting material is fed and, to a lesser extent, when the finished batch is discharged from the coater. Furthermore, uncoated and coated particles rub against each other during their movement in the different phases (heating, spraying, drying, and cooling) of the entire coating process. Together with the spray-dried droplets, the abraded dust streams into the filter, pneumatically transported by the fluidizing air. Another possibility is partial absorption by the spray and the redeposition on particle surfaces.

In addition, superficial active material can also be abraded. In some cases, as for enzymes, the abrasion of the core material should not reach the protective layer. Fluidizing air transports the dust away out of the spray zone. In sensitive individuals, the enzyme dust possibly causes sensibilization from exposure via the lung or through direct contact. Sensibilization is a preliminary stage and triggers allergy during a second contact. In a plant, the operator must avoid deposits of enzyme dust (see Chapter 10). This can be achieved by modifying the construction of the fluidized bed and operating conditions.

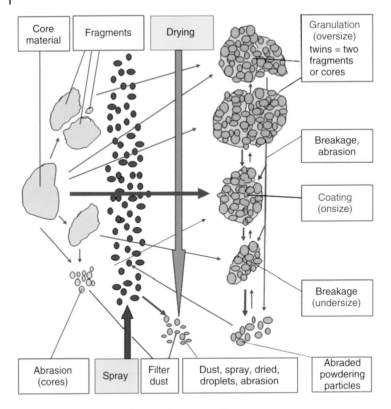

Figure 6.7 Processes in large-scale coating plants.

Note, that some particles (uncoated, partially and fully coated) break or crumble during the coating process as a result of pressure and impact. The tendency to fracture should be minimized through the optimization of particle composition and coating formulation. The proportion of fragments decreases significantly by using coatings with elastic properties. Broken particles agglomerate with the core material or other fragments and get a new coating. Some granules are then of the right size, while others are too big, and a few are too small.

A downstream screening/sifting of product separates the oversize particles, especially twins, triplets, and the undersize ones such as fragments from the onsize particles. The usual way is to mill and recycle the sieved parts, whereby the formulation of the product can change slightly. In pharmaceutics, such deviations are unacceptable; over- and undersized particles are discarded as garbage.

Some yields from experiments for coatings of sieved particles that originated from an extrusion process are as follows: dust 1–3% (attrition and spray-drying); fines 3–6% (from particle production and breakage); coarse fraction 2–6% (predominantly granulation); this resulted in an onsize particle yield of 85–92%. The percentages not only depend on the process but also on definition of the width for the onsize range.

Working with a *melt* [13] leads to considerable difficulties. A seamless trace heating from the feed tank up to all nozzles is needed in order to ensure temperatures that are well above the melting point in all the regions. The starting material in the coater should be preheated to temperatures just below melting temperature, so that molten droplets are not allowed to solidify on the way to the particles, but rather impinge as liquid, spread out, and run into each other before the melt freezes. An incorrect temperature profile impedes the droplets from adhering on the particle surface. The use of a surfactant in small quantities promotes an immediate spreading, even for a melt.

The spraying of oil/water-emulsions causes repulsion on hydrophilic surfaces. After an impingement, the aqueous phase penetrates into the water-soluble particles, while (silicone) oil drops remain adsorbed on the surface. The principle is simple and works well, especially when *nanoemulsions* are used. A nanoemulsion includes two surfactants, so that impinging droplets adhere and spread out in order to unite subsequently with neighboring droplets, forming a closed layer. Uniform film thickness is achievable by spraying extremely small droplets, preferably using twin-fluid nozzles.

The coating usually takes place in the presence of organic polymers. Of late, some manufacturers of sophisticated or delicate products (e.g., enzymes and sodium percarbonate) have shown a preference for inorganic materials (salts) for safe wrapping of particles. In some cases, they use several differently composed layers. Possible coating solutions for applying the layers are displayed in Table 6.6.

Table 6.6 Typical fluids for particle coatings in the chemistry.

Coating fluids (without and with suspended solids)

Use of (colored) coating liquids
    Salts
    Sugar
    Polymers, natural, modified, or synthetic

Use of a solution with suspended particles
    Pigments
    Mica
    Talc
    Supersaturated solutions
    Insoluble salts (calcium carbonate and fumed silica)

Utilization of an (nano-) emulsion
    Oils, especially silicone oil, possibly mixed with polymers
    Fatty compounds (alcohols, acids, etc.)

Use of pure organics, especially a melting
    Monomers and polymers
    Saccharides
    Surfactants
    Fatty compounds
    Waxes and fats
    Glycols

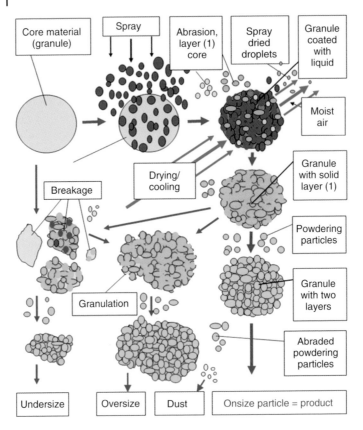

Figure 6.8 Double coating (solution/powdering) of a core material with granulation, breakage, and abrasion.

Figure 6.8 shows the principle of a double coating in chemistry. In this case, particles are first coated with an aqueous solution and thereafter powdered with fine particles. In a fluidized bed coater, powdering is not possible. Therefore, the second coating requires an additional step in a mixer or spheronizer. In fluidized beds, the air flow (fluidizing and drying air) transports the powder particles into the filter. The separation into two steps is also necessary for the recirculation of filter dusts. After the end of a complete coating, there always ensues a screening of oversized particles, for a high quality product as also for the separation of the undersize particles.

In chemistry, a common technique consists in powdering before applying the film. Stickiness is removed by upstream powdering. This ensures free-flowing and swirling particles in the coater. Powdering in pharmaceuticals (such as talc, powdered sugar, calcium carbonate, calcium sulfate, starch, and fumed silica) offers the possibility to form a separating layer between two coatings.

In the manufacture of tablets, the application of double or multiple layers is state of the art [14]. A core tablet is often coated with active substances that are dissolved in organic solvents or in water. This coating represents the first layer. In a closed coating system, two or more films are applied successively under drying conditions. In tablets, the layer thicknesses are usually in the range 30–100 µm, and in special cases up to several millimeters. These values are significantly higher than typical layer thicknesses in chemistry. A multicoating on tablets is formed by the following steps: step 1: a base film for protection against moisture by the addition of a solution followed by drying; step 2: separation of the layer by powdering, possibly with subsequent drying; step 3: stomach acid-resistant film by a solution or by drying; step 4: a final layer for the aesthetics (color, gloss, and smooth surface) by a solution with subsequent drying/cooling, possibly integrated in step 3.

Coating of particles with *reactive substances*, briefly mentioned here, leads to mostly water-insoluble films that are more stable and dense. There are many examples in the field of polymer chemistry. Suitable polymers, which are sprayed as melt, react and cure by UV irradiation. Another possibility is to spray on monomers, add a catalyst, and/or increase the temperature, which triggers the polymerization or cross-linking reactions. A good, but extravagant and expensive method for uniform coating of particles is represented by the evaporation process (usually in vacuum), in which ultrathin (metal) layers are formed. Further processes use hydrolyzable compounds (e.g., titanium tetrachloride). These substances are applied as liquid on support material (alumina and silica) and then converted at high temperatures to the oxide (titanium dioxide).

6.5
Coating Technologies

Typical coating processes are shown in Figure 6.9. In industry, the powdering takes place predominantly in continuous mixers or spheronizers, while typical coatings with solutions are batch processes in fluidized beds. In almost all procedures, the atomization takes place with two-fluid nozzles in order to produce very fine

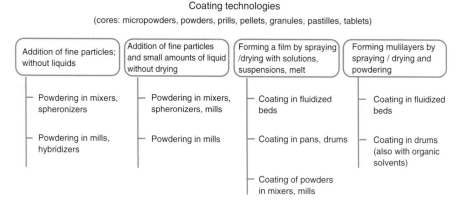

Figure 6.9 Overview of the standard coating technologies.

droplets. Powder coatings can be carried out in mixers with horizontal and vertical shaft or in mills preferably conveyed pneumatically. Mills have powders, which are used as grinding aids, coatings with melt at moderate temperatures, and solutions under conditions of grinding and drying.

The construction of fluidized beds differs in the supply of hot air, in the arrangement of the nozzles (top, bottom, and tangentially), and in the execution of the bottom plate and product discharge. The bottom plates of these fluidized beds have holes of different sizes or slits. Furthermore, bottoms with movable slats or fixed rings exist, wherein the rings are mounted one above the other in stages, with annular air guiding slots (Innojet®/Hüttlin). In addition, in the pharmaceutical industry, two coaters are often used, namely, pan and drum coaters.

An interesting method of powdering with hard materials uses mechanical incorporation of fine dry solids on the surface of the core particles. A tackiness of the particle surfaces is not required. By high mechanical energy input, hammering (mechanofusion) of powdering particles into the surface is carried out. The machine, called the *hybridizer*, acts like a mill with large tools. The intensive particle "merger" on the surface, displayed in Figure 6.10, generates an interesting morphology (surface design). The mechanical and chemical properties change significantly with this process. The particles show improved application

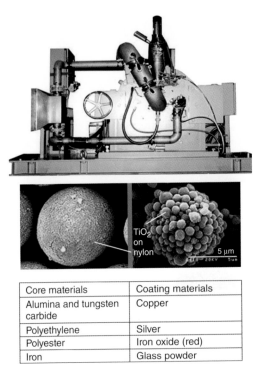

Core materials	Coating materials
Alumina and tungsten carbide	Copper
Polyethylene	Silver
Polyester	Iron oxide (red)
Iron	Glass powder

Figure 6.10 Hybridizer, dry-coated small particle with tiny powdering particles, and selected examples (Courtesy of Nara Machinery®, Europe).

characteristics, or even give rise to new products such as catalysts. Owing to high energy input, however, the hybridizer is not suitable for all particulate materials, especially soft particles. The application examples depict the variety of product design, such as coating of tungsten carbide with copper or polyethylene with silver. This technology allows combinations, especially with metals or glass powders that are not achievable in other ways. The process extends the possibilities of product design for disperse products.

Usually, powdering takes place in less energy-intensive processes. The suitable core particles should be slightly sticky. Small amounts of fine powder particles adhere on surfaces, if the surface texture of the particles permits. Powdering should not tend to stick and be hygroscopic. Applied mixers run with small Froude numbers and simpler mechanics. In the mixer, the particles should execute a rolling motion, as in a spheronizer at reduced rotational velocities. A powdering is possible for certain materials in mills, where the powder immediately covers new sticky surfaces. Here, the powdering supports the milling and the transport through the machinery. In most cases, the target of a powdering is to improve flowability by reducing adhesion.

Many coatings in the chemical industry are carried out in specially designed fluidized beds or mixers, in exceptional cases, in mills. In the usual method, the spraying of a solution takes place simultaneously with drying. The narrow size distributions of the core particles promote the quality of coating, because then the secondary process of "granulation" barely occurs. The pharmaceutical industry utilizes drums, balling discs (pans), and fluidized beds, such as the "Wurster" coater [15, 16]. Current coating technologies are depicted in Figure 6.11.

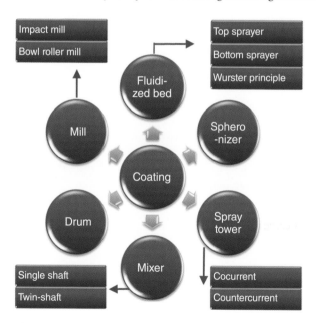

Figure 6.11 Coating technologies.

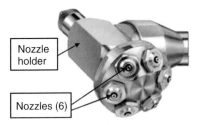

Figure 6.12 Nozzle holder with six twin-fluid nozzles (Courtesy of Glatt).

In addition to the preferred batch fluidized beds, some continuous variants for coatings also exist [17]. Theoretical principles are described in the literature [17, 18]. Batch fluidized beds have proved to be the top sprayers in the field of biotechnology for coating of enzymes.

In fluidized bed processes, the nozzles are arranged either directly in or above the expanded fluidized bed (top sprayers), less frequently in the fluidized bottom plate (bottom sprayer). Atomization of the liquid phase usually occurs in twin-fluid nozzles, as shown in Figure 6.12. These nozzles produce very fine droplets up to the micron range. Nozzle holders of this type are used in top, bottom, and tangential spray systems. A uniform coating allows many nozzles, preferably arranged on one or more nozzle holders. For symmetrically covering the entire spraying area, several nozzle holders are necessary in large plants. The use of a few relatively large nozzles, which additionally cover the spraying area asymmetrically with droplets, leads to poor coatings.

Coating facilities are constructed mostly in cylindrical design (Figure 6.13), but for continuous operations there is an oblong version (Figure 6.14). The hot air flows through slots or holes in the bottom plate of the apparatus and fluidizes the particles. The air is cooled down by water evaporation, and it exits the fluidized bed via a filter. The airstream transports the dust from abrasion and dried spray droplets to a filter fabric. Cleaning of exhaust air preferably takes place through an integrated filter, arranged above the fluidized bed. When a top sprayer is used, the jet cleaned air may carry fine particles back into the spray zone.

Depending on the preferred product discharge technique in the batch procedure (Figure 6.15), the hot fluidizing air comes in from below or from above before flowing through the bottom screen plate. The coated material exits the coater through a lateral or central discharge tube. Another possibility is an opening by rotary motion of bottom slats or the whole bottom plate. Fluidized beds, used for coating, are also suitable for granulations or spray agglomerations. In continuous operations, both the top and bottom sprayers (ProCell® technology) are run, either as coater or granulator.

The coating, also referred as *lacquer*, is formed on small tablets at high throughput preferably in a device with tubes inside, called the *Wurster coater* [20]. The two-fluid nozzle is arranged centrally at the lower end of a tube. The particles are conveyed through the tube inserts into the spray cone. Small coaters have one tube insert, whereas large coaters have up to 100. The momentum of air spray accelerates

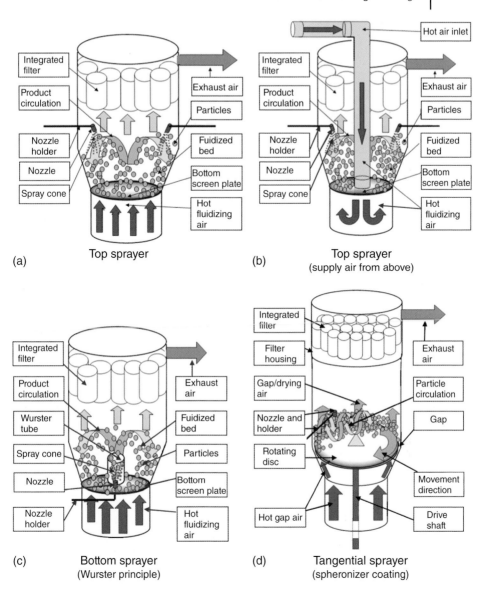

Figure 6.13 Discontinuous coating processes in fluidized beds (a) top sprayer, cylindrical apparatuses with integrated filters; (b) top sprayer with air supply from above; (c) bottom sprayer with the Wurster tube; and (d) tangential sprayer.

the particles in the tube, so that they achieve high speeds during the spraying. If particles leave the pipe, they fall back on the bottom plate and flow once more into the tube, where further acceleration starts. The Wurster coater generates uniform films. One of the reasons for high quality is the very short residence times in the spray cone. However, the complete coating time remains the same in comparison

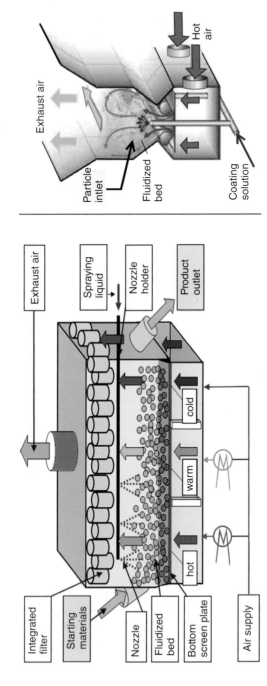

Figure 6.14 Continuous coating processes in fluidized beds (a) stretched version, shown as top sprayer, with three temperature zones and (b) bottom sprayer ProCell® for larger particles without the bottom plate [19]. (Courtesy of Glatt.)

Figure 6.15 Industrial versions of coaters (a) cylindrical top sprayer, lower part with the fluidization chamber and bottom outlet and (b) spheronizer with product outlet at the side and with melt coating or powdering of particles.

to other batch processes. The particle sizes (tablets) are limited in contrast to the drum coater. The Wurster coater is used mainly in the pharmaceutical industry for large batches, in which the chemistry dominates the top sprayer.

Coating in a spheronizer represents a newer procedure, where the droplets tangentially impinge on the torus of the circulating particle. With the use of a rotating disc or pot with tip speeds of about 20 m/s, the particles move at high speeds along the wall and around each other. The high speed ensures uniform covering of the individual particles (short residence time). Warm air across the gap assists in slow drying, in addition to slight fluidization of the particle torus. Further, the air prevents individual particles from falling through the gap. The amount of gap air is much smaller than for fluidization in a top sprayer. Therefore, this process is particularly suitable for application of a melt coating. In a practical comparison test with the top and Wurster coater, the spheronizer yielded the better results, that is, the most uniform melt coating for enzymes.

The pharmaceutical industry preferably uses the typical pan and drum coaters for the coating of tablets, as shown in Figures 6.16 and 6.17. A drum coater gently moves the tablets, and so practically generates no breakage or abrasion. Because of lower airflow, less liquid evaporates in drum coaters. Therefore, individual batches take longer. The pharmaceutical industry prefers this coater because of the uniformity of the layers and the possibility of applying multiple layers sequentially. A major advantage is the processing of solvents as coating solutions/suspensions with solvent recovery. Even small batches can be handled in the machine.

Judged by the weight, thick layers may represent the largest part of a tablet, and, therefore, they can be the reason for significant raw material and manufacturing costs. For a spherical particle of 500 μm diameter, it is estimated that a 25 μm coating (= particle diameter + 10%) results in an additional weight of 16% compared to the uncoated particle at comparable density. For a 50 μm layer (+20%), the weight of

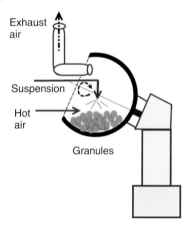

Figure 6.16 Pan coater for tablets [3, 21].

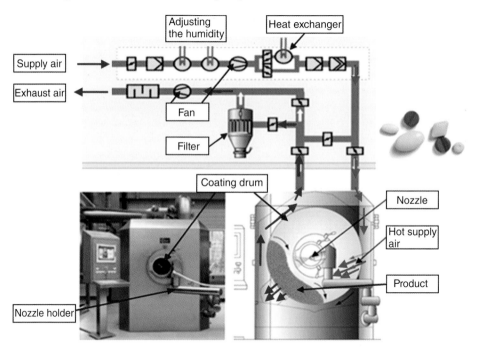

Figure 6.17 Drum coater in the pharmaceutical version for single or multiple coatings of tablets, evaporation of water, or solvents (Courtesy of Glatt).

layer is already almost 50% higher, and 100 μm layer (+40%) results in more than double the weight of the uncoated particle.

Powder coatings can be realized in suitable mixers, possibly in combination with a fluid bed drying. One of the technical principles is explained in Figure 6.18. The chopper is only needed when the powder forms lumps. The spraying takes place

Option 1:
Addition of the liquid through
a lance or an injector

Option 2:
Addition of the liquid by atomization
(or by a pressureless feed)

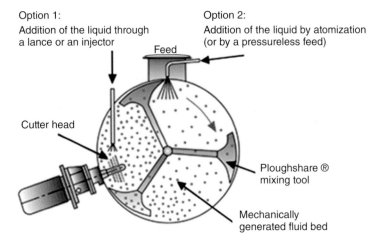

Figure 6.18 Coating process in a mechanically generated fluidized bed (Courtesy of Lödige).

via twin- fluid nozzles. Stable particles are mechanically fluidized and sprayed with the coating solution. If particles tolerate the moisture, then a drying step is unnecessary. This version represents an economically interesting alternative. A pharmaceutical mixer and some products from chemistry, coated in mixers, are depicted in Figure 6.19. The chemical industry uses simpler mixer versions, where the only wetted parts consist of polished stainless steel. Another principle is the intensive mixing in a batch mixer with a vessel rotating in the opposite direction as shown in Figure 6.20. There are also some product examples. The rotating vessel promotes both the granulation as well the coating.

A mill drying plant, which transports the crushed product pneumatically, is also useful for a coating operation. For this application, tangentially mounted nozzles allow to spray a solution or suspension or melt in the direction of the airstream. Secondly, it is possible to add substances that melt under milling conditions and thus provide a coating.

The spray tower allows producing particles economically with a thin coating layer. The components of a droplet can segregate in the drying process in a known manner. When drying an aqueous droplet, the hydrophobic constituents prefer places near and on the surface (low energy state). With this knowledge, it is possible to build up a coating layer with spraying procedures. In the presence of surfactants, emulsified oils as well as small amounts of organic substances (for example, fatty alcohols or suitable polymers) congregate at the surface, and constitute parts of the outer layer after drying. In the drop, the hydrophobic portion is located in the outer edge, while the hydrophilic substances prefer the interior.

A high degree of layer uniformity characterizes a good coating. The film should adhere firmly. Further, the outer shell should show certain elasticity to mechanical stress and not chip off easily. Measurements using the scanning electron microscope (SEM) demonstrate the quality of a coating. Figure 6.21 presents three

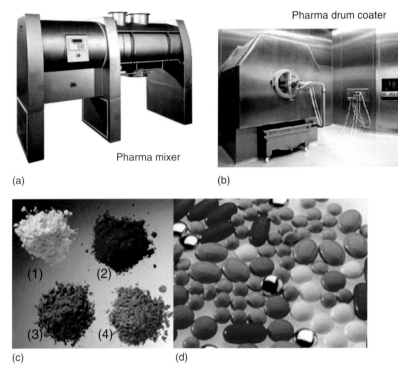

Pharma drum coater

Pharma mixer

(a)   (b)

(1)   (2)

(3)   (4)

(c)   (d)

Figure 6.19 Pharmaceutical coating equipment and coated products (a) good manufacturing practice (GMP) mixer, (b) drum coater, (c) powders, coated in the mixer: 1: hydrophobizing of salt with fat; 2: instantizing of cocoa beverage powder with lecithin; 3: silanization of fire extinguishing powder; and 4: pigments on lacquer polymers; and (d) products (tablets, seeds, and catalysts), coated in a drum (Courtesy of Lödige).

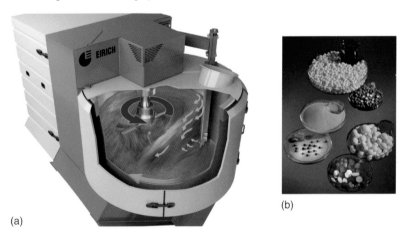

(a)

(b)

Figure 6.20 (a,b) Intensive mixer for granulation and coating with examples for coatings of products (Courtesy of Eirich).

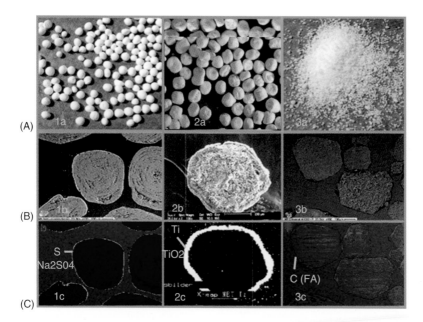

Figure 6.21 Examples for industrial coated particles: 1. percarbonate coated with sodium sulfate/sodium carbonate; 2. enzymes (proteases) coated with polymer and titanium dioxide; 3. citric acid coated with fatty acid (FA); (A) top row: images of product surfaces; (B) middle row: SEM top view of Epon-embedded, cut particles in resin; and (C) bottom row: energy dispersive X-ray microanalysis (EDX) images.

practical examples, namely, the view of the particles' surface (A), intersected particles embedded in a resin (Epon™; B), which were examined with the energy dispersive X-ray microanalysis (EDX, C). All coating layers are clearly visible at the edges [1] of the particles. The sodium percarbonate spheres originated from fluidized bed spray granulation, enzyme particles from granulation/extrusion, and fine citric acid particles from crystallization. Thereafter a process of coating in a fluidized bed took place.

6.6
Learnings

√ Coating is an essential method to determine and optimize product design. Layering is carried out preferably in fluidized beds, but also in mixers and coating drums.

√ The solid protective layer originates from powdering or spraying of a solution, suspension, or melt followed by drying/cooling. The operating conditions lie between those for spray-drying and granulation.

√ The coating process allows the adjustment of product properties, particularly the prevention of contact, the stability, flowability, and aesthetics, as well as control of dissolution.

√ Medical tablets show time- and pH-controlled dissolution (controlled release) by one or more coating layers.

√ The layer masks unpleasant tastes of tablets.

√ An insecticide and fungicide layer protects seeds and developing plants.

√ Sweets are coated for improving taste and aesthetics.

√ Enzymes need a protection layer to prevent contact and abrasion.

√ Suitable coatings of enzymes consist mostly of a salt mixture and/or polymers, often in combination with pigments.

References

1. Ghosh, S.K. (ed) (2006) *Functional Coatings: By Polymer Microencapsulation, Wiley-VCH Verlag GmbH*, Weinheim.
2. Mollet, H. and Grubenmann, A. (2009) *Formulierungstechnik, Emulsionen, Suspensionen, Feste, Formen, Wiley-VCH Verlag GmbH*, Weinheim.
3. M. Kumpugdee-Vollrath, *and* J.-P. Krause *(eds.), (2011) Easy Coating, Vieweg + Teubener Verlag*, Wiesbaden.
4. *Evonik Industries Eudragit*® *Application Guidelines, 12th edn, Evonik Industries, www.eudragit.com (accessed 26 November 2013).*
5. Bron, J.S. et al. (1993) Byk Nederland BV, DE 43 13 726 A1, Apr. 27, 1993.
6. Jakob, H., Zimmermann, K., Overdick, R., and Leonhardt, W. (2007) Degussa; DE 603 09 070 T2, May 16, 2007.
7. Rähse, W. (2007) Henkel AG & Co. KGaA, DE 10 2005 036 346 A1 Feb. 01, 2007.
8. Locquenghien, K.H., Engelhardt, K., Kleinbach, E., and Müller, M.W. (1995) BASF DE 195 21 502 A1, Jun. 13, 1995.
9. Bürger, H. et al. (1994) Aglukon Spezialdünger GmbH, DE 44 14 724 A1, Apr. 25, 1994.
10. Rähse, W., Baur, D., and Pichler, W. (2001) Henkel AG & Co. KGaA, DE 101 42 124 A 1, Aug. 30, 2001.
11. Rähse, W. (2004) *Produktdesign – Möglichkeiten der Produktgestaltung.* Chem. Ing. Tech., **76** (8), 1051–1064.
12. Rähse, W., Kühne, N., Jung, D., Fues, J.-F., and Sandkühler, P. (1995) Henkel AG & Co. KGaA, DE 195 47 457 A1, Dec. 19, 1995.
13. Paatz, K., Rähse, W., Pichler, W., Kühne, N., and Upadeck, H., Cognis, DE 43 22 229 A1, May 7, 1993.
14. B. Skalsky, *and* H.-U. Petereit, *(2008) in Aqueous Polymeric Coatings for Pharmaceutical Dosage Forms, 3rd edn, Chapter 9 (eds J. McGinity, and L. A. Felton), Informa Healthcare*, New York, p. 267 ff.
15. Pietsch, W. (2005) *Agglomeration in Industry, Vols.* **1 and 2**, *Wiley-VCH Verlag GmbH*, Weinheim.
16. E. Tsotsas, *and* A.S. Mujumdar *(eds) (2011) Modern Drying Technology, Product Quality and Formulation, Vol.* **3**, *Wiley-VCH Verlag GmbH*, Weinheim.
17. Porter, S.C. (2001) in *Pharmaceutical Process Scale Up* (ed M. Levin), *Chapter 9, Marcel Dekker*, New York, p. 259 ff.
18. Jacob, M., Piskova, E., Mörl, L., Krüger, G., Heinrich, S., Peglow, M., and Rümpler, K.-H. (2005) Proceedings Nr. 081–003, 1–13, 7th World Congress of Chemical Engineering, Glasgow, Scotland, July 10–14.
19. Glatt Innovative Technologies for Granules and Pellets, download from Glatt, www.glatt.com (accessed 26 November 2013).
20. F. N. Christensen, *and* P. Bertelsen, *(1997) Drug Dev. Ind. Pharm.*, **23** *(5)*, 451.
21. E. Teunou, *and* D. Poncelet, *(2005) Encapsulated and Powdered Foods, Chapter 7 (ed. C. Onwulata), CRC Press*, p. 180 ff.

7
Product Design Out of Disperse and Continuous Phases by Crushing

Summary

Nearly all manufacturing processes in the chemical, food, and pharmaceutical industries include steps for size reduction of substances. By crushing, atomization or (under water) granulation, disperse materials (solids, suspensions, pastes, and emulsions) and continua (liquids, melt, plasticized materials, and gases) are broken up for setting the required particle size and size distribution, and also to improve the particle form. For the products, a size range between about 500 nm and a few millimeters may be desired. Performed in solid or liquid phase, these energy-consuming actions form particles out of larger units, characterized by a significant increase of the specific surface area. In practice, the application or acceleration of subsequent chemical reactions, or the intensification of physical processes requires finely divided material. Occasionally, other operations such as drying, cooling, coating, or reacting can take place during crushing. The physicochemical and application properties are adjusted in the laboratory and in pilot plants. The resulting measurement ranges (specification list) constitute the basis for a global reproduction of the products. Partially complex settings and coordinated steps in production define the product design. A discussion with examples ensues, explaining the procedures.

7.1
Breaking Up of Materials

Typical breaking-up processes are the crushing of disperse materials into fine powders and of continuous phases (liquid) into droplets. Solid, liquid, and gaseous phases disintegrate in appropriate devices (Figure 7.1). Large solid units (continuum) must first be pulverized. Disperse solids constitute the feedstock for extremely fine grindings. Thus, while solids usually exist in disperse form as starting material, liquids (pure substances, suspension, pastes, or emulsions) as well as gases start as a continuum. Out of each breaking-up process, disperse solids (Figure 7.2) result, namely, micropowders, suspended micropowders, powders, or grains. An exception is emulsification, wherein a two-phase system originates

Industrial Product Design of Solids and Liquids: A Practical Guide, First Edition. Wilfried Rähse.
© 2014 Wiley-VCH Verlag GmbH & Co. KGaA. Published 2014 by Wiley-VCH Verlag GmbH & Co. KGaA.

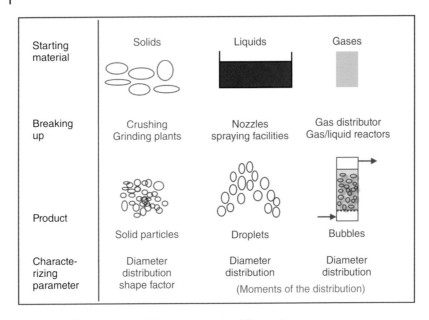

Figure 7.1 Breaking-up (crushing) processes for different phases.

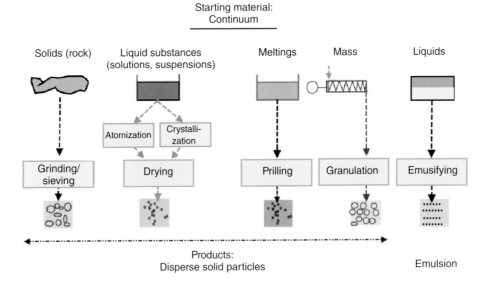

Figure 7.2 Targeted transfer of continuous phases into disperse products.

from two liquids or a three-phase system, with additionally dispersed fine particles. An emulsion is formed from two immiscible liquids with the aid of surfactants, whereby small droplets are dispersed in the continuous phase.

7.2
Importance of Crushing Processes

According to Rumpf (M. Pahl, private communication, Universität Paderborn, 2010), the breaking up of solids, liquids, and gases leads to "particles." The breaking down determines the average size and size distribution of the resulting particles, and partly also their form and surface chemistry. First, the fineness of the particles is necessary for their use in the application [1]. Second, chemical reactions or physical processes require an enlarged surface area. In addition to usual sieving, further operations take place during fragmentation. Simultaneous procedures may be drying (drying and grinding), cooling, mixing, reacting, activating, mechanofusion [2, 3], extracting and adsorption, or deodorizing as well as coating, including dyeing and perfuming. The operations take place in one step or in several circle runs. Processes with two or more operations at the same time can be described as hybrid procedures and are economically highly interesting, especially when running in the circulation mode. Complex processes today provide customized particles (product design), in which the crushing process represents an essential part.

Crushing plays a dominant role in the extraction of raw materials such as minerals [4, 5] and ores (Figure 7.3), coal, and renewable resources, as also in their recovery from waste material. In addition, for building materials, food, and

Figure 7.3 Ore mining (Courtesy of Netzsch).

in the chemical, pharmaceutical and plastics industries, grinding is an important operation. The discussion here considers fine crushing into particle sizes in the range from about 1 µm to a few millimeters.

Natural raw materials constitute the basis for production of solids. They are either directly or indirectly extracted by chemical, biochemical, and physical processes. The processing requires chemical treatment, isolation, and purification. Necessary procedures may include stages in solid and liquid or gas phase.

Grinding, agglomerations drying, reaction, and extraction are carried out on solids. Often, a path to the pure product leads through liquid phases by dissolving, followed by evaporation, precipitation, and crystallization. Reactions in the gas phase with subsequent sublimations also allow the manufacturing of pure solids. Precipitations and crystallizations include additional purification steps, which are desirable for the preparation of solid particles. Depending on the operating conditions, crystallization and recrystallization adjust the particle sizes. Only very fine particles require a further size reduction. Thus, crystallization represents an important technology in product design.

In addition, homogeneously dissolved solids exist, which are in the form of melts at elevated temperature [6] or are heterogeneously dispersed as suspension. Further, the formulation of a solid mixture or a solid-containing emulsion is possible. In all cases, the material properties and intended use determine the product design. Extensive laboratory tests set possible ranges of main parameters for the determination of a specification list and of application test methods. The measured values of relevant physicochemical, technical, and application tests describe the product and ensure a possible reproduction anywhere in the world.

Particle size enlargement [7] and reduction processes [8] are essential product design technologies for solids and multiphase systems. The preceding chapters treated the agglomeration of particles (granulation, Chapter 5) and their shaping by coating the surface (coating, Chapter 6). In the following, the third important procedure for adjusting the product design, namely, the targeted breaking up of larger grains and continuous phases into small particles in millimeter, micron-, and nanoscale sizes is discussed. Figure 7.4 displays several examples for the methods applied. Division ensues under high-energy input, resulting in the formation of dispersed materials from larger units with a significant increase in the specific surface area.

7.3
Particle Properties by Breaking Up

In many cases, it is not sufficient to execute a simple crushing for the desired product design. The application (toner, inkjet inks, metal powders, and pigments) requires mostly minimum and maximum particle sizes as well as compliance with other specified parameters. The targeted crushing includes the entire process, in particular, the type of mill, the starting material, the degree of comminution, as well as the operating conditions and the interior or exterior separation of outsize

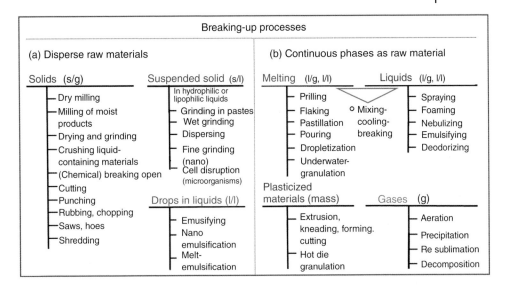

Figure 7.4 Industrial processes of crushing to form disperse materials.

particles. In some cases, several mill passages or a connection of two or more mills is necessary. Table 7.1 gives an idea of the product designs realized by crushing procedures.

The breaking up must be executed in a way that the particles meet all requirements in the value ranges of a specification list. Not only particle sizes and distributions but also, in particular, the adjustment of essential performance properties, such as moisture content, solubility, processing, flow properties, bulk density, dust and heavy metal content, shape, and tableting behavior are specified.

Some examples should describe the product design through crushing processes in more detail. For preparation of cellulose ethers, the chemical industry needs pure cellulose, originated from wood or cotton. Chemical pulp, pressed into rolls or plates, is cut to fiber length approximately in the range of (d_{10}, d_{90}) 100–400 µm in cutting mills with an inside sieve. During cutting, the operator pays attention to temperature and, in particular, to lower particle sizes, in order to avoid polymer degradation (reduced viscosity). Significantly, too long fibers however impede the homogeneous alkalization and etherification, so product quality (solubility) decreases [10].

In the food industry, a targeted size reduction of spices is reached to enhance the digestibility, solubility, drug release, and extractability [11, 12]. Here, the required fineness lies generally in the range of <150 to <600 µm (95%). Powdered sugar (<100 µm, classification of the coarse fraction) is available for use in some desserts and icings. For decorating a cake with icing sugar, an addition of rice starch to sugar improves the flow properties (moisture control).

Titanium dioxide is recognized as the best white pigment. Whitening power depends on the modification (rutile or anatase), particle sizes (10 nm to about 10 µm), and their distributions. Up to an average particle size of 1 µm, dry grinding

Table 7.1 Design of solid particles after a problem-tailored crushing process [9] (limits of particle sizes min. d_{10}, max. d_{90}).

Product	Crushing process	Application	Particle size distribution	Special features
Chocolate and cocoa	Rolling or impact mills and agitated ball mill	Food	>5 μm and <20 μm	In mouth not perceptible as particle
Spinach	Cutting mill	Food	95% <4 mm	Gentle milling
Toner	Opposed jet mill with a double sifting	Copier and printer	>3 μm and <15 μm, $d_{50} = 7$ μm	Combined classification and spheronization, pressure-shock-resistant design (10 bar)
Emulsion-polyvinyl chloride (E-PVC)	Spray-drying and air classifier mill	Different products	Quality 1: 99.9% <40 μm; Quality 2: 99.9% <20 μm	Soft substance
Polyethylene	Pin mill and sieve	Powder from granules	<150 μm	Circulation and cooling with liquid nitrogen
Powder coating and thin film	Opposed jet mill and cyclone	Automobile, Office furniture, Shelves, Domestic appliances	Quality 1: <12 μm Quality 2: 12–25 μm Quality 3: 25–45 μm Quality 4: 45–63 μm	Without oversize particles; pressure-shock-resistant system (10 bar)
Kaolin, gypsum	Agitated ball mill and sieve	Fillers and pigments for the paper industry	80%, <2 μm	With additives of up to 70% solids content
Magnesium powder	Jet mill and classifier	Chemical industry, pyrotechnics, and metallurgy	>10 μm and <63 μm	Circulation with nitrogen and oxygen monitoring, pressure-shock-resistant design

in a jet mill suffices. Further crushing takes place preferably in an agitated bead mill (nano mill) by wet grinding up to about 50 nm. According to application, the particle sizes of white pigments can be adapted and additionally improved with inorganic or organic coatings. With silica or alumina hydrate via precipitation methods, coated pigments are used in emulsion paints. Deposition of organic material [13] is carried out to improve the processing of plastics (with silicone oil) and gloss of lacquers (polyalcohols and triethanolamine). The cosmetic industry uses titania (0.2–0.4 μm) coated with alkylsilane for sunscreens.

In addition to solid grinding, spraying represents an important breaking-up procedure for continuous phases. There are at least two unit operations combined in the process: spraying (i.e., shaping) with drying and agglomeration or coating and spraying with cooling. For the production of powders from solutions or suspensions, several spray-drying methods (see Chapter 8) are used by the industry worldwide. There, the choice of the type and size of the nozzle, and partly the operating conditions, allow a rough adjustment of particle size distribution. The fine-tuning ensues by milling/sieving to remove coarse fractions and to recycle undersize particles.

"Crushing" of a melt represents a technically important molding process. Examples are presented in Table 7.2. From the melt, it is possible to produce any desired particles and moldings by shaping and cooling (solidifying). For particle formation, four processes exist, namely, prilling (spray/cooling), dropletization in a cooling channel, flaking, and several pastillation methods (Figure 7.5). An extruder allows the production of needles, spaghetti, hollow bodies, granules, and any other

Table 7.2 Shaped products from melts in the chemical and food industries [14].

Substance	Technology	Product shape
Urea	Prilling	Powder
	Pastillation	Pastilles
Stearate	Dropletization	Spheres
	Flaking (roller)	Flakes
	Pastillation	Micropastilles
Water glass	Pouring, break	Piece glass
Paraffins	Pressurizing	Candles
Polyvinylalcohol	Extrusion	Granulates
	Pouring or extruding	Foil
	Injection molding	Shaped bodies
	Blow molding	Bottles
Polyester	Extrusion	Fibers, films, tapes, and shaped bodies
Sulfur	Pastillation	Pastilles
		Flakes
Goody	Pouring	Ellipsoid
Chocolate	Pouring	Chocolate tablet

Figure 7.5 Breaking up of melts (a) chocolate, pastilles; (b) sulfur, flakes; (c) bitumen, flakes; (d) lauric acid diethanolamide, micropastilles; (e) polyethylene glycol, flakes from roller; and (f) fatty alcohol, spheres from dropletization (Courtesy of Sandvik (a–c) and COGNIS (d–f)).

shaped bodies by forming a melt or by plastication of masses. The shaping takes place simultaneously in the extrusion head. Recycling of faulty materials in the melting tank is possible by remelting.

7.4
Variants of Crushing

The crushing of solid materials is achieved by several mechanisms of stress that are partially coupled. Stress mechanisms such as collision, impact, beating, pressing, shearing, cutting, and rubbing are implemented in various mills (Table 7.3), often through rotor-stator machines constructed for the purpose.

The way in which grinding stress affects the particles depends on the construction of the mill, tools, and operating conditions. In mills, multiple processes take place simultaneously, whereby one of the processes is predominant. In impact mills, particles are accelerated and thrown against solid walls, especially against the stator. Collisions take place not only with the wall, but also with tools and preferably with other particles. This principle is also used in the air turbulence mill, supported by induced vortex and a specially constructed wall. These mills run with a tip speed of more than $100\,\mathrm{m\,s^{-1}}$. In an opposing jet mill, the particles are fired against each other with such high speeds that they shatter on impact. The required energy supplies compressed air at 6 or 10 bar that is reduced to normal pressure with

Table 7.3 Fine crushing of dry solid materials.

Mechanism of crushing	Typical machines	Examples of different products	Some principles
Impact/collision	Pinmill (left), turbo, air turbulence, and jet mill (right)	Fertilizers, dyes, powder coatings, bentonite, gypsum, clay, quartz, feldspar, toner, cereals, sugar, spices, and tea	
Beating	Ball, vibratory, and rod mill		
Pressing/shear stress (squeezing)	Roller, table roller, and pan mill	Cellulose ethers, coal, cement, coke, limestone, gypsum, and agglomerates	
Cutting/tear	Cutting mill	Wood cellulose, linters, rubber, leather, soft metals, films, fibers, wood, paper, and plastics	
Rubbing	Grater and corundum stone mill	Agglomerates and food	—

subsequent acceleration of the particles. Another very effective mill operates with superheated steam.

Compared to impact mills, in a rolling ball mill, the particles move only with low speed. Balls and particles run up the mill wall, then fall down together. Particles crushed by shocks between the balls as well as between the ball and the wall. In roller mills with adjustable gap and shear field, the movement of a particle is not of importance. Here, the particles or agglomerates are squeezed and crushed. A similar mechanism takes place in the table roller mill (also called the *bowl roller*). The squeezing and crushing and further additional shearing off take place under the influence of pressure forces between a moving ribbed roller and ripped bowl with significantly longer residence times.

Granulators that are often fitted with an internal sieve shred fibrous materials and films. The cutting process takes place between the rapidly rotating blade of the rotor and just a few stator blades, wherein the gap width is only 0.3–0.6 mm. Up to an installed electric power of 500 kW, this machine is used widely for recycling of waste, such as plastics (films, profile scraps, and residues), cable scraps, rubber, paper, and leather, and for the grinding of chemical pulp.

Particles in suspensions and pastes can be crushed effectively in stirred bead mills, which exist in different forms. The industrially used horizontal or vertical stirred ball mills crush particles from 10 to 300 µm down to below 5 µm, and in special cases, also below 1 µm into the nano range. Proven in the pilot scale at high energy consumptions, these mills with very small beads allow the production of colloidal particles below 100 nm from feed materials that are several microns in size.

There are different processes, which are functions of the starting materials (type and size), for crushing in gases or liquids. The areas of application of conventional mills, depending on the desired particle sizes, are shown in Figure 7.6A,B for fine crushing. Some product examples are depicted in Figure 7.7.

7.4.1
Grinding of Solids

The selection of a suitable mill takes place depending on the material and desired particle design, and based on economic criteria. In plants for grinding, homogenization, and grinding/drying, the particle sizes of solids are reduced by impacts, shocks, cutting, and pressing with or without shear forces. The comminution of very soft, usually moist materials (food industry) is carried out by

Figure 7.6 (A) Typical areas for dry and wet grinding of particles. (B) Simplified mill selection with dependence on particle size of the final product: (a) eight-stage high-pressure homogenizer for the production of nanoemulsions; (b) stirred bead mill for wet milling; (c) vertical stirred bead mill with disks; (d) corundum stone mill for pastes; (e) turbo mill; (f) universal impact mill; (g) cutting mill; (h) ball mill; (i) roller mill; and (j) steam jet mill for extremely fine dry grinding (Courtesy of GEA Niro Soavi (a), Netzsch (b, c, e, f, and j), FrymaKoruma (d), Pallmann (g), Hosokawa Alpine (h), and Neuhaus Neotec (i)).

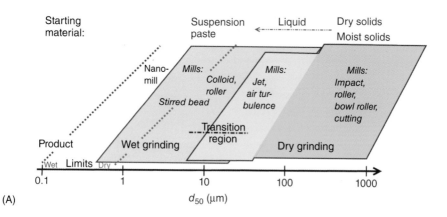

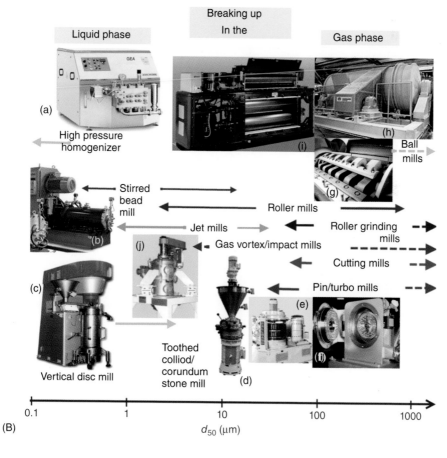

(A) d_{50} (µm)

(B) d_{50} (µm)

Figure 7.7 Typical products before and after grinding: From top to bottom: Plastics, initial chemical products 1, food and spices, initial products 2, dyes, and cosmetics (Courtesy of Pallmann, Jäckering, Netzsch, and FrymaKoruma).

cutting, rubbing, or pressing. The industries grind dry or moist solids preferably in gas phase with simultaneous drying, if necessary. The diameters of the produced particles (d_{50}) are frequently more than $100\,\mu m$, whereas in some cases they are down to about $50\,\mu m$. Usual grinding with a cycle to reduce the oversized particles achieve these requiremnets. In jet and vortex mills dry fine crushing is possible up to about $1\,\mu m$.

The tasks of these crushing processes can focus on the following:

1) Setting the required particle sizes and particle size distributions in suitable mill/sieve or mill/sifter procedures. Some examples are building materials, especially cement clinker, further colors (pigments) and fillers, powder coatings, iron oxide, toner, and food (spices, malt, and sugar).
2) Crushing for fast dissolution and performing chemical reactions. The extraction, dissolution, or reaction takes place after the particles sizes fall below a limit. Examples are oil seeds, coffee, salts, magnesium, coal gasification, roasting of metal sulfides, the calcination of limestone, and extraction of cellulose fibers from wood pulp.
3) Treatment of heterogeneous composite materials prior to separation. Besides cereals (wheat, rye, and barley), mineral and fossil resources as well as heterogeneous composite products (packaging, waste) often need grinding before separation, for example, ore/gangue, coal/rock, grain/shell/seedling, and recyclables/waste composites.
4) Uniformly shaped products from special crushing. These methods are generally found in the food industry, for example, almonds in the form of pins, thin slices, chopped almonds, nuts, diced carrots, French fries, chips, paper, and sheet metal punching.
5) Conversion of solids into fine powder as starting materials for targeted granulations, for example, detergent granules.

7.4.2
Deagglomeration

In mechanically operating procedures such as granulation, pelletization, roller compaction, extrusion, and coating, often undesirable large particles arise that are out of spec. On sieving these particles, a coarse fraction is obtained, which should be recycled after grinding (deagglomeration). The amount depends on the size distribution and mesh sizes of the sieve. To fulfill the task, crushing of this proportion takes place either in an impact mill for sharp grinding or in a roller mill for gentle deagglomeration (see Chapters 5 and 11). The ground material flows back to the agglomeration. Furthermore, integrated or external crushing processes allow the recycling of materials from the start-up and shutdown as well as products that do not meet all points of the specification (faulty batches).

An unwanted buildup of particles occurs during thermal processes, in particular, in case of drying from solutions/suspensions or filter cakes. The process is characterized by the formation of oversize particles, clumps, and large chunks. This necessitates recirculation after dry crushing back on the sieve or dissolution in the suspension tank [15]. Drying of wet filter cakes or pastes in a grinding and drying plant, where grinding tools prevent each agglomeration is recommended. The product accrues as fine powder under minimal energy consumption, particularly when superheated steam is used as the drying gas with energy recovery.

7.4.3
Split Up Sensitive Materials in a Cold Milling Process

Heat generated during grinding causes many organic products to lose some valuable contents. In the case of spices, these are precious essential oils. Fats can melt and smear the screens in the mill. Feedstock sticks together when the ingredients are melted. For such products, it is advisable to use a cold or cryogenic milling [12], preferably with recycling of the inert gas. The process requires a cooling with liquid nitrogen of the starting materials in the reservoir, feed, and circulation lines as well as the crushing machine. For plastics, the cooling should fall below the glass transition temperature, to embrittle the material. By evaporation of a controlled amount of liquid gas, the temperature is kept constant in spite of heat generation during grinding. Depending on product properties, the temperature of crushing lies between -50 and $15\,°C$, in exceptional cases down to $-80\,°C$ [9]. Temperature-sensitive and/or soft products break up in the cold atmosphere at about twice the rate. These include thermoplastics and elastomers, as well as waxes and paint additives, as well as active ingredients of the pharmaceutical and chemical industry and additives for food (proteins).

7.4.4
Milling of Suspended Solids

Fine crushing of solids to particle sizes below $25\,\mu m$, in particular, by intensive grinding with recirculation for particle sizes even below $1\,\mu m$, that is, into the nanoscale, runs better in suspension. A special case of wet grinding causes cell disruption of microorganisms in bead mills to break up the cell walls by introduction of energy. By this measure, proteins are able to exit the cell. So they can then be isolated from the solution. The tasks of wet grinding may be stated as follows:

1) Fine crushing to unify the particle size distribution, some with a breaking of grains (food), occur in liquids with low, medium, and high solids concentration. There are numerous applications in various fields, for example, ointments, lotions, creams, lipstick, toothpaste, mascara, color pigments, ketchup, almond paste, mustard, cocoa, fish, pasta, pet food, putty and joint compound, polymer dispersions, ceramic masses, and agrochemicals.

2) Destruction of cells ($0.1–10\,\mu m$) allows the isolation of biotechnologically produced, intracellular substances, particularly medically active substances.

3) In agitated bead mills, the conversion of preground, suspended solids into the micro- and nanoscale happens, for example, talc, calcium carbonate (paper), pens (paint), emulsion paints, titanium dioxide, pigments, printing inks,

technical ceramic compounds, herbicides, fungicides, chip polish, LCD color pigments, and inkjet ink.

7.4.5
Breaking Down and Transforming of Liquids into Dispersed Solids

Particulate solids originate from solid-containing liquids such as solutions, suspensions, emulsions, and melts by an atomization with simultaneous drying or cooling [15]. Agglomerations occur owing to the presence of moisture and stickiness. Therefore, these processes contain integrated classification and deagglomeration steps.

1) Atomization of solutions, emulsions, and solutions/suspensions provides flowable powder in the spray tower procedure. Examples are milk, egg and tomato powder, washing powder and ingredients, dyes, pigments, ceramic compounds, pesticides, fertilizers, biochemical and pharmaceutical intermediates, and plastics.
2) Melts solidify after spraying by an air cooling (prilling) process, followed by dropletization in a cooling channel and deposition of the droplets on water-cooled steel tapes or cooled double rolls. These are solidified as the output of an extruder with underwater pelletizing. Examples are fatty alcohols, fatty acids, polyglycols, waxes, metals, alloys, and thermoplastics.

7.4.6
Splitting with Simultaneous Absorbing/Reaction

During breaking up, starting materials react with suitable gases or liquids and undergo a coloring and scenting or coating with simultaneous removal of moisture.

7.5
Processes for Crushing of Materials

Apart from the central grinding machine, crushing plants comprise some other components, such as storage silos, mixing and feed tanks, metering units, classifications, filters, pumps, fans, and the filling station within the warehouse. Their design depends on the capacity and the required product quality, because circulating streams influence extremely the necessary equipment sizes, and thus ultimately determine the investment and production costs. Wetted parts of plants consist mainly of stainless steel in the chemical, food, and pharmaceutical industry in known qualities (AISI 304-V2A, 304L, 316Ti-V4A, 316, 316L, and 321; see Chapter 16).

Other qualities for heavy usage are constituted by the special expertise of the machine manufacturer. Extremely hard products, such as minerals, require wear-resistant materials for the machinery, which vary with the type of mill. Roller grinding mills contain chrome cast iron or cast composites, while crusher and mills use air- or water-quenched and tempered metals, such as high-wear-resistance steel. These quality requirements can be met by stainless steels with the German material no.1.87xx, 1.8704, 1.8709, 1.8710, and 1.8714, or similar special structural steels with Brinell hardnesses from 300 to 600 HBW (ISO 6506). For special tasks, ball mills are constructed with grinding bowls and balls of tungsten carbide, zirconia, or ceramics. These and other linings in the mills guarantee iron-free products.

For the demanding grinding applications in the biotechnology, food, and pharmaceutical industries, a smoothing of the stainless steel surfaces by pickling, electropolishing, sanding, and polishing is carried out (Chapter 16). By direct installation, an automated cleaning system (Clean in Place: CIP) provides the optimal cleaning. In addition, disinfection and sterilization (saturated steam at 121 °C) are possible. This requires finely polished stainless steel surfaces, dead space-free valves and outlets, and adapted seals and filter materials. The shaft bushings necessitate special seals. The sterilization is automatized (sterilization-in-place: SIP).

For dust-explosive powders, the grinding occurs preferably in pressure-shock-resistant equipment (10 bar) with explosion protection valves (ATEX 94/9/EC). They separate the hazard area in the complex. Processing with explosion protection valves in combination with zero-delay rupture disks is an alternative. In all cases, the chemical engineer arranges extinguishant containers with automatic quick release at various points. In addition, for problematic products, an inertization takes place with monitoring of the oxygen concentration.

7.5.1
Equipment for Grinding of Disperse Dry Raw Materials

Grinding plants consist of a crushing machine, often equipped with an internal sieve or classifier, and a screening plant with lines and receivers for recycling materials. Sieving occurs with single covering or with multistage screens. In the grinding process, solids flow from a silo into the mill, metered on a belt, screw, or vibrating chute. The crushing takes place there, possibly in combination with classifications. An airstream simultaneously transports and cools the milled product. In the grinding and drying, the hot inlet air supplies the heat of evaporation for the wet materials and takes moisture away. Downstream filters or cyclones separate the particles, which fall into the receiver, while the gas flows into the exhaust or into circulation. Some typical machines for grinding of dry solids are indicated in Figure 7.8.

In various grinding processes, the ground material is fed either mechanically or pneumatically by metering into the mill or into the airstream. The same happens with the removal of ground particles that fall directly out of the mill or flow with the conveying air. Table 7.4 demonstrates some examples. The pneumatic transport of milled particles leads to an internal or external classifier. The high airflow ensures an efficient transfer of milling heat. Usually, a screen separates larger particles.

Because of explosion hazard from fine organic particles, separation from the airstream is not always easy. This takes place in cyclones and filters, secured

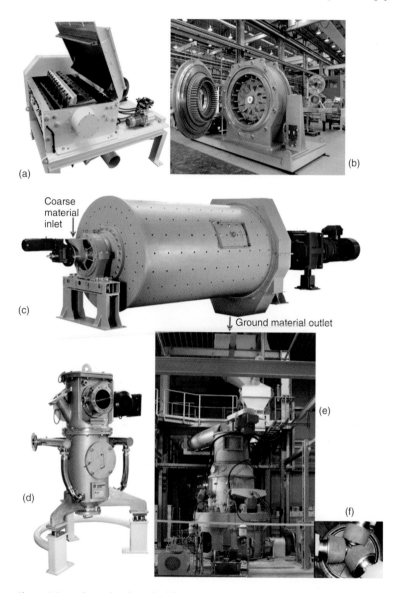

Figure 7.8 Industrial mills with different main crushing principles; (a) cutting mill; (b) fine impact mill; (c) ball mill, (d) turbo twin opposed jet mill; (e) table roller mill; and (f) the rollers inside (Courtesy of Hosokawa Alpine).

by relief valves or by a pressure-shock-resistant design. Using a filter allows to suppress the dust load in the exhaust below a value of $10\,\mathrm{mg\,m^{-3}}$ (usually not possible in a cyclone). The fine product is removed via rotary valves and bagged.

Many manufacturing and recycling processes involve grinding/screening stages, in which a series of tasks are completed. Besides setting the correct particle size

Table 7.4 Supply and removal of materials in crushing processes.

Mill	Supply of starting material	Discharge of product
Impact	Pneumatic conveying	Pneumatic conveying
Ball	Dosing device	Free discharge
Table roller	Dosing device	Pneumatic conveying
Cutting	Mechanical device	Pneumatic conveying

distribution, for pneumatically fed mills, the airflow and temperature adjust the residual moisture content and flowability, as well as influence the smell within limits. In conventional processes, air is at an ambient temperature. Depending on the task, an adjustment of supply air ensues in the range of about 10–350 °C. A fine crushing plant with pneumatic product transport through the mill is shown in Figure 7.9a. This mill type is able to form swirls at the flappers. The starting material is metered either into the airstream or directly into the grinding chamber out of container located above the chamber.

In combination with sieving, roller and rod and ball mills work without air support, which may be advantageous for some products. Roller mills crush dry materials down to about 100 μm and particles in pastes down to about 10 μm. Pneumatic particle transport of fine material out of the grinding zone supports further grinding because the fines hinder the comminution and this leads to a disproportionately high energy input. The separation of fine particles (<100 μm) ensues advantageously with a classifier, which needs the use of pneumatic transport.

Inert processing occurs mostly in nitrogen atmosphere, but also in carbon dioxide or in superheated steam. Thereby, the gas is recirculated. Size reduction of temperature-sensitive and soft or tough materials takes place preferably at low working temperatures. Liquid nitrogen or carbon dioxide effects the cooling and guarantees an inerting at the same time. For the grinding and drying of temperature-stable, water-containing raw materials in an inert atmosphere, superheated steam is advisable as drying gas. The energy of crushing is utilized for evaporation of wet materials. The moisture is removed from the circulation in the form of excess steam (see Chapter 8).

Figure 7.9 demonstrates different examples for fine crushing. The first one is a vertical impact mill with internal classifier for grinding of organic materials dosed into the mill by a screw. The fed air transports the explosive fines through the classifier into the filter. Special valves and rupture disks protect the grinding chamber and filter. The second process uses a ball mill with an external classifier, reachable by conveying air, for separating and recycling of oversize particles. An impact mill, running at lower temperatures with cooled nitrogen in circulation is depicted in the third image. Cooling occurs by liquid nitrogen supplied at different places to control the temperature. The last picture explains the principle of mill drying with superheated steam in circulation. The crushing ensues in a table roller mill.

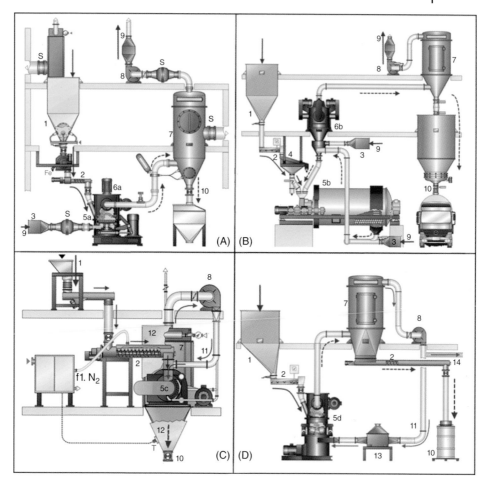

Figure 7.9 Process flow diagrams of grinding plants (A) Ex-proof system with pressure relief, (B) grinding plant with external classifier, (C) cryogenic crushing in nitrogen, and (D) grinding and drying with superheated steam. 1: Container for raw materials; 2: dosed feeding of solids; 3: filter; 4: sieve; 5a: Ex-proof impact mill, 5b: ball mill, 5c: impact mill with the filter behind, 5d: Table roller mill; 6a: integrated classifier, 6b: external classifier; 7: product filter; 8: fan; 9: airflow; 10: container for ground product; 11: circulation of inert gases; 12: insulation; 13: heat exchanger for superheating the steam; 14: discharge of excess steam; and S: explosion protection valves, bursting disks with blowout (Courtesy of Hosokawa Alpine).

7.5.2
Equipment for Grinding of Disperse Raw Materials in Liquid Phases

Some fine particles are only available suspended in a liquid. In these cases, and if the solid particles are below 25 μm after reduction, wet grinding is commonly the preferred process. There, the comminution of suspended particles takes place in water or an organic solvent. Suitable mills for wet grinding are the three-stage rolling mill, the corundum stone mill, and the agitated ball mill. They allow the fine crushing of particles in viscous suspensions (pastes). When using precomminuted, suspended matter in water (d_{90} <100 µm), the nanoscale (<1 µm) is reachable easily by agitated bead mills. On a drive shaft, disks are arranged, or disks with pins and at the opposite side stator pins or displacement body, which forms a ring gap with the wall of the mill. Energy is transferred to the crushed particles moving about the hard grinding media (mostly beads with diameters <2 mm). Fine grinding of precomminuted, suspended particles is shown in Figure 7.10. A grinding and drying process converts the ground material into dry powder. In other cases, the product is a suspension for which the viscosity is adjusted.

Except for ordinary products, the suspension is pumped several times through the stirred ball mill. All grinding beads remain in the machine, usually retained through slots/sieves at the outlet. The longer the grinding takes, the more problematic are the abrasions of the beads and the grinding heat. In an optimized procedure (machine, grinding media, residence time, and liquid), high energy inputs allow

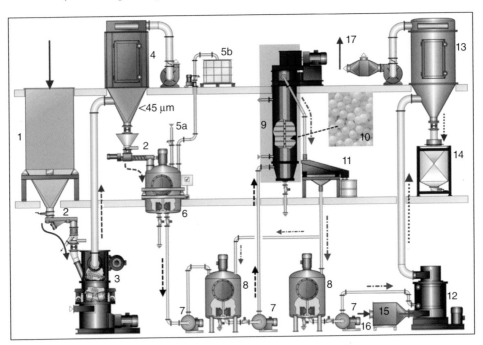

Figure 7.10 Wet grinding of dry precomminuted particles in a stirred ball mill with subsequent mill drying (calcium carbonate); 1: silo for starting material; 2: dosage of particles; 3: bowl roller mill for dry grinding; 4: filter; 5a: process water, 5b: additives; 6: stirred slurry container on load cells; 7: slurry pumps; 8: stirred suspension container; 9: vertical stirred ball (highlighted); 10: grinding beads of yttria-stabilized zirconia, image magnified insert, 11: sieve; 12: impact mill for grinding and drying; 13: filter for fine particles; 14: product container; 15: heat exchanger for heating the drying air; 16: supply air; and 17: exhaust with blower/silencer (Courtesy of Hosokawa Alpine).

particle sizes significantly below 500 nm. Furthermore, the product is normally processed as a suspension. Otherwise, drying should ensue under conditions of deagglomeration in a mill that dries and crushes developing agglomerates.

For preparation of stable suspensions and pastes, solids must fall below a certain size so that the particles do not separate. This is important as storage stability depends particularly on particle size distribution and viscosity.

Paints are an example for small particles in liquids. Extremely fine pigments are invisible as particles in the application. If possible, a direct production of micro- or nanoparticles takes place. Otherwise, particles originate from a grinding in the liquid phase. One form of design is presented in Figure 7.11, which shows the annular gap ball mill. This mill, suitable for extremely high energy inputs of 50–600 kWh m^{-3}, demonstrates optimal heat removal through internal and external surfaces for heat exchange. These machines may be cleaned by CIP. In fine grinding, beads are in the range 0.5–5 mm, and in specific cases and in special machines, down to 50 μm; they are moved by rotating disks or paddles. The beads consist of ceramic materials, mainly zirconia and alumina, or glass. The walls of stirred mills may be lined with polyurethane, rubber, ceramics, and cast iron or manufactured directly in special steels. The complete removal of the beads from paste is a difficult task, for which different solutions exist.

The bead mill (Figure 7.12) with disks or pins represents another form of construction that is preferred for large plants. Depending on the operating conditions, wet grinding takes place here for particle sizes below 10 μm, usually less than 1 μm, and several cases, even in nano ranges. Examples of these are coating materials (<10 μm), gravure ink concentrates (<3 μm), and calcium carbonate (less than 1 μm for food colors), provitamin A in oil, as well as aluminum oxide and silicon carbide,

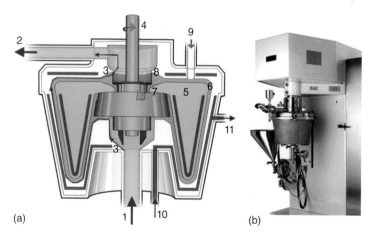

(a)   (b)

Figure 7.11 (a,b) Annular gap bead mill: principle of the mill and a small production plant. 1: suspension inlet; 2: product outlet; 3: suspension way through the mill; 4: drive shaft; 5: rotor; 6: grinding gap with product and beads; 7: bead recirculation channel; 8: bead separation; 9: bead supply; and 10, 11: cooling water in and out (Courtesy of FrymaKoruma).

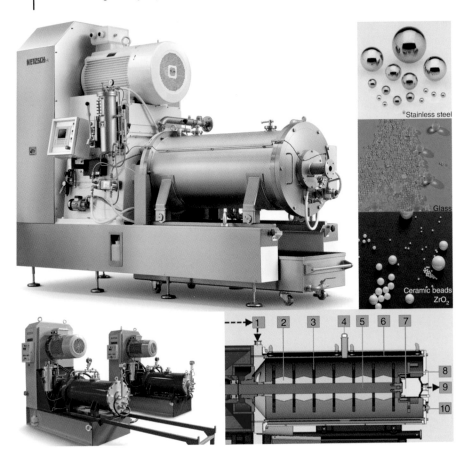

Figure 7.12 Horizontal disk bead mills and typical beads; representation of the functional principle. 1: tangential inlet of suspension; 2: double-conical spacer bushings; 3: grinding disks; 4: centered grinding media fill port; 5: agitator shaft; 6: intensive grinding chamber cooling; 7: separator sieve; 8: separator rotor; 9: product (paste) outlet; and 10: drain plug (Courtesy of Netzsch).

finally titanium dioxide (less than 0.5 µm, i.e., ≤ 500 nm). Multiple runs through the machine or longer milling times are the demanding tasks that are required. Stirred bead mills need an input of about $0.1-1\,\mathrm{kW\,l^{-1}}$ of milling space, or from 0.01 to $0.4\,\mathrm{kWh\,l^{-1}}$ of product, also determined by the material. Usually machines are available with milling chambers up to 1.000 l.

For resources from nature such as minerals and ores, there are specialized bead mills (Figure 7.13) with volumes up to 46 m³ and 8 MW installed electrical power. The largest machine has an unladen weight of 126 t. By using high rotation speeds (tip speeds up to $22\,\mathrm{m\,s^{-1}}$) and small beads, large specific energies are transferred to the ground material. For example, the coarse feed shows sizes of about 250 µm and is milled to a diameter below 10 µm. Typical materials range from lead and zinc sulfides, platinum concentrates, and industrial minerals to iron oxide and

Figure 7.13 Grinding plants in production (a) milling of calcium carbonate in two 500 l mills; and (b) large bead mill for minerals and ores (Courtesy of Netzsch).

refractory gold concentrate. This new technology has been embraced for achieving better results in metal recovery.

Other machines for fine crushing of particles in suspensions and pastes operate according to the rotor/stator principle. Perforated disk mills, toothed colloids, and corundum stones mills (Figure 7.14) proved their suitability in several applications, especially for food and cosmetics. In theory, a fruit or spice shall pass successively through the three mills in the given order (Figure 7.14). In the first machine, the fruit pieces are reduced to the size of a few millimeters. With each subsequent stage, the comminution makes progress by ~ one order of magnitude. The grinding success is based on the material, operating conditions, and especially the adjustable gap. The product then leaves the last stage, the corundum mill, with particle sizes of about 50 μm.

The required particle sizes determine the selected machinery, with possibly suitable alternatives. Customers should use the experience of renowned machine manufacturers. Their crushing machines would have run in different industrial applications for years. Additionally, various substances pass through the mills in the manufacturer's pilot plant. Therefore, in addition to experience with machine design, these manufacturers also have extensive experience with materials and optimal operating conditions. The customer can benefit from such experience. Various applications of crushing in liquids are shown in Table 7.5.

Emulsions represent an interesting case. An emulsion consists of an insoluble liquid in the form of droplets in a continuous phase. The small droplet size is achieved via high energy dissipation. By applying appropriate surfactants in high pressure homogenizers, the nanometer size range is achievable (see Chapter 13). The droplets can be formed at elevated temperatures from a melt. Fine solid particles arise from the melt emulsion processes [16] after they cool down. For preparation of narrow distributions, multiple passes through the homogenizer with pressure in the range above 500 bar are necessary. Additives in the continuous phase (such as polyvinyl alcohol) allow a coating of the nanoparticles formed, which

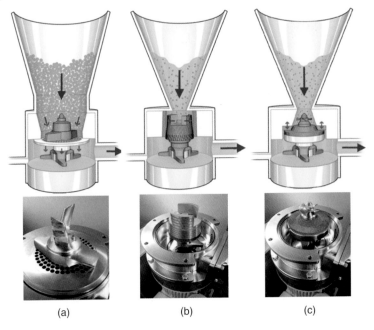

(a)   (b)   (c)

Figure 7.14 Crushing of liquid-containing solids and solids in liquids; schematic illustrations of principles and grinding tools (a) Perforated disk mill for liquid-containing solids (vegetables, fruits, fruits and juices, and animal feed); (b) Toothed colloid mill (juices, soy milk, nuts, paints, and varnishes); and (c) Corundum stone mill (mustard, chocolate, nut pastes, animal feed, makeup, and creams) (Courtesy of FrymaKoruma).

are separated centrifugally. Water-insoluble substances, whose melting points are at atmospheric pressure between 30 and 90 °C, are suitable for comminution in liquid phase before solidifying by cooling. In other cases, at temperatures up to 250 °C, the whole process takes place under pressure.

7.5.3
Breaking Up of Materials in Combination with Drying Methods

Chapter 8 describes the process of spray-drying in detail. The discussion here compares spray drying with the mill drying, where mill drying also may be an alternative to spray-drying for some products. This implies that the suspension is present in high concentration. The production of free-flowing powders from solutions or suspensions takes place worldwide in spray towers [15, 17, 18], where a partial agglomeration to larger particles occurs. Usually, these coarse components of the powder are separated on a sieve. This portion is deagglomerated in a mill and then recycled on the sieve or in the spraying zone, or alternatively transferred without milling into the suspension tank for dissolving/dispersing (Figure 7.15). Particle sizes and distributions can be controlled using various parameters. When

Table 7.5 Liquid (multiphase) products out of comminution devices.

Industry	Products
Pharmaceuticals	Eye drops
	Geles
	Ointments
Cosmetic	Creams
	Mascra
	Lotions
	Make-up
	Hair care
	Toothpastes
Food	Sauces
	Juices
	Pastes
	Jam
	Marzipan
	Mayonnaise
Chemistry	Agrochemicals
	Paints and varnishes
	Silicone emulsions
	Filler and sealer
	Adhesives
	Furniture polish
	Paper coatings

Courtesy of FrymaKoruma.

fine powders are desired, the procedure must prevent agglomeration in the manner in which this is done during grinding and drying.

Mill drying, such as that performed in impact mills, long gap or air turbulence mills, ball mills, and bowl roller mills, allows the drying of particles with simultaneous comminution. The process is suitable for moist solids, and also for pastes and suspensions with high solid contents. Usually, single-fluid nozzles spray the liquids on the particles. Spraying occurs either in front of the mill or, in most cases, directly in the grinding chamber (Figure 7.16). The quantity of water evaporated is limited by maximal heat input of the air and the additional crushing heat. Dry micropowder arises out of a mill drying, which is supplied wet ground material from a stirred bead mill.

While mill drying allows the production of fine powders from pastes, spray agglomeration in fluidized bed needs suspensions with lower solid content for forming large, spherical particles. The optimal process, selected from spray drying, spray agglomeration, and mill drying, depends on the concentration of solids in the suspension and the particle size distribution required. In some cases, two procedures are suitable and there is an overlapping area (Figure 7.17).

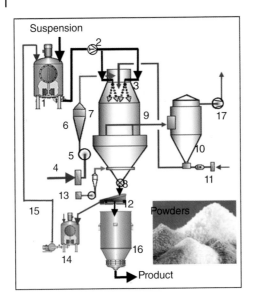

Figure 7.15 Plant for drying of suspensions in spray tower with recycling of coarse and fine fractions; 1: container for suspension; 2: high pressure pump; 3: holder with pressure nozzles; 4: supply air; 5: supply air fan; 6: heat exchanger for heating the air; 7: cocurrent tower; 8: powder discharge; 9: dusty, moist air; 10: filter; 11: airlift for filter dust transport into the spraying zone; 12: sieve; 13: additional air for drying; 14: container for dissolving the screened coarse fraction; 15: line for recirculated solution/suspension; 16: silo for powder; and 17: exhaust fan (Tower and powder images: courtesy of GEA/Niro).

7.6
Energy Requirements

Schönert [19] describes the mechanisms of breakage and micro processes of crushing in solids. Stated simply, comminution takes place by the action of forces on the particles, whereby stresses occur. Stress spikes lead to cracks at existing structural defects in real solids. The creation of new surfaces begins at crystal defects. Two parameters dominate the required energy, namely, the properties of the material and the mean diameter of the crushed material. Besides the degree of comminution (d_{before}/d_{after}), the machine and operating parameters influence the energy input. In extreme cases, as for mica (layered silicate), crushing needs more than 40 times the specific energy compared to that needed by glass, and about 10 times in relation to many common substances. The measured Bond values [20] document this dependence of material. In addition, the smaller the particles are, the fewer defects they show, and the more difficult it is to carry out further grinding. In general, grinding of dry solids succeeds only up to the single-digit micron range. In suspension, the particles achieve sizes that are up to one magnitude lower.

Because of the extreme material dependency, the choice of mill [21] and scale-up take place in pilot-scale trials under any probable operating conditions. The trial and production mills should be geometrically similar. The measured variables are

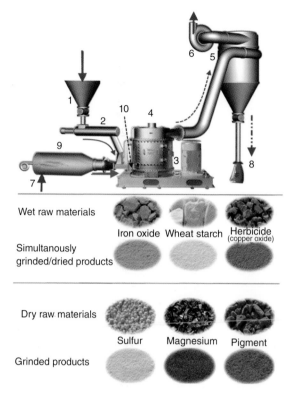

Wet raw materials

Iron oxide Wheat starch Herbicide (copper oxide)

Simultanously grinded/dried products

Dry raw materials

Sulfur Magnesium Pigment

Grinded products

Figure 7.16 Plant for grinding, mill drying of moist solids and of pastes; 1: container for (moist) raw materials; 2: dosing screw for feeding the solids; 3: air turbulence mill; 4: integrated classifier; 5: product filter; 6: blower, 7: airflow; 8: container for the ground product; 9: gas burner for heating the drying air; and 10: spraying a suspension (if required) (Courtesy of Jäckering).

the particle size distributions as function of operating conditions, in particular, the energy input, solids throughput as well as the gas stream and temperature. The final product quality, required safety measures, and necessary materials form the basis for selection of the process. In some cases, the product may not contain heavy metals or the material to be crushed is very hard and produces severe abrasion.

For fine crushing, the average particle diameter is a function of specific energy. An approach similar to that of Bond and Wang [described in 22] describes the dependency and allows a rough estimation. Each two parallel lines cover large areas, which include many measurements with different solids and liquids (Figure 7.18). They are meant to be an orientation. The figure shows the energy input per unit volume for splitting of liquid phases during the formation of emulsions in comparison to that in solid crushing. This actually inadmissible comparison shall show that the absolute slope of the straight line in the double-log plot depends on the task of crushing: the starting point is the breaking up of atomized liquid

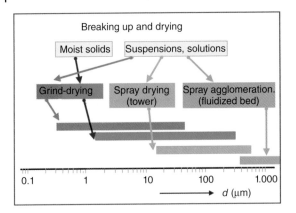

Figure 7.17 Breaking up under drying conditions (suspension may be precomminuted in bead mills).

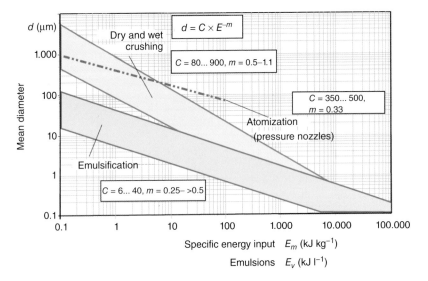

Figure 7.18 Ranges of the average diameter d_{50} as a function of specific energy input; for crushed solids and atomized suspensions (detergent slurries) in (kJ/kg); for emulsions in (kJ/l) with d_{Sauter}.

(continuum; $m = 0.33$), followed by the formation of coalescence-free emulsions (about 0.25–0.4 [23]), and then the coalescing of emulsions into solids (up to about 1). For the solids, the measured values in Figure 7.18 show sizes from few millimeters [24] up to the nanoscale. These small sizes are achievable in technical systems with bead mills [25–28]. For emulsions [29], the mean droplet sizes are in the upper range at <100 µm and at lower end about 100 nm. Owing to differences in density and phase proportions as well as methods for determining the diameter, an exact comparison of energy inputs, of course, not possible.

Schuchmann describes [30] droplet breakup and the concept of energy density concept in mechanical emulsifications. In addition to specific energy input and residence time as well as tools in the machine, for emulsification, it is necessary to decide particularly on the choice and concentration of appropriate surfactants, the stability, and the sizes of droplets and their distribution [31]. Optimized surfactants should minimize the coalescence. The establishment of precise boundary conditions and additional experience during the process of emulsification is required for the measurement of the necessary power input required. With other emulsifiers, and a high pressure [32] or similar dispersing machines, the drops can divide down to the nano range (up to mean diameters of about 100 nm).

The dashed line in Figure 7.18 is for orientation. It originates from atomization of detergent suspensions with single-fluid nozzles. The use of two-fluid nozzles would move the estimated course by about one order of magnitude down and to the right. The equation obtained (P. Walzel, private communication, Technische Universität Dortmund, 2010) is based on observations taken from practice.

7.7
Determination of Product Design via Specifications

The characteristics of the desired product design should be determined before starting development work. For doing so, specific values arising from physico-chemical and application measurements during development should be included as elements of the specification list. Extensive experience in applications in the laboratory helps by establishing allowable limits for the values. The specification lists describe many chemical and disperse requirements. This means that the disperse properties of the particles, as discussed by Rumpf [33], cover only part of the specifications. Usually, the chemistry represents an essential part of the application. The following example from the industry demonstrates this.

The addition of cellulose ether to mortar prevents too rapid a drying of the plaster on fresh masonry walls. The mortar is sprayed on the walls by special machines. Even small quantities cause significant improvements of the adhesion strength of plasters. These properties of cellulose ethers can be proved by measuring the water retention (Figure 7.19), preferably as a function of temperature.

Determination of water retention is carried out under defined conditions. For measurement in the laboratory, a fixed amount of water is added to freshly prepared mortar of standard consistency. Then the mortar is poured into a filter, equipped with weighted filter paper. Under vacuum, a part of the water leaves the mortar and flows into the filter paper. The water retention capacity (WRC) of mortar represents the amount of water held in the mortar in relation to the initial water content (German cement industry, instructions EN 413–2).

Systematic studies show that two parameters increase the WRC significantly, namely, the size distribution of the particles [34] and the etherifying agents; furthermore, the realized degrees of substitution also increase. The degree of

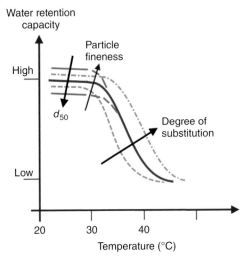

Figure 7.19 Schematic course of the water retention capacity (WRC) of cellulose ethers in mortar as a function of temperature; high WRC: fine grinding and/or higher degrees of substitution.

substitution [35] specifies how many OH groups of an anhydroglucose unit (max. 3, [10]) on average are etherified. Optimizations of the effect in mortars require consideration of both effects, that is, the chemical, but also disperse properties of the cellulose ethers, especially in the lower temperature range.

7.8
Design of Products by Breaking-Up Processes

The breaking up of disperse and continuous phases have the following common features:

1) The particles produced show distributed diameters, and thus differ significantly in particle volumes.
2) Same size particles are usually desirable.
3) Narrow diameter distributions require additional expenditure on equipment and energy.
4) The smaller the desired diameters, the more energy is required.
5) Particles tend to grow during crushing, because fine solids stick together or agglomerate with bigger ones; droplets or gas bubbles coalesce.
6) Particle shape, flowability, and bulk densities are controllable in some cases during the processing.
7) Coatings allow a precise adjustment of the surface properties of ground material.

7.8.1
Determination of Particle Properties for Solids

The target of comminution of solids is primarily an adjustment of the required particle sizes. In some cases, a suitable grinder solves this problem. Much more difficult is the setting of a defined particle size distribution, which represents an important quality criterion. Normally, particles should not exceed a fixed, maximum size. In the case of larger particles (>500 µm), the separation of the coarse fraction takes place through a sieve. Finer particles (<200 µm) need an air classifier. Many mills are equipped with one of these separation units. Removal of undersize particles with an exact separation size occurs through an effective sifter. In modern classifiers, fine particles in the range of <20 µm are separated in an optimal manner with low axial flow rates, using multiple, high speed rotating, small classifier wheels (Figure 7.20). To realize high rotating speeds, the wheels are

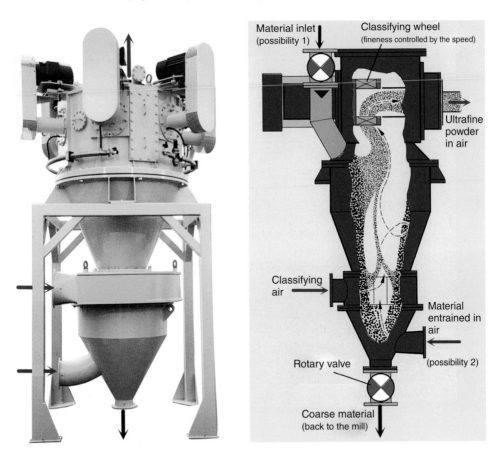

Figure 7.20 External wheel classifier for ultrafine classifying operations (up to $d_{50} = 0,3$ µm) (a) large four-wheel classifier; and (b) one-wheel classifier, showing the operating principle (Courtesy of Hosokawa Alpine).

not only small but also horizontally arranged and equipped with their own drives. In less demanding situations, it is enough to blow the dust with the conveying air into the filter, as it occurs frequently at a spray tower. Dusts interfere in application as well as in processing and reduce the product quality.

Owing to high adhesive forces, fine particles (<50 μm) tend to form clumps. Therefore, a uniform removal from a silo fails and the dosage is difficult because the material "shoots." In these cases, an added special product, which improves the grinding as well as the flowability and storage properties, represents aspects of product design. As grinding aids [36], many substances are suitable in amounts of 0.5–2%. The dry fine excipients help in powdering/moisture absorption. Examples include lactose, and fumed silica (Aerosil®); further, ethylene glycols, amines, and fatty acids and their salts (e.g., stearic acid and magnesium stearate), alcohols, and ligninsulfonates. Extensive studies [37–39] demonstrate the use of excipients in wet grinding.

In consumer goods, bulk densities play a major role. By technical measures, the values of powders are adjustable within limits. Bulk densities rise after compaction of materials. Double rollers are suitable for this, as also ball or rod mills. The process runs with an increase in bulk density in two stages: the introduction of pressure forces are effective, either for compression or for pregrinding between rollers, followed by fine grinding. In the fine-grinding step, product moisture gets adjusted owing to the humidity of air, temperature, and throughput. The achievable bulk density depends strongly on the water content of material. Injection of water is more effective. Then dissolution at the particle surface starts, and under action of high shear forces (special mixer), the voidage reduces. After the necessary drying, this method brings the greatest gain in bulk density.

With appropriate effort it is possible to meet the requirements of disperse properties. Stable incorporation of the fine particles in a matrix may require the application of a layer on the particle surface (coating). Activity losses (percarbonate) or degradation of polymers (cellulose ether), controllable via the process, temperatures, and residence times occur in crushing processes. Undesirable side reactions may be restricted by inertizing, lower temperatures, and shorter residence times. The heavy metal content before and after grinding (machine material; grinding aids) should always be measured, and where appropriate, the mill should be lined. Next to mechanical abrasion, chemical reactions may change the surfaces.

7.8.2
Solids from Melting

The breakup of liquids takes place under targeted conditions. The solids present in solution and suspension or paste and emulsion are fine as in milk or coarse as in slurries. Their drying in spray towers is controllable by several parameters (see Chapter 8).

In contrast, molten substances require appropriate technologies at higher temperatures. The processing of melt is not very easy as it depends on temperatures and material properties. The same technology suffices in the production of particles from melts in all temperature ranges. Substances such as paraffin (alkanes) and fatty acids as well as fatty alcohols characterize the lower region (melting points from 24 °C to about 100 °C). Plastics, waxes, sulfur as well as lead and tin fall into the lower middle range (100 °C $<T_F<$ 400 °C). Mineral and metallic melts, such as glasses, slags, and ceramics as well as iron, silicon, or copper constitute the uppermost region ($T_F<$ 2000 °C).

Breaking up of nonmetal melting (minerals) preferably takes place in two-fluid nozzles with gases or with water (also oil) as liquid propellant, in order to ensure rapid cooling [40]. In this process, irregular, sometimes sharp-edged powders are formed. To avoid oxidation, in the case of molten metals, the propellant consists of an inert gas (mostly nitrogen and sometimes argon) for the atomization in two-fluid nozzles. These droplets cool down much more slowly, so that the capacity of plant drops by almost an order of magnitude.

Gas-atomized molten metals (high surface tension and low viscosity) form beads, mainly with average diameters (d_{50}) of about 100 μm. By changing the operating conditions, the size can be set from about 50 μm to 400 μm. For blast furnace slag and other mineral melts, a sandy structure (foundry sand) or fibrous shape results from the gas atomization, depending on the operating conditions and nozzle geometry. In these ways, the product design is controllable within wide ranges.

7.9
Product Design Out of Multistep Processes (Examples)

Some complex examples demonstrate the influence of breakup on the product design, where, in addition to fragmentation, further process steps occur.

7.9.1
Powders from Molten Metals

Metals can be melted and atomized in various ways [40, 41]. The atomization of a melt sometimes takes place on a rotating disk or turbine. Here, the liquid may arise by melting metal wires by means of a laser. The process operates in vacuum and is characterized by high quality of the product (no oxidation). A significantly higher performance is achieved in towers with twin-fluid nozzles, especially with water as propellant. For reasons of quality, the use of an inert gas, mostly nitrogen, is preferred. This gas prevents oxidation as well as explosions from occurring. The nozzles, constructed analogous to the twin-fluid nozzles with outside mixing, are described in Section 8.2.1. Melting occurs in the upper part of the tower under vacuum or in the presence of an inert gas, less frequently in the presence of air.

In plasma atomization, a metal wire is melted in plasma flames and atomized with the gas. This relatively new technique produces high quality and perfectly spherical particles. The particle sizes are very small (<100 μm and up to 200 μm). The procedure is limited to alloys that can be formed into wire feedstock. Another modern method of the metal atomization is represented by the EIGA method (electrode induction melting gas atomization). In this process, the melt is produced from a rotating metal bar with heat supply from the noncontact induction coil. The molten film flows downward into the gas stream of the nozzle for atomization. Here, the combination of ceramic-free melting (all alloys, titanium, zirconium, aluminum, copper, gold, nickel, platinum, and vanadium) with spraying under protective gas, preferably argon, occurs in an economic way. The particles solidify in the spray tower, resulting in pure, fine metal powders. This spray process allows the setting of an average particle size ranging from 20–60 µm, in some cases up to 500 µm. The morphology is similar to gas-atomized particles. Figure 7.21

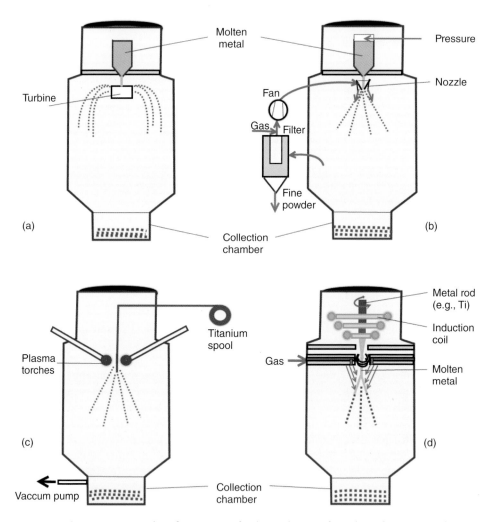

Figure 7.21 Principles of atomization for the production of metal powders (a) centrifugal atomization; (b) atomization in twin-fluid nozzles with inert gases or water; (c) plasma atomization; and (d) electrode induction melting gas atomization.

depicts the principles of metal atomization. The metal powder can be pressed into complex work pieces before sintering. This method offers an economic production of difficult machine parts made up of various metals and alloys.

7.9.2
Powdered Metals for Metallic Paints

Metallic paints have a special glow and typical reflection. These effects are brought about by incorporating metallic pigments (platelet) in paint and varnish layers. Chemical additives provide orientation and distribution of the platelets for gloss and glitter effects (brilliance and brightness) and affect the color and opacity. Pigments (aluminum, brass, and copper; see Figure 7.22) generate extreme silver and gold shine in many variants. In addition, optimal use of pigments generates a fine appearance of the surfaces. In particular, the paints for cars, but also inks, coatings, and laminations of plastic films benefit from these metallic effects.

Metal platelets for automotive coatings require not only a specific geometry (size, thickness, shape, and surface roughness) but also an orientation parallel to the surface and a uniform distribution in the paint. Many demands on the product design, described in a specification list, can be met through an optimized combination of chemistry and chemical engineering. The production begins with a pure melt, for example, an aluminum melt ($T > 660\,°C$, at least 99.5% Al). The atomization takes place in a spray tower under nitrogen [40]. The resulting particles, known as *metal grit*, show sizes of about 100 μm. After screening of the oversize particles, this powder represents a starting material for the grinding process that follows.

Because fine aluminum particles react with the moisture in air with the formation of the hydroxide and hydrogen, all further comminutions ensue in a solvent (white spirit). Simultaneous crushing and flattening of the suspended grit is carried out in a ball mill. Depending on the equipment and operating conditions of the mill, a surface structure is formed that resembles "cornflakes" or a superior quality "silver dollar" with smooth surfaces. Particle sizes can be set exactly by targeted grinding. On the basis of the average diameter d_{50}, the different batches show platelet-sizes in the 5 μm range. These mainly lie in the range 4–35 μm, and in some grades up to about 100 μm or more (up to 350 μm in aluminum plates from foils). Platelets

(a)   (b)   (c)

Figure 7.22 Metal-effect pigments (a) aluminum pigment powder, (b) copper pigment, and (c) gold bronze pigment.

have a desired l/d ratio of 50 : 1 up to 100 : 1. Accordingly, they are <0.1 to 1 µm thick. The more finely ground the flakes, the higher is the opacity (in cm^2 g^{-1}).

After grinding, the metal platelets are separated from gasoline in a filter presses. The described method, working in a solvent, is known as the "*Hall*" process. Lately, it has been possible to crush and to shape the Al-grit, just like the brass, in dry nitrogen atmosphere, wherein the ball mill operates with a well-adjusted classifier in the circuit.

A cold welding of particles during grinding prevents separating agents. Stearic acid for manufacture of floating "leafing" flakes and oleic acid for "non-leafing" pigments are in use, and are distributed uniformly in an organic medium. Non-leafing products are suitable for car paints, whereas leafing pastes show advantages in corrosion protection (Zn). The classified products are available as powders or pastes with metal content of 60–87% in various petroleum solvents.

In contrast to the reactive aluminum, size reduction of gold-colored pigments (see Table 7.6) takes place more easily by dry grinding in a ball mill. These pigments are made of brass (copper-zinc alloy) with different Cu contents. In one or more passes through the mill, sprayed grit is ground to the desired fineness (the "Hametag" process). The size of the grinding media (balls) allows adapting to the quality required. During this grinding, the balls hit the grits to form the desired platelets. In these processes, the focus is not just on the degree of grinding. A further essential task is the creation of almost round and smooth plates, which cause an increase in brilliance of the paint. This is an impressive example for the importance of a grinding process in product design.

7.9.3
Reinforcing Materials and Fillers for Polymers

Inorganic compounds allow a filling and reinforcing of polymers. For example, calcium carbonate, silica, talc, and kaolin (Mg- and Al-silicates) are suitable fillers. Uniform and durable incorporation requires the aid of adhesion promoters. For

Table 7.6 Metallic pigments from ball mills.

Pigment	Color	Manufacturer	Platelet diameter d_{50}
Aluminum	Silver	Wet grinding in mineral spirits	7–150 µm, depending on the quality
Brass (copper-zinc alloy)	Rich gold to copper gold, depending on the alloy (Zn 10–30%)	Dry milling	3–42 µm, depending on quality
Copper	Copper colored	Dry milling	3–40 µm, depending on quality
Zinc	Silver and anticorrosion	Wet milling	10–30 µm

synthetic silica and talc and kaolin, some silane compounds are used as promoters. According to mechanistic ideas, just formed OH-groups of the silicon atom react with the inorganic compound. A reactive group, situated on the other side of the molecule, connects with the polymeric matrix when incorporated in the melt [42].

Naturally mined calcium carbonate is widely used as filler. This is because stearic acid helps calcium carbonate as adhesion promoter, wherein the acid group probably reacts with the inorganic material at elevated temperatures. The lipophilic group causes solubility in the polymer melt. The coating of powder with the adhesion promoter takes place in a mill during grinding, to prevent the otherwise inevitable agglomerations. In the first step of the process, stearic acid ($T_F = 69\,°C$) is melted in a container by supplying heat. The atomization of melt then occurs in a spray tower with the use of twin-fluid nozzles. The resulting fine particles of stearic acid are mixed in amounts of 0.5–2% with the calcium carbonate. Subsequently, the preheated mixture flows into an impact mill. There, crushing heat is used to melt the stearic acid, which then covers the particles uniformly. The resulting product is widely used as filler that is needed mainly in polyvinyl chloride (PVC) and polyethylene products. The coated particles of calcium carbonate show a d_{50} of about 5–8 µm (precipitated: <2 µm). On coating of fillers with adhesion promoters, the product properties change and it is converted into a reactive form. These interesting examples of product design could be potentially applied to other materials.

7.10
Consequences

The basic operations for setting the product design are the breaking up of disperse and continuous phases, the agglomeration, crystallization, and coating. Many raw materials and products require crushing procedures and demonstrate thereafter particle diameter in the range from 0.25 µm up to a few millimeters. The breaking up always takes place in combination with classification and powder separation. In some cases, further crushing and drying, coating, agglomeration, or cooling may be integrated in the process, which affects the product design significantly.

The characteristic feature of breaking up is to create new surfaces. Depending on the material, energy demand increases with decreasing particle diameters and requires longer residence times. The product requirements, fixed in a specification list, define suitable, often multistep processes. The right choice of additives (chemistry) and interconnected steps of process (technology) determine the product design. This refers to particle sizes, their distributions (narrow, free of dust, and oversize), solubility, and reactivity, and in some cases, to particle shape and color, surface composition, and rate of dissolution. In product application, the chemistry is responsible for performance, while the technology sets the dispersed parameters, influences the product performance to some extent, and shows responsibility for the product handling (convenience) and aesthetics.

7.11
Learnings

√ Crushing constitutes the basis for extraction of minerals, ores, and coals as well as for processing of all solids and for generating fine solid particles.

√ All breaking-up processes start from larger particles or continuous phases (rocks and liquids).

√ In single or multistage crushing in combination with sieving and sifting, particles of the desired size and distribution are obtained.

√ Simultaneously, further controlled operations take place, such as drying, cooling, mixing, reaction, and coating.

√ In addition, specially designed procedures allow the adjustment of particle shape, in particular for molten feedstock, which means inorganic and organic compounds, and also metals and thermoplastics.

√ Organic products and metals as starting materials need an inert atmosphere during crushing. In most cases, the process takes place in nitrogen, carbon dioxide, argon, and superheated steam in circulation mode.

√ Wet milling processes allow particle sizes below $10\,\mu m$, especially down to the nano range.

√ Grinding and drying is optimal to produce finest particles form wet ground pastes.

√ Product design (size distribution, solubility, bulk density, and flowability) can be controlled within limits.

√ Energy expenditure increases with decreasing diameter. For the grinding of solids and emulsification of immiscible liquids, the energy consumption is comparable in the range of $0.8–3\,\mu m$.

√ By using about 80–95% of crushed components, the granulations require less time and allow the manufacture of spherical particles.

References

1. Teipel, U. (2003) Partikeltechnologie: gestaltung partikulärer Produkte und disperser Systeme. *Chem. Ing. Tech.*, **75** (6), 679–684.
2. Heegn, H.-P. (2001) Mühlen als Mechanoreaktoren. *Chem. Ing. Tech.*, **73** (12), 1529–1539.
3. Teunou, E. and Poncelet, D. (2005) Dry coating, in *Encapsulated and Powdered Foods* (ed C. Onwulata), Chapter 7, CRC Press, p. 180 ff.
4. Kawatra, S.K. (ed) (2006) *Advances in Comminution*, Society of Mining, Metallurgy and Exploration, Littleton, part 1, p. 1–128.
5. Gupta, A. and Yan, D.S. (2006) *Mineral Processing Design and Operation: An Introduction*, Elsevier, Amsterdam.
6. Rähse, W. (2010) Produktdesign von Kunststoffen für die Waschmittelindustrie. *Chem. Ing. Tech.*, **82** (12), 2073–2088.
7. Pietsch, W. (2005) *Agglomeration in Industry*, Vols. **1 and 2**, Wiley-VCH Verlag GmbH, Weinheim.
8. Salman, A.D., Ghadiri, M., and Hounslow, M. (eds) (2007) *Particle Breakage, Handbook of Powder Technology*, vol. **12**, Elsevier, Amsterdam.

9. *Handbuch mechanische Verfahrenstechnik* (2006) 2nd edn., published by Hosokawa Alpine.

10. Rähse, W. and Dicoi, O. (2010) Produktdesign disperser Rohstoffe durch chemische Reaktionen. *Chem. Ing. Tech.,* **82** (10), 1655–1670.

11. Gerhardt, U. (1994) *Gewürze in der Lebensmittelindustrie,* 2nd, Chapter 7.6 edn, Behr's Verlag, Hamburg, pp. 216–226.

12. Jacob, S., Kasthurirengan, S., Karunanithi, R., and Behera, U. (2000) in *Advances in Cryogenic Engineering* (eds Q.-S. Shu, P. Kittel, D. Glaister, J. Hull, *et al.*), Vol. **45** (Parts A and B), Kluwer Academic, New York, p. 1731 ff.

13. Müller, B. (2009) *Additive kompakt (Farbe und Lacke Edition),* Vincentz Network, Hannover, p. 32 ff.

14. Rähse, W. (2007) *Produktdesign in der chemischen Industrie,* Springer-Verlag, Berlin.

15. Rähse, W. and Dicoi, O. (2009) Produktdesign disperser Stoffe: Industrielle Sprühtrocknung. *Chem. Ing. Tech.,* **81** (6), 699–716.

16. Stang, M. and Wolf, H. (2005) in *Emulgiertechnik* (ed H. Schubert), Chapter 18, B. Behrs Verlag, Hamburg, pp. 547–555.

17. Walzel, P. (1990) Zerstäuben von Flüssigkeiten. *Chem. Ing. Tech.,* **62** (12), 983–994.

18. Walzel, P. (2009) *Ullmann's Encyclopedia of Industrial Chemistry, Reprint,* 7th edn, Wiley-VCH Verlag GmbH, Weinheim, pp. 1–20.

19. Schönert, K. (2003) in *Handbuch der mechanischen Verfahrenstechnik,* vol. 1 (ed H. Schubert), Wiley-VCH Verlag GmbH, Weinheim, pp. 183–213.

20. Zogg, M. (1993) *Einführung in die mechanische Verfahrenstechnik,* Teubner-Verlag, Stuttgart, p. 56.

21. Redeker, D. (2003) Zerkleinerung – Eine uralte Technologie und immer noch eine Herausforderung. *Chem. Ing. Tech.,* **75** (10), 1438–1442.

22. Vauck, W.R.A. and Müller, H.A. (1966) *Grundoperationen Chemischer Verfahrenstechnik,* Verlag Theodor Steinkopff, Dresden und Leipzig, p. 197.

23. Karbstein, H. (1994) Untersuchungen zum Herstellen und Stabilisieren von Öl-in-Wasser-Emulsionen. Dissertation. Universität Karlsruhe.

24. Furchner, B. (2009) Fine grinding with impact mills. *Chem. Eng.,* **116** (8), 26–33.

25. Breitung-Faes, S. and Kwade, A. (2007) Einsatz unterschiedlicher Rührwerkskugelmühlen für die Erzeugung von Nanopartikeln. *Chem. Ing. Tech.,* **79** (3), 241–248.

26. Mende, S., Stenger, F., Peukert, W., and Schwedes, J. (2002) Mechanische Erzeugung und Stabilisierung von Nanopartikeln in Rührwerkskugelmühlen. *Chem. Ing. Tech.,* **74** (7), 994–1000.

27. Stenger, F. and Peukert, W. (2001) The role of particle- particle interactions in submicron grinding in stirred ball mills. *Miner. Proc.,* **41** (10), 477–486.

28. Lietzow, R. (2006) *Herstellung von Nanosuspensionen mittels Entspannung überkritischer Fluide (RESSAS),* Cuvillier-Verlag, Göttingen.

29. Schuchmann, H.P. (2007) in *Product Design and Engineering,* vol. **1**, Chapter 4 (eds U. Bröckel, W. Meier, and G. Wagner), Wiley-VCH Verlag GmbH, Weinheim, pp. 63–93.

30. Schuchmann, H.-P. (2005) in *Emulgiertechnik* (ed H. Schubert), Chapter 7, B. Behrs Verlag, Hamburg, pp. 171–205.

31. Rähse, W. and Dicoi, O. (2009) Produktdesign disperser Stoffe: Emulsionen für die kosmetische Industrie. *Chem. Ing. Tech.,* **81** (9), 1369–1383.

32. Kempa, L., Schuchmann, H.P., and Schubert, H. (2006) Tropfenzerkleinerung und Tropfenkoaleszenz beim mechanischen Emulgieren mit Hochdruckhomogenisatoren. *Chem. Ing. Tech.,* **78** (6), 765–768.

33. Rumpf, H. (1967) Über die Eigenschaften von Nutzstäuben. *Staub-Reinhaltung der Luft,* **27** (1), 3–13.

34. Baumann, R., Scharlemann, S., and Neubauer, J. (2010) Regelung der Anwendungseigenschaften von Zementputzen durch Celluloseether. *ZKG Int.,* **63** (4), 68–75.

35. Kessel, H. (1985) Bestimmung der funktionellen Gruppe und des durchschnittlichen Substitutionsgrades von Carboxymethylstärke. *Starch- Stärke*, **37** (10), 334–336.

36. Schubert, H. (1988) Zum Einfluss des Mediums und von Zusatzstoffen auf Mahlprozesse. *Aufbereitungstechnik*, **29** (3), 115–120.

37. Breitung-Faes, S. and Kwade, A. (2009) Produktgestaltung bei der Nanozerkleinerung durch Einsatz kleinster Mahlkörper. *Chem. Ing. Tech.*, **81** (6), 767–774.

38. Reinsch, E. (2003) Der Einfluss von Zusatzstoffen auf die nasse Feinstbreakmahlung in Rührwerkskugelmühlen. Dissertation. TU Bergakademie Freiberg.

39. Klimpel, R.R. (1999) The selection of wet grinding chemical additives based on slurry rheology. *Powder Technol.*, **105**, 430–435.

40. Schönert, K. and Bauckhage, K. (2003) in *Handbuch der mechanischen Verfahrenstechnik*, vol. **1**, Chapter 5 (ed H. Schubert), Wiley-VCH Verlag GmbH, Weinheim, pp. 299–431.

41. Capus, J. (2005) *Metal Powders: A Global Survey of Production: Applications and Markets 2001–2010*, Elsevier Advanced Technology.

42. Domininghaus, H., Eyerer, P., Elsner, P., and Hirth, T. (eds) (2005) *Die Kunststoffe und ihre Eigenschaften*, 6th edn, Springer-Verlag, Berlin, p. 162.

8
Product Design Out of Continuous Phases by Spray Drying and Crystallization

Summary

Drying in spray towers is the standard method for manufacturing of powders from liquids (solutions, suspensions, and emulsions). This seemingly simple process offers many parameters to optimize. Thus, the industry discussed regularly such established procedures, to improve the capacity, product quality and plant safety, furthermore to reduce energy consumption and exhaust pollution. Capacity increases are achievable by use of much more nozzles with correspondingly smaller boreholes. Then, the possible increased amount of sprayed liquid per cubic meter gas/h reduces the specific energy consumption. Material-dependent structure formation during the drying process determines the particle design (shape, bulk density, and flowability). Depending on material properties and solids content, dried particles are greater or smaller, in comparison to sprayed droplets. Compared to droplet size of the spray, particles at the cone of tower exhibit significantly higher values for the particle diameter, due to agglomerations. For safe operation of a spray drying tower, ignition sources must be avoided. Many organic-based formulations require a drying in nitrogen. An interesting alternative is the inert drying with superheated steam, because no exhaust gas escaped and energy can be recovered substantially. Usual scale-up from laboratory to pilot plant and to production is not possible. Starting with the selection of a nozzle, the way of scale-up will be described in detail.

8.1
Importance of Spray Drying

Drying of atomized liquids with hot air is an old method. Since about 1850, milk is dried in the technical scale [1]. In 2006, estimated more than 25 000 towers were worldwide in operation for the targeted production of flowable powders, especially for dairy products [1]. Hundreds of products emerge by the spray-drying process: not only almost all washing powders and many ingredients, but also food products (milk, egg, and tomato powder), beverages (instant coffee and tea), chemicals and macromolecules, ceramic materials, fine chemicals, dyes and pigments, pesticides,

Industrial Product Design of Solids and Liquids: A Practical Guide, First Edition. Wilfried Rähse.
© 2014 Wiley-VCH Verlag GmbH & Co. KGaA. Published 2014 by Wiley-VCH Verlag GmbH & Co. KGaA.

fertilizers as well as biochemical and pharmaceutical products and intermediates, to name a few examples. The spray-drying process represents the most widely used method for powder manufacturing out of liquids.

Starting materials are solutions, suspensions, and solids containing emulsions. Because of large specific surface area of the droplets as well of significant differences in temperatures between the drop and gas, the drying expires quickly. Originated from the spray mist, the particles are, therefore, only a short temperature stress exposed, so that even temperature-sensitive materials dry without appearance of side reactions.

Spray drying belongs to the few processes (Table 8.1), in which two operations, namely, in this case the drying and shaping, happen simultaneously. The installations are simple in design, suitable for many materials and for all capacities up to about 15 t of evaporated water per hour in one production line. In exceptional cases, $50\,t\,h^{-1}$ of water are evaporated [1]. For the manufacture of powders from liquids, the drying in spray towers shows the lowest costs. A cost comparison with contact dryers (rolls and thin film dryers) results advantages for the spray drying. In particular, by the simultaneous formation of the free-flowing powders arise after screening already salable products. The superior quality comes from the lower load of temperature.

In industry, improvements of the spray drying deal with capacity increases, energy consumption, exhaust pollution, caking and incrustations as well product quality. That means the economy and ecology, further plant safety and particle design represent central points of discussions [2]. The following text refers to these points. Discussions of extensive sub-processes, two-phase flow with momentum, heat and mass transport, measurement and control technology, problems with the formation of deposits and chemical reactions coupled with the drying process could

Table 8.1 Examples of product design technologies: drying and shaping in one step.

Starting material	Technology	Process	Product
Liquid (solution, suspension, and emulsion)	Generation of particles	*Spray drying*	Powder
Liquid (suspension and paste)	Generation of particles	*Thin-layer drying*	Powder
Liquid (solution, suspension, and emulsion) on disperse solids	Particles growth in layers, formation of new particles	*Spray agglomeration in fluidized beds*	Granules
Solids (granules and powder)	Shaping of plastificable materials	*Spheronization*	Granules (powder)
Solids (lumpy material, granules, and powder)	Particle reduction	*Mill drying*	Powder (granules)
Solids (granules and powder) and applied *Liquids* (solution, melt, suspension, and emulsion)	Surface coating	*Fluidized bed coating*	Granules (powder)

not take place here, for reasons of space. The majority of results presented refer to the "detergent-counterflow tower" without downstream dryer/granulator [3].

8.2
Basics of Atomization and Drying

8.2.1
Atomizing with Nozzles

Generation of liquid droplets occurs in atomization installations [4]. In addition to the rotary atomizer (spinning disc), the industry uses nozzles of various designs [5]. The elementary processes of droplet formation affect the product properties [6]. To increase the efficiency, concentrated liquids are preferred, minimizing the evaporated liquid per kilogram of solids. The concentrating leads to highly viscous suspensions, which only can be sprayed into fine droplet with high efforts. Under no circumstances, the nozzle should clog by fine solids. Otherwise, the suspension must be ground in a device for wet milling prior to atomization. The solids content of suspension is a compromise between water content, viscosity, and sprayability.

Classified according to the shape of spray, there are hollow and full cone nozzles, flat spray, and fan jet nozzles. At single-fluid pressure nozzles, the spray arises by converting the pressure into kinetic energy. Pressure energy accelerates the fluid in the nozzle, generates rotation and forms at the nozzle orifice a thin liquid film, which tears off and disintegrates into droplets. Pressure nozzles work under a pressure of 5–150 bar. Figure 8.1 shows the two best-known types of nozzles in simplified depiction. Detailed representations of nozzles and nozzle assemblies, further the calculation of drop sizes is described in the references [1, 4, 7, 8].

In pneumatic nozzle, a pressurized gas (air, nitrogen, carbon dioxide, or steam, 1–10 bar) causes the breakup of the liquid. The supply of compressed gas for the pneumatic nozzles is expensive owing to the high energy consumption. In exchange, the twin-fluid nozzles create small droplet sizes, which are about 1 order of magnitude below those of single-fluid nozzles. In the field of pneumatic atomizers, twin-, triple-, and fourfold fluid nozzle exist. In chemical industry (coating; spray agglomerations in fluidized beds [9]), twin-fluid nozzles are common with full cone (Figure 8.1b). These are designed with external mixing in case of low tendency to clog, and internal mixing at high solids contents. Triple-fluid nozzles (g/l/g or l/g/l) and fourfold-fluid nozzles are special designs, for example, to get a homogenous spray jet with simultaneous mixing of liquids.

8.2.2
Description of Drying and Structure Formation

When drying a spray, the free surface water evaporates initially in drying phase 1 with constant drying rate. During this process, the droplets disappear immediately, forming wet particles. With decreasing evaporation rates, further drying occurs in

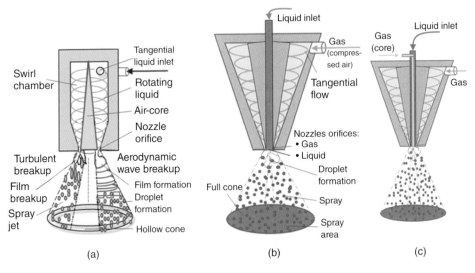

Figure 8.1 Mode of operation for frequently used nozzles in schematic representations (a) hollow-cone nozzle with pressure on the liquid; (b) twin-fluid nozzle with external mixing, compressed air as propellant gas; and (c) triple-fluid nozzle (gas/liquid/gas) with external mixing.

drying phase 2. There, water vapor diffuses from deeper layers and voids to surface, especially via pores. In drying phase 3, the diffusion of vapor out of the inner parts is rate limiting, and the drying rate shows a further descending course. In many cases, the particles are already sufficiently dry in drying phase 2.

During evaporation, complicated composed droplets (dissolved/undissolved inorganic and organic substances) either form agglomerates out of the suspended solids. Or throughout drying, the local concentrations of the solutes increase and supersaturated. Thereby, solid structures and/or crystals arise. The presence of other substances (e.g., surfactants) influences the formation of particles and crystal structures. A representation of resulting material-dependent structures illustrates schematically Figure 8.2. Corresponding images of different spray-dried substances are in Section 8.5.2.

Detergent formulations form particles with open structures, so with cavities and pores. In principle, particles resulting from spray drying, may be smaller, of same size or larger than the droplets (Figure 8.3). The generation of particle with larger diameters requires a high solid content in the droplets. Depending on voidage of the particles and concentration of the suspension, individual droplets during drying grow or shrink. Frequently, they grow in the case of solids contents of about >60%, below this value, they shrink usually. These observed and calculated effects are based on measurements by single-droplets. In practice, formed wet particles may agglomerate on the drying way to discharge at the tower cone. Owing to the superimposition with the agglomeration, the measured particle diameters are larger than the droplet diameters.

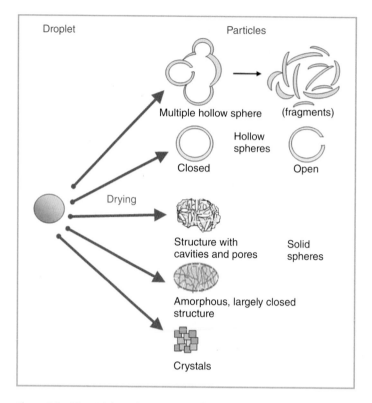

Figure 8.2 Material-dependent structure formation in the spray drying of droplets.

Multiple hollow spheres arise in the drying process of water glass. By moving the particles, these structures break. Resulting material shows extremely low bulk densities with insufficient flowability. By downstream milling succeeds the production of salable powders. Open and closed hollow spheres may originate by spraying of inorganic suspensions.

High bulk densities arise from starting substances with high densities [10]. Thereby the solid density is increased, while the void volume remains almost unchanged at 0.4 to 0.5. By this measure, the mean particle diameter decreases. The same effect results from dilution of the mixture with water. During drying, initially the dissolved salts concentrated and supersaturated before the crystallization starts. Some organic compounds, in particular polymers, dry under formation of amorphous particles. They show pores, but practically no major cavities.

8.2.3
Industrial Spray Drying

Measurements of particle moisture along the height of a production spray tower show that about 50 cm below the nozzle area (4 mm-hollow cone nozzles and 75 bar), the transition from droplets to particles occurs. Approximately 1 m below

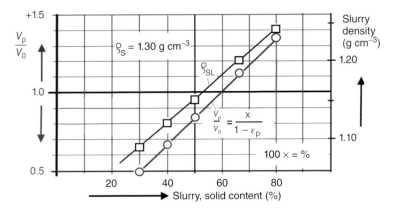

Figure 8.3 Calculated changes in the particle volume V_p, related to the drop volume V_0 at the formation, in dependence of the solids concentration x in the droplet (measurements for detergent powder: $x = 0.67 \leftrightarrow 67\%$ solids: slurry density $\rho_{SL} = 1200\,g\,l^{-1}$; bulk density $\gamma = 390\,g\,l^{-1} \leftrightarrow 300\,ml\,l^{-1}$; calculated values: porosity $\epsilon_p = 0.4 \leftrightarrow 200\,ml\,l^{-1}$; voidage $\epsilon_v = 0.5 \leftrightarrow 500\,ml\,l^{-1}$).

the nozzles, the particle moisture is then halved, with values of 15–20%, according to the formulation. In the spray space (area × 1 m depth) of a countercurrent tower, hot air vaporizes about 66% of the total water volume. A calculation reveals that evaporation capacity of spray space lies around 200 kg $H_2O\,m^{-3}$ air.

Optimum drying can only be achieved by fine spraying with full coverage of the spray area. Therefore, the maximum possible number of nozzles is required, which then have correspondingly smaller nozzle openings. For doubling or tripling the number of nozzles in a high spray tower, spraying ensues in two or three spray levels [2, 11]. Each spray level should operate always with all nozzles. A change of single-fluid to twin-fluid nozzles for finer atomization is no alternative for economic reasons, because the large amounts of compressed air are expensive. The mean residence time of particles depends on the height of tower. In small cocurrent towers (2.7 m), the residence time amounts to about 10 s [12], and in large towers (22 and 35 m) with counterflow to 25–50 s.

The viscosity of suspension plays a major role in optimal spraying, and is adjustable by temperature, shear rate, and concentration. It should be noted that by a dilution the viscosity decreases, but the cost of water evaporation increases significantly.

8.3
Spray Drying Plants and Plant Safety

Some conventional spray-drying methods [13, 14], in which the product dries in one step up to the residual moisture content, illustrates Figure 8.4. (A possible residual drying in an integrated/flanged fluidized bed or belt dryer is described in

Ref. [15, 16]). The used drying gas comes directly out of a burner or from a heat exchanger, where ambient air is indirectly heated. The gas streams in co- or countercurrent manner through the tower in comparison to the spray direction (from above to bottom). The dried powder, originated from the atomization, is separated in the cone, cyclone, or filter. By particle separation in the tower cone, the dust filtering may ensue in an integrated filter on the tower top. Alternatively, the air conveys the dust together with the powder into an external filter.

In the case of cocurrent towers, there are three possibilities for removal of the drying gas: firstly the conduction of the air with the entire powder via the cone into an external filter (good solution), secondly the conduction of the air with dust through a centrally mounted discharge pipe (bad solution and vacuum cleaner effect) or through a ring channel (neutral solution). The vacuum cleaner effect leads to a removal of less dried particles out of the drying chamber.

In hot air towers, an energy recovery is possible to a limited extent. First, the heat of exhaust gas may be used via exchangers to heat water (Figure 8.4b). This results in little over 10% increase in the calculated efficiency. Caution, because excessive cooling of the exhaust gas leads to a falling below the dew point and its starts to rain in the area of tower. Second, it is advantageous to recycle a part of hot exhaust air and make a mixture with the hot flue gas (Figure 8.4c). A bottom open tower allows the inflow of cold ambient air, which cools the escaping powder.

During atomization of dissolved organic materials (dust-ex.), a responsible engineer avoids all ignition sources and works additionally under inert conditions. As inert gas is nitrogen in circulation (Figure 8.5) standard. Inerting with nitrogen requires the dehumidification of recycle gas, which usually ensues in a scrubber by decrease of temperature [16, 17]. A strong dehumidification needs a cool washing liquid. Owing to the low level of water temperature, an energy recovery is not worth. The compensation of gas losses through discharge of the powder requires from time to time a feeding of further nitrogen.

Alternatively, it is possible to inertize the drying air with flue gas. For this purpose, the cycle gas streams through the burner, where either natural gas or oil reacts with oxygen from the air to carbon dioxide and water. This operation also demands the dehumidification of the recycle gas. By permanent addition of air/combustion air, a part of the circulating gas is suitably removed accordingly to the mass balance.

Figure 8.4 Drying of suspensions in spray towers with hot air (a) pressure nozzle – countercurrent tower with two spray levels and integrated filter; (b) the same tower with energy recovery from the exhaust, numbers are examples; (c) same tower with energy recovery by recirculation of a portion of exhaust air; (d) small cocurrent tower with pressure nozzles and external particle separation; (e) wide cocurrent tower with pressure nozzles, rotating disc, or pneumatic atomization; (f) fontaine tower with pressure or compressed-air nozzle and with external powder separation; (g) tower with integrated fluidized bed for drying and agglomerating; supply a part of slurry at the bottom possible; (h) tower with flanged fluidized bed for drying and agglomerating; and (i) tower with flanged belt dryer for drying of pasty products.

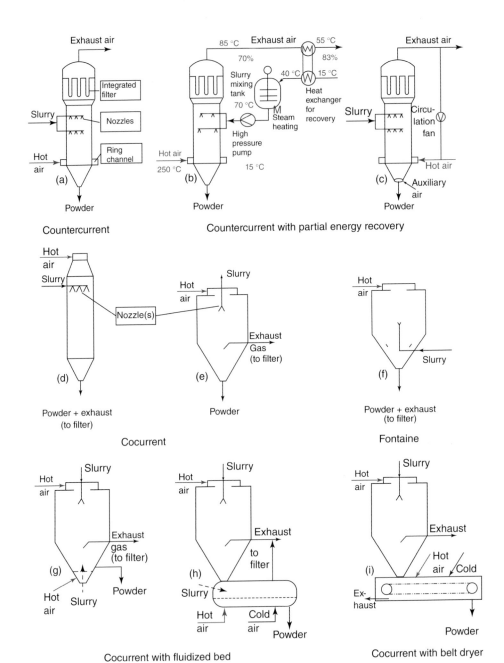

Countercurrent

Countercurrent with partial energy recovery

Cocurrent

Fontaine

Cocurrent with fluidized bed

Cocurrent with belt dryer

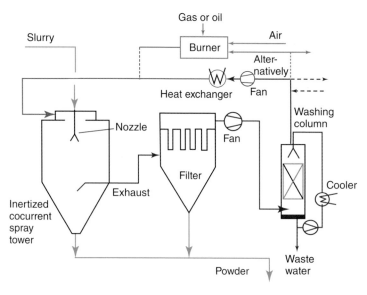

Figure 8.5 Drying of suspensions with hot nitrogen or alternatively with air/flue gas in the circulation mode: moisture removal by temperature decrease in the scrubber.

Figure 8.6 depicts three economic process alternatives for spray drying in superheated steam, which means under inert conditions. Use of superheated steam [18–20] is possible for all products that tolerate temporarily a temperature of about 110 °C. From 1994 to 2012, the slurry drying for detergents runs in a 28 m high tower in the countercurrent mode. The tower, implemented by Henkel Austria in Vienna, realized a water evaporation of $6 \, th^{-1}$. During production, the oxygen content in steam indicates usually values lower than 1%. The operation of tower takes place without exhaust gas, but with energy recovery via condensation of the excess steam. A few years ago, Gea/Niro presented a cocurrent spray tower with integrated filter out of metal mesh. Using superheated steam, this drying tower is suitable for sterile work. Systems with external filters may also operate with steam, as own large-scale pilot experiments showed.

According to Table 8.2, drying with superheated steam demonstrates a number of environmental and economic benefits [20]. Steam drying means, drying gas and evaporated liquid are identical. The steam gets the energy via an indirect heat exchanging. By a superheater, the steam inlet temperature rises up to 250–450 °C. Thereafter, the hot gas flows into the tower tangential, often through a ring channel. The superheated steam causes the evaporation of water from droplets, whereby the relative humidity (water activity) increases to about 80%. In this process, the gas cools down from 350 to 120 °C. Same amount of water vapor, which evaporates from the droplets, streams through a pipe to the heat recovery plant. There, by

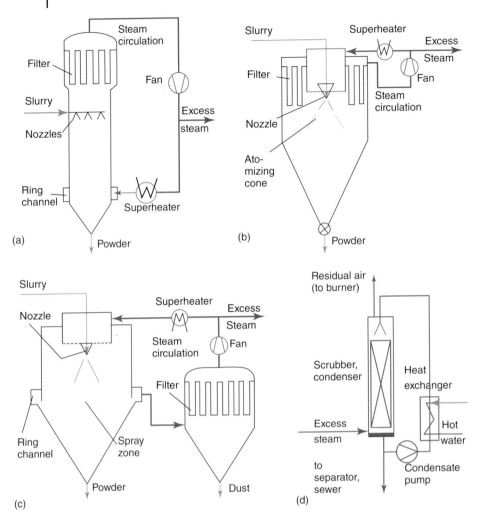

Figure 8.6 Drying of suspensions in spray towers with superheated steam (a,b) counter-current and cocurrent tower with integrated filter, (c) cocurrent tower with external filter, and (d) condensing the excess steam for recovering the energy.

generation of hot water (80 °C) ensues the heat recovery with a gain of 65–75% of the used energy.

Because the temperatures at steam outlet and before fan (imaginary outside) are the same, thermal efficiency is 1. (The burner for supplying the superheater shows an efficiency of 85–90%, but this is separately considered as a rule). The thermal efficiency results from ratio of heat of vaporization to heat supplied by the steam. Steam drying requires the least energy of all drying variants taking into account the energy recovery.

Table 8.2 Advantages and disadvantages of drying with superheated steam in spray towers.

	Pros	Cons
Economics	High efficiency	Thermal stress of products, not suitable for all
	Energy recovery	Extended startup and shutdown
	Sterility	Time-consuming product changes (in the case of a tower cleaning)
		Higher investment
Ecology	Drying without exhaust	No
	Simple condensate treatment	
	Energy recovery	
Product quality	Color (no oxidations)	Increased stickiness of some products at the tower cone
	Sterility	
Plant safety	Inert atmosphere, no risk of explosion or fire	No
Process plant	Closed plant	Increased effort due to recycling, pressure protection needed, and product discharge difficult

8.4
Improved Capacity and Energy Consumption

An increase of capacity in a hot air-spray tower takes place on three ways. First, higher inlet temperatures improve the thermal efficiency. The product and/or exhaust quality or the plant safety set the upper temperature limit. The second possibility consists in an increase of gas stream. Average axial gas velocity in a countercurrent tower, equipped with single-fluid nozzles, may rise to values of 1.8–2 m/s. For realizing this velocity in a "normal" tower, the engineer must install a fan with higher power, a filter with increased area (attention: pressure loss) and possibly also a more powerful burner. The third method for improving the capacity is the use of many nozzles with small borings instead of a few with larger borings. The finer spray, also reachable with increased spraying pressures, leads to higher drying rates and improved energy efficiency.

The economics of methods depend on the energy consumption. Mass balances of tower include the ingoing (slurry and hot air) and outgoing streams (powder and exhaust gas). The heat balance is made up of the supplied energy for heating the slurry (about 60–80 °C) and the drying gas (supply of evaporation energy and heat losses), as well as for the electrical drives. The feed pumps and in particular the fans require motors and therefore electric energy. The atomization needs in the case of single-fluid nozzles a high-pressure pump, for the rotating disc a large drive or for twin-fluid nozzles a compressor, which generates compressed air. Power consumption of the fans for gas supply into the burner and for exhaust gas or circulation should be included also. An open system loses a large part of the supplied heat by the exhaust, and a small part over the discharged powder.

About the temperature levels (in-out and thermal efficiency), nozzles and their arrangement, amount and flow of the gas as well as heat recovery from exhaust gas, the energy consumption in the spray tower sets together and is influenceable (Table 8.3). The following examples demonstrate some experiments and measurements, which originate mainly from two different spray towers. The first one shows a height of 25 m with a diameter of 4.5 m and equipped with an integrated filter, the second tower a height of 45 m with 6.5 m diameter and an external filter. Both are countercurrent production towers. The experiments took place always during production. Production means the drying of detergent suspensions with solid contents of about 65% and water evaporation between 3 and 12 t h^{-1}.

A controlled flow of gas provides the possibility of reducing the energy consumption in spray towers, especially in countercurrent operation mode. There are three characteristics of drying gas to observe, namely, the formation of wisps and swirls and the central flow. In large towers ($d > 4$ m), the mixing of centrally flowing hot gases with colder, near-wall streams does not happen by itself. Therefore, air flows in different wisps through the tower. This can be demonstrated through measurable hot air wisps, which are detectable above the spray area and even in the exhaust pipe.

The swirls arise from changes in the flow direction and in particular by a strong contraction at the gas outlet. The gas outlet, located at the edge of the tower, forces the flow to a 90° deflection under strong contraction. This causes, noticeably, an asymmetrical flow profile in the upper part of tower and the swirls. Oscillations

Table 8.3 Ways for reducing the energy use at a spray tower.

Method	Execution
Reduction of the central flow of gas	Flow control
	Arrangement of the nozzles
	Gas feeding
Extending the residence time	Height of tower
	Tangential swirl flow
Increasing the number of droplets	Higher amount of suspension per cubic meter drying gas
	More (finer) nozzles
	Higher pressure
Improving the efficiency	See points 3 and 5
	Higher flow velocity
	Flash drying, use of a suspension above 100 °C
	Drying with superheated steam
	Higher inlet temperature
Heat recovery	Heat of exhaust for preheating of slurry water
	Recycling of a part of exhaust gas into the burner
	Condensation of excess steam for superheated steam drying

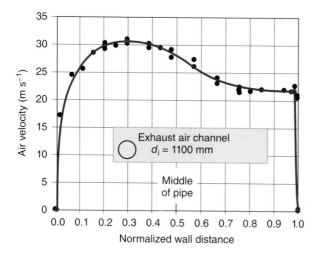

Figure 8.7 Measured local gas velocities in the exhaust gas channel.

in the tower induce some opposite running, extensive vortex that not only affect the spraying, but also increase the amount of dust discharged into the filter. Improvements would be a tangential, or still better, a central outlet in the flow direction with gradual constriction. These appropriate modifications of installation reduce the irregular high flow turbulences. The asymmetry of the flow continues in the exhaust duct, as Figure 8.7 shows, as differences in velocities and temperatures. A more uniform gas exit minimizes all turbulences. The tangential gas inflow and/or the radial-tangential arrangement of the nozzles as well as the addition of swirl air, generates a powerful rotational flow over the whole tower area. This rotational flow influences the gas flow through the tower, contributes to a more uniform flow and to a better mixing of cold and hot air wisps.

The central flow of air shows Figure 8.8, which reproduces experimentally detected values during production. The actual flow rate depends on the position and temperature. The flow varies not only along the tower, but also along the diameter. Near to the outer ring channel for supplying the hot air, maximum velocities occur, measured in a distance of 0.35 d away from the tower wall. Close to the wall, there is a downward flow. Furthermore, a not fully symmetrical flow profile arises, particularly in the cone. Where the hot gas flows, should be the atomization cones, distributed symmetrically over the entire spray space (area of the transition from liquid to solid state). By the length of nozzle fittings and their inclination, most of the sprays should be focused on the central region. This arrangement can eliminate largely hot wisps of the gas [21].

Figure 8.9 displays measurements of the tangential flow in a tower under production conditions. Maximum values appear at 0.35–0.40 and from 0.65 to 0.70 d, zero for the center and rather low values near the wall. A marked asymmetry in the cone has been observed, probably due to the specific tower geometry.

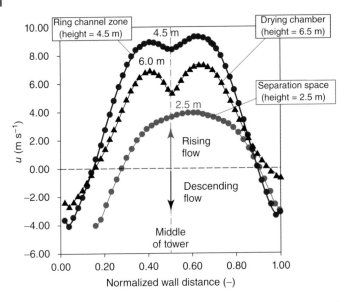

Figure 8.8 Actual axial flow velocities in a large counterflow tower, operating with swirl air, in dependence of the position along the tower diameter: ($d = 4.5$ m, height 25 m; periphery: 0 and 1, center: 0.5; parameters: tower heights, 4.5 m near the gas supply, 6.0 m in the drying chamber, and 2.5 m in the cone; slurry flow rate $9\,t\,h^{-1}$, just as in Figure 8.9).

To extend the residence time and to increase the mixing of drying air with the air inside, a tangential flow is applied, called *swirl air*. This occurs by a special supply of 10–30% of drying air through a separate pipe, preferably added tangentially in the ring channel. The swirl air causes a stabilization of the flow through the tower and a deflection of spray cones from radial into the tangential direction in accordance with Figure 8.10. The images made with an IR camera, show the nozzles and the flow of the droplets. It can be seen that in the region of maximum tangential speed arises a rotation of the atomized droplets around the tower axis. Also, two images prove hot air wisps, which move around the center, indicating reduced energy efficiency (heat losses). The resulting particles move downward helically on the tower wall in straggly clusters. Depending on strength of swirl flow, a straggly movement of particles begins ~1 m below the spray level, and leads to the desired extension of residence time. On the way down, residual drying happens of about 12–15% down to 3–5%. There the particles agglomerate, partly also dust from abrasion. Measurements of particle density in a running tower as function of location with conventional methods and with laser reflection revealed that below the spray space almost no particles exist in the tower center. They are all located close to the wall and directly on the wall.

Numerous experiments (more than 50) at the production tower ($d = 4.5$ m and hot air) verify the dependency, depicted in Figure 8.11. Namely, that the specific

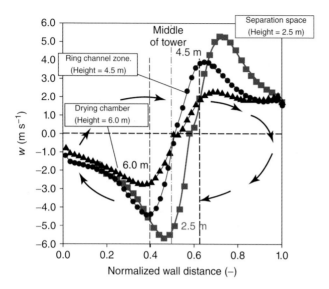

Figure 8.9 Tangential flow velocities in a counterflow tower, operating with swirl air, in dependence of the position along the tower diameter. Positive velocities: arrows point from left to right and negative velocities: arrows from right to left.

energy consumption decrease by increasing the spray density (= droplet density). That means an increasing throughput of suspension per cubic meter per hour drying gas, with identical moisture content of product up to a limit, reduces the specific energy requirement. Presumably, the local turbulence (momentum exchange) improves significantly. This measure works in many recipes, requires adjustments to the inlet temperature and/or pressure as well as especially to nozzle sizes, number, and location, and brings a significant increase in performance of spray towers. Spray drying with air allowed in detergent counterflow towers a maximum thermal efficiency of about 74%, the max. values in cocurrent towers are 70%. Limitations are the safety and the quality.

The efficiency improvement by a flash drying [22, 23] is a development, done at the beginning of the 2000. This technology is worthwhile for temperature-insensitive materials for spraying with single-fluid pressure nozzles. Implementing this technology ensues by installation of a suitable heat exchanger (with baffles) for the suspension. Heating up happens under pressure in the temperature region from 120 to over 200 °C. While spraying, the so-supplied energy causes lossless water evaporation. The residual moisture must be removed conventionally with the hot drying gas. The drying gas is in almost all cases ambient air, heated up directly in a burner. Another positive aspect applies the increase of throughput, reached in this way at the same facility. An example displays Figure 8.12.

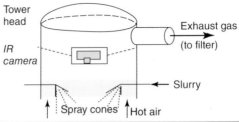

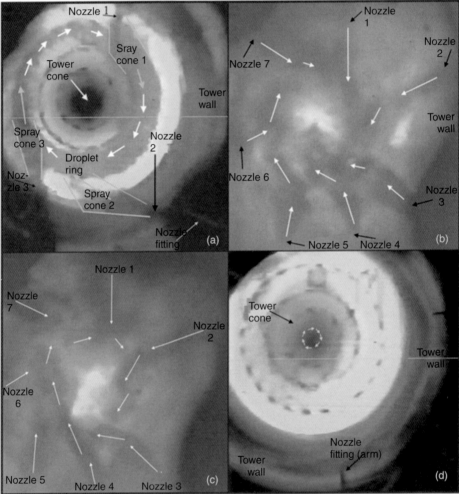

Figure 8.10 View with an infrared video camera into the spray area of a countercurrent tower (camera is located above the nozzle level, see the depiction above). Air inlet temperature 250°; operation with swirl air, tower equipped with seven nozzle fittings through the tower wall (a) startup with three nozzles; (b,c) operation with seven nozzles; white spots in the middle means the breakthrough of hot air wisps, the real flight directions of sprayed droplets are indicated in (b,c) with white arrows, and (d) shutdown, no active nozzle, most recently sprayed particles rotate still in the hot cone.

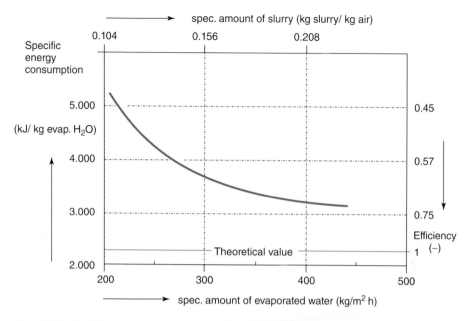

Figure 8.11 Specific energy consumption and thermal efficiency as function of evaporated water per square meter tower area and hour (measured with detergent suspensions).

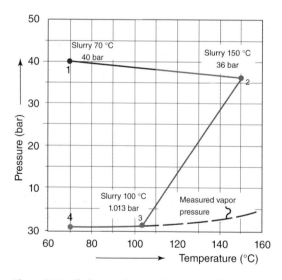

Figure 8.12 Flash spray drying with increased efficiency: (in this example: energy saving > 25%), route 1–2: heating, 2–3: adiabatic relaxation during spraying, and 3–4: evaporative cooling to drying temperature of the particles.

8.5
Influencing the Product Design

Fixing of product design in the spray tower first requires detailed knowledge of all processes in the spray tower. These can be obtained through sensors (Figure 8.13) arranged in the spray tower (Figure 8.14). The sensors help to control the procedure, and facilitate the reproducibility of products.

For adjustment of major particle properties (Figure 8.15), there are a number of setscrews. The performance, esthetic and dispersive properties must fulfill the specifications. Controlling the particle design is possible to limited extent. First, some critical properties depend strongly on materials and not so much on spray conditions, like the bulk density. Secondly, a complex structure formation takes place during the solidification, practically unalterable. Thirdly, it is possible to adjust in most cases the particle size distribution, the residual moisture content and some other properties within limits. Therefore, the knowledge of spray drying fundamentals and chemical substance properties as well as targeted pilot tests is essential for the optimization.

Nevertheless, a careful selection of suitable spray equipment and of operating conditions (Table 8.4) facilitates the settings. If the sprayed powder with respect to particle size, shape, flowability, moist content, solubility, or bulk density does

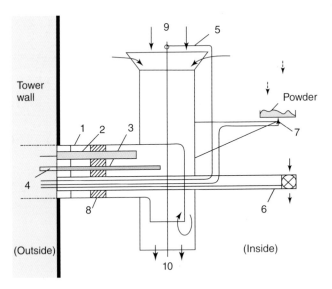

Figure 8.13 Smart sensors for measuring the gas velocity, temperature, humidity, pressure difference, and powder weight 1: measuring device; 2: relative humidity and temperature; 3: pressure difference; 4: Polytetrafluoroethylene (PTFE) tube for pressure transducers; 5: second temperature measurement; 6: vane anemometer; 7: weigh cell; 8: PTFE seal; 9: drying air in; and 10: drying air out (Courtesy of Ovid Dicoi).

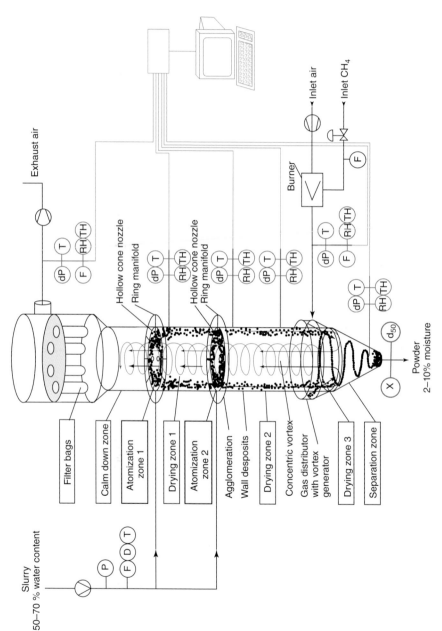

Figure 8.14 Countercurrent two level spraying tower for detergent powders, equipped with smart sensors: *P*, *dP*: pressure and pressure differences; *F*: flow; *D*: density, solid content, and viscosity; *T*: temperature; *RH*: relative humidity with temperature *TH*; *X*: water content in the powder; and *d₅₀*: particle size distribution (Courtesy of Ovid Dicoi).

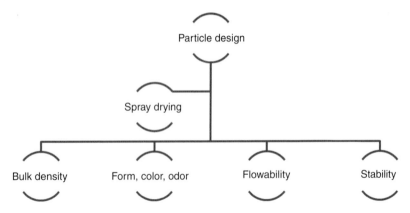

Figure 8.15 Main parameters of the product design for spray-dried particles.

Table 8.4 Options for controlling the product design in spray drying.

Influenceable operation parameter	Manufacturing variants
Drying gas	Air (dry and moist)
	Air/flue gas
	Nitrogen
	Superheated steam
Spray conditions	Nozzles (type, number, levels, and arrangement)
	Spray pressure
	Local pressures and temperatures in the tower
	Gas velocity
	Slurry concentration, viscosity, density, and surface tension
	Sizes of suspended particle
Tower geometry	In- and outlet of the gas
	Diameter ratios: spray cone to tower and diameter to height
	Cone
	Filter integrated or external
	Powder discharge
Mode of operation	Cocurrent without/with swirl flow
	Countercurrent without/with swirl flow
	Fontaine
Follow-up treatment	Airlift
	Sieve and sifter
	Fluidized bed
	Coater
	Belt dryer

not fulfill all requirements, the properties may be adjust by grinding, compacting/grinding, granulating [24], granulating/drying in a fluidized bed [25], and/or surface coating [26].

8.5.1
Choice of Atomizer and of Operating Mode

The choice of atomizer and type of atomization determine the particle size as well as the distribution (Figure 8.16). The atomization of a liquid with pressurized gases (compressed air, nitrogen, steam, or carbon dioxide) generates in the so-called twin-fluid nozzles extremely fine droplets, especially utilizing liquid carbon dioxide in the spraying process. Liquid carbon dioxide works if the product is soluble in the system and decreases the viscosity of liquid [27–29]. Therefore, twin-fluid nozzles are used in coating plants [26] and fluidized bed agglomerations [9, 25]. Very fine droplets arise with several small nozzles. An increase in the spray pressure also results in finer droplets. Advantageously, both methods can be applied.

The atomization pressure in twin-fluid nozzles (range 1–10 bar, standard 2–6 bar) significantly affects the mean particle size. High pressures promote the formation of fine droplets, but worsen the profitability. In all atomization procedures, the viscosities of fluids and thus their concentration limit the desired droplet formation. The limits are often in the range of 35% solids (milk) up to 70% by weight (simple detergent formulation).

Centrifugal atomizer cannot clog. They work in the cocurrent mode and manage well liquids with relatively high solids contents. The liquid is supplied to a rapidly rotating disc (30–75 m s^{-1} tip speed), where the liquid film breaks off by centrifugal forces at the edge and constitutes fine droplets. Disc atomizers run in wide towers to minimize the caking. Due to moving parts (drive, disc as ignition sources) arises an explosion hazard. Drying of organic matter requires an inerting with nitrogen.

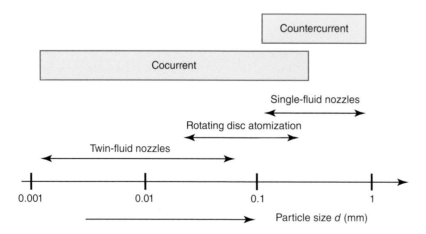

Figure 8.16 Particle sizes depending on the atomizer and gas guidance.

Depending on material, the choosing of wheel diameter and the rotational speed enables an adjustment of average particle sizes in the range of 30–250 μm.

Relatively coarse particles in broad distribution are producible with single-fluid nozzles in a countercurrent tower. To achieve the objective, the operator uses nozzles with large borings and atomizes with low pressures (5–25 bar). With use of towers with pressure nozzles, a piston pump builds up the pressure of atomization (5–150 bar, preferably 25–70 bar) in the supply line. The suspensions atomize commonly in several hollow cone nozzles. Choice of smaller holes in the nozzles, of higher pressures and larger spray angles lead to finer particles. Fine particles in relatively narrow distribution with improved form factors result from countercurrent drying with flash support. However, here nearly dry particles pass through the hot gas inlet zone. The thermal stress is high. Possibly, in the drying process occur tacky particles, depending on the material. Particles may adhere on the hot tower wall and build up cacking, a big problem in spray drying.

Processing of temperature sensitive formulations requires the use of a cocurrent spray tower. Here, the efficiency is lower. In cocurrent mode, hot drying air meets on the just formed droplets, so initially the water evaporates without strong increase of the particle temperature. In this phase, gas temperatures decrease significantly. The drying is gentler. Dried particles from the cocurrent tower, compared to counterflow, are finer and lighter, and therefore significantly more particles fly into the filter. Fontaine atomizing runs in rather small towers and represents a hybrid between counter- and cocurrent. Accordingly, with pressure nozzles emerge particle sizes in the range from 100 to 500 μm because of the overlapping of two flow types.

The particle size distribution can be classified in desired dimensions by sieving the over- and undersize. As dust removal is costly, in many productions only screening of the coarse grains takes place.

8.5.2
Material Dependence of Particle Design

To investigate the material-depending behavior of particle formation and design, numerous attempts have been carried out in a small drying channel. In this laboratory plant, made of glass, also the drying kinetics was measured (Figure 8.17). For measurement, an about 2 mm large droplet is placed on a platinum loop, nearly in contact with the tip of a thermocouple. A microbalance offers a connection to the platinum loop, which registered the weight of the drops as function of drying time and separately the temperature. The drying process was filmed by a video camera.

Particle design depends, largely unalterable, on the chemistry of atomized material. The substance-dependent structures arise during the drying process and determine particle shape and porosity, flowability, and bulk density. Probative images depict Figure 8.18, in which the drying of three drops was monitored by a video camera. The drops were hung up on a platinum loop at the end of a hot air channel.

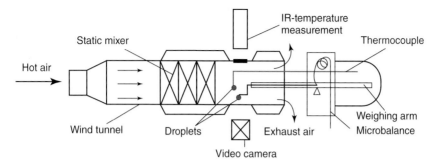

Figure 8.17 Laboratory facility for measuring the drying kinetics of different droplets (Courtesy of Ovid Dicoi).

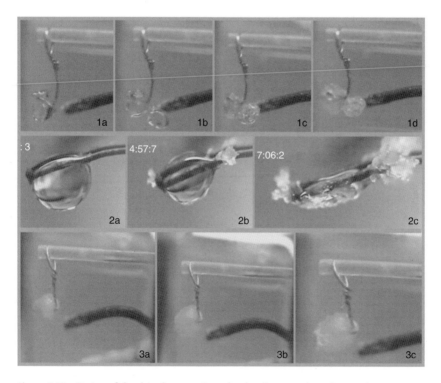

Figure 8.18 Drying of droplets, hung up in a platinum loop, with hot air in a laboratory flow channel; measurement of droplet weight, the time and the temperature, at the same time video recording for the documentation: Drop in a platinum loop on weighing arm; lower wire right: thermocouple; droplet diameter about 2 mm; air temperature around 200 °C; flow rate of 1 m s⁻¹. Product 1: water glass solution 37/40, 2: sodium sulfate solution; 3: detergent suspension; (a) immediately after start of drying, and (1d, 2c, and 3c): drying end, images are taken from the video recordings.

The top row of the images shows a clear translucent drop, which arose from water glass solution. This droplet swells in contact with hot air (200 °C) powerful, and forms more bubbles. The dried product, consisting of thin, milky fragments of the hollow spheres, had an extremely low bulk density between 15 and 150 g l^{-1}.

The middle row illustrates the drying of dissolved sodium sulfate. In hot air, the droplets evaporate water until the crystallization at the edges occurs. Finally, free water disappears, and later, also the crystal water. This results in a total shrinkage of the drops. Crystals grow up on the loop and form new shapes. A crystal formation point to a rather high bulk density (about 800 g l^{-1}).

Images in the lower part of figure explain the drying of a slurry droplet (solid content: 65%), derived from detergents. This swells up visibly after the initial contact with hot air, but then remains constant in size until the end of drying. Crystallization cannot be seen; the particles exhibit amorphous structures. This behavior is typical for powder with bulk densities ranging from 300 to 600 g l^{-1}.

An example for drying of an aqueous polymer solution illustrates in Figure 8.19a. The evaporated water leaves behind a slightly shrunken, amorphous particle having a smooth surface, which, however, has some cavities (craters). The right picture displays the dried droplets of an anionic surfactant. During drying, the size increases and some craters arise. The five examples demonstrate that every substance dries material-dependent differently. Based on physical and chemical properties, the drying behavior cannot be predicted, also not the characteristics of resulting particles.

Another example of material dependency represents sodium carbonate with the big changes of density. During the drying of a solution, short-term originates the decahydrate from the supersaturated solution at temperatures <32.5 °C with a density of 1.44 g cm^{-3}, afterwards the heptahydrate with 1.51 g cm^{-3}, subsequently at temperatures >35.4 °C the monohydrate with 2.25 g cm^{-3} and finally the anhydrous product at the conversion point of >107 °C with 2.53 g cm^{-3}. It follows, that the density almost doubles during drying (same mass – half volume). Conversely, the

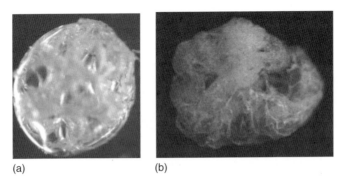

(a)   (b)

Figure 8.19 Droplets to particle by drying in a laboratory hot air channel, hung up on a platinum loop. (a) sodium salt of an acrylic acid copolymer (~2 mm droplet diameter at the start, dried about 1.8 mm, 150 °C inlet temperature, 1 m s^{-1} air speed), and (b) a particle of the sodium salt from the alkyl benzene sulfonic acid (150 °C, 1 m s^{-1}).

solid structures change appreciably by absorbing moisture and increasing volume, even if particles contain sodium carbonate only in subordinate amounts.

The observation of drying in the laboratory flow channel provides indications for the crystallization and structure formation in the drying process as well as for the kinetics. The behavior of sprayed droplets, which are an order of magnitude smaller, may differ due to higher drying speed (larger specific surface). However, all dryable 2-mm drops in laboratory are in any case also dryable in spray tower. According to product images, the powders differ not only in particle size and distribution, but also in shape as well as in their agglomeration behavior.

A look at different products indicates that inorganic materials frequently form a spherical shape and rarely agglomerate, in line with Figure 8.20. With some exceptions, these substances dry easily with high tower inlet temperatures and increased throughput in co- or countercurrent towers. In contrary, drying of organic materials do not allow high product temperatures, otherwise chemical reactions occur (discolorations, decomposition, and oxidations). Organic materials and organic/inorganic mixed products tend to agglomerate, especially if sticky, plastic phases appear during drying. Formed from the primary particles, the agglomerates differ significantly in size and shape and lead to a broad particle size distribution.

Shortly after beginning the drying process, droplets of some organic substances or formulations form impermeable, more or less dry and ductile layers at the surface. These layers impede the removal of moisture. Further drying is controlled by diffusion with low mass transfer rates. In this case, finer spraying assists the drying, possibly after changing the atomization equipment. Recommended is the use of a disc or better of two-fluid nozzles, and to dry faster (cocurrent and high input temperature). Other hand, drying takes place in combination of a spray tower with a fluidized bed or a belt dryer. Alternatively, in difficult cases, the use of a thin film dryer (centrifugal dryers) often provides good results, but the dry particles become very fine.

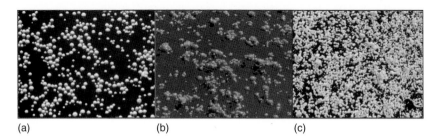

(a)   (b)   (c)

Figure 8.20 Product images from cocurrent towers after nozzle spray drying (a) inorganic material: aluminum oxide; (b) organic matter: dye; and (c) titanium dioxide (rotating disc) (Courtesy of GEA/Niro).

8.5.3
Agglomeration of Droplets and Particles

When spraying a suspension in usual concentration, that means for detergents a solid content of about 50–70%, the produced particle size distributions are not identical to the size distribution of droplets. They differ due to agglomerations during the drying process. It follows, that the distributions of dried powders are wider with increased mean diameter, in comparison to the droplet distributions. Shrinking of droplets, proved in many cases at individual droplets, is not observed for the particle collective after measurements at the outlet of spray tower (Figure 8.21a). The agglomeration is stronger than the effects of drying. About the process, it is possible to shift the curves of the cumulative distributions for the droplets as well as for the particles. Changing of operating conditions and/or of equipment lead to modifications of sizes in limits, but the differences remain for substantially same formulations.

Two different mechanisms effect the particle enlargement. Firstly, the plurality of used nozzles with the partially superimposed spray cones increases the probability of droplet coalescence or fusion of a droplet with a particle. This influences the agglomeration only in small extent, because the droplets collide elastically without flowing into one another. Secondly, during the drying with still high residual moistures, particle agglomerations play an essential role. The evaluation of agglomeration behavior reveals that the fine particles below 100 μm remain in same size, because they dry very quickly. However, the big particles enlarge even more in volume (up to a factor 6); preferably if more large drops are present and the particles contain high moisture contents. Agglomerations occur frequently by stickiness of particles during the drying process. This can be reduced by injecting dry powders (flow aids). In the theoretical case, that drop coalescence/agglomeration do not depend on size, then, the two curves in Figure 8.21a would run parallel and not move further and further apart with increasing diameters.

The shape factor, derived from the ratio of largest length to smallest diameter, depends only a little on particle size, indicates more or less constant values. It shows in the range of < 100 μm minimum values, increases to 200 μm to a maximum (0.73), and then remains constant from 500–900 μm at just over 0.7. The agglomeration does not change the particle shape perceptibly.

8.5.4
Recirculation of Coarse Particles and Dust

The particle size distribution of tower powder is set by the screen mesh size for over- and undersize. Figure 8.22 displays the possibilities of recycling the materials. In the detergent industry, in usual manner a sieving of 2–5% oversize particles takes place, additionally to small amounts of incrustations, spalled from the tower wall. Some wet particles adhere at the hot tower wall during drying and constitute very hard and discolored caking. The amounts depend on the stickiness of moist particles and process control, but also on the wall material.

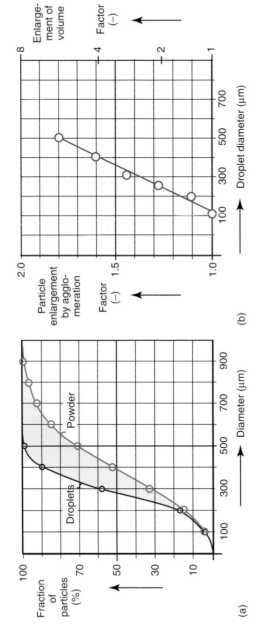

Figure 8.21 Agglomeration of particles during spray drying in a countercurrent tower (a) cumulative distribution for the diameters of atomized droplets from a detergent suspension (75 bar, hollow cone nozzles) and the obtained there from particles and (b) agglomeration depending on the droplet size.

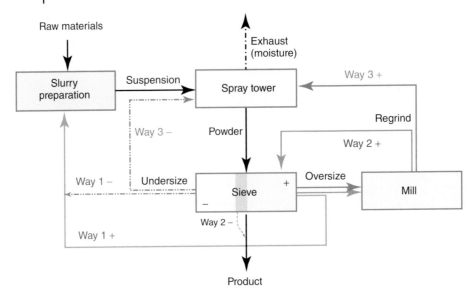

Figure 8.22 Alternative pathways for processing of over and undersized particles from the spray tower.

The oversize goes back either into the slurry-mixing tank to dissolve or into a mill. From there, the regrind flows after sieving directly back into the process. According to the setting of mill, the distribution may change adversely by increasing amounts of dust. The dissolution of outsize ensue in the slurry tank, and however, leads advantageously to no change in particle size distribution and therefore also not in the aesthetics, but costs more money for additionally water evaporation. Alternatively, conveying air carries the undersize and ground material direct into the spraying zone. Fine particles remain often in the product although they worsen the quality.

In the case of integrated filter, after jet cleaning of filter bags the airborne dust moves back into the spraying zone. Here, dust may adhere at the droplets. In all other cases, which means with external exhaust purification, the conveying air blows the separated dust directly into the spray zone. This statement is valid for both the co- and the countercurrent tower. Most favorable for addition of dust is the spraying space, in which the droplets convert into moist particles.

A trouble-free application of product requires the previous separation of dust. The term "dust" is not defined exactly, but it represents a negative characteristic. Product-dependent, dust means particles smaller than 50 or 80 or 150 μm. Additional removals of fine particles with a sieve or a sifter is complex, but leads in most cases to significant product advantages (dissolving, incorporating, dosing, conveying, etc.). Again, the two alternatives exist, such as the recycling in the spray zone or in the mixing tank. The properties of produced products help to determine the better method.

8.6
Scale-Up of Spray Dryers

A scientifically based scale-up from laboratory to industrial scale is not possible in the spray drying. Laboratory nozzles differ distinctly in size (geometry, nozzle diameter, type, spray cone, and power requirement) from the nozzles, used in production. In a laboratory spray tower, many suspensions form powders, which are not representative of the large scale. The resulting material is suitable for a pre-selection of application properties. Important characteristics such as the particle size distribution, bulk density, and flow behavior, furthermore solubility and residual moisture do not represent the product out of a production plant. Proceeds the drying in the laboratory system not completely, this does not mean that it will be the same in the production plant. While conversely, positive laboratory tests likely make a practicality in production.

For production, the correct nozzle choice is the key for a trouble-free atomization and for the manufacturing of free-flowing powders. If unknown, the responsible engineer selects an appropriate nozzle in cooperation with a competent manufacturer. The choice takes into account the geometry of production tower and the required throughput, further a possible arrangement of nozzles/nozzle assemblies as well as the desired particle size distribution. An optimal nozzle allows a uniform atomization into fine droplets, at a throughput, which is 3–33% of the required tower capacity. One of these suitable nozzles (exceptionally smaller), checked on a test stand with the representative suspension, operates for many tests in a pilot plant. The pilot tower is geometrically similar to the production, has a minimal diameter of about 2 m (often larger). This size is necessary, due to the capacity of selected nozzle with the necessary amount of gas for the evaporation of water, furthermore owing to the width of spray cones.

Number and arrangement of selected nozzles and required amount of process gas for the supply of evaporation heat and losses determine the diameter of process plant. The developer calculated the minimum diameter from three parameters: first, the quantity of evaporated water, secondly the thermal efficiency limited by maximum inlet temperature and thirdly taking into account the known or to be determined maximum gas velocity in the tower. The height of the tower or the fall distance, estimated from the particle fall velocity and from the required residence time for drying, is determined and verified experimentally.

The pilot plant, designed as replication of production plant in all relevant ways, comprises several functions: first, to examine the operating parameters regarding to the atomization and guidance of gas as well of the formed and recycled particles. Secondly, the pilot plant enables an optimization toward the desired product properties. This basis permits a designing of the production plant. Thirdly, particles from the pilot plant, produced under adapted production conditions, correspond to particles out of the production tower in many parameters. This allows producing the required product samples for necessary application tests. Some points for scale-up display Table 8.5.

Table 8.5 Scale-up parameters for spray towers.

Parameter	Pilot scale	Production plant

Tower
  Height h
  Diameter d
  Supply and discharge of
    the gas
  Cone and product
    discharge
	Calculation of h via the residence time; of d via the diameter of spray cone taking account the average flow velocity	Geometrically similar to the pilot plant, including the inlet and outlet lines

Nozzles
  Type, size, number, and
    arrangement
  Pressure of liquid or
    propellant
  Throughput
  Viscosity
  Specific flow density
  Temperatures
	Selection on theoretical considerations and experiments, pilot studies with a production nozzle (identical in type and size); centrally deployed; experimental determination of optimal settings	About 3–35 nozzles, depending on the scale-up factor; located near by the center, spraying in the direction of center, using the determined settings from pilot scale

Gas
  Velocities (axial and
    tangential)
  Air, inert atmosphere:
    nitrogen, steam with
    recirculation
  Co- or countercurrent
  Temperatures
  Airlift (cooling)
  Exhaust purification
  Heat recovery
	Calculate the amount of gas, based on the amount of evaporated water, maximum temperatures, and gas velocities; measurement of exhaust contaminations (actions necessary?); determine the gas flow in dependencies of plant safety, product properties and caking; influence the gas flow by a controlling the swirl flow	Extrapolation of measured values and gas runs comparable

Product
  Properties
  Application and dispersity
  Residual moisture
  Possibly further treatment
    in a granulator,
    fluidized bed, mill,
    or coater
  Separation, discharge, and
    caking
  Recycling of over- and
    undersize
  Blowing in of adjuvants
    and dust
	Focus in the pilot tests on desired/required product properties, evaluated according to the application and physical-chemical parameters; decision on installation of downstream processes; process optimization for high product quality under safe operation conditions; fixing of optimal parameters for the product and operations	Optimized and extrapolated settings are taken

A major problem during the spray drying represents the materials, which may clog the nozzle or enlarge abrasively the nozzle boring. They may be responsible for caking on the tower wall and influence the drying characteristics. Caking depend on the recipe, of blowing in of adjuvants, operating conditions and choice of tower materials. Ultimately, caking constitutes an optimization problem – they are not entirely preventable. Therefore, scale-up experiments are carried out carefully over a long period, also with the objective of minimizing hard films on the tower walls.

8.7
Exhausted Air and Waste Water

Not only water evaporates from the atomized droplets, but also other substances with significant vapor pressure enter the gas phase. Sampling with gas mice from the exhaust are complicated, without heat tracing of the lines. Because then the measurements exhibit values, that are an order of magnitude less than the actual load. These substances are mostly steam volatile organic compounds. In accordance with the German exhaust regulation (TA-Luft) of 2002 [30], the load of exhaust air with organic, non-classified compounds (organic C) may not exceed $50 \, \text{mg m}^{-3}$. Additionally, the exhaust often also contains aerosols and small amounts of fine dust, which passed the filter bag. The authorized dust load of $20 \, \text{mg m}^{-3}$ for not classified solids can easily fall below with modern filter fabrics, but not reliable with cyclones during changing operating conditions.

When exceeding the limits of organic C, various actions (Table 8.6) are available. The lowering of tower inlet temperature of process gas is only a small setscrew for the regulation. This measure worsens the thermal efficiency. Meaningful is a modification of the formulation and process. That means, a partially or completely removal of the responsible organic substance ensues from the suspension. The

Table 8.6 Ways to reduce and to prevent the organic C-load in exhaust gas.

Ranges of measured values (mg m^{-3})	Measures
<100	Lowering the tower inlet temperature
<250	• Shift the slurry formulation
	• Gas scrubbing
<300	Partially recirculation of the gas
	• Air
	• Flue gas/air
All ranges	Completely recirculation of the gas (no exhaust)
	• Nitrogen
	Recirculation of the main stream, discharge of excess steam (no exhaust)
	• Superheated steam

necessary addition of these substances occurs downstream, by spraying on the dried powder.

Another option represents a scrubbing of the hot exhaust air (70–95 °C). Therewith the needed process plants not become too large, scrubbing runs with cold water drops and high gas velocities. The aerosols, formed by cooling in the scrubber, can be dejected with the help of water droplets. Unfortunately, the separation is just in the interesting range of 0.5–5 µm particularly poor, so it is not simply to press a value of about 300 mg m^{-3} reliable to less than 50 mg m^{-3}.

At high loads of the gas with organic C, only the circulation mode prevents exhaust gases. The use of nitrogen requires the scrubbing of entire cycle gas, that is, a large scrubber. Working with superheated steam, the mainstream flow in the cycle, while only the excess steam condenses, which evaporates from the droplets. For energy recovery, the condensate of excess steam flows in a circulation from the scrubber to heat exchanger and back. Excess condensate runs into a tank, where the deposited organic compounds can be skimmed off. The organic C-load of the aqueous phase is often lower than municipal wastewater; otherwise, it makes sense to clean the water with a membrane unit.

Also spray towers with hot air in open passage operate normally without wastewater. The wash water from the tower cleaning runs back into the slurry tank, so that a completely recycling is guaranteed.

8.8
Spray Agglomeration

A special case of spray drying represents the fluidized bed spray granulation. Starting materials in the fluidized-bed are some granules from the last batch. The valuable substance is dissolved or suspended in a liquid. After spraying, water evaporates during the drying process in contact with hot air. The very fine atomization (about 10–20 µm) is generated in twin-fluid nozzles. Unlike the coating process, droplets impinge on particles of same composition. Due to constant spraying of new layers and simultaneous drying, granules arise with an onion structure.

The agglomerations of two or more particles, forming large structures, are undesirable. Also, breaking of particles, abrasion and complete drying of sprayed droplets before impact should be avoided (Figure 8.23). In fluidized bed arising dust (abrasion and dried spray) streams with the air into the filter and returns from there back into the spray zone by a jet cleaning of filter bags. There exist descriptions of fluidized bed spray granulation with theoretical aspects and practical treatments [9, 25]. The modern method is suitable for difficult drying substances as well as for preparation of an optically beautiful particle with a dense structure and relatively high bulk density. Interestingly, the procedure forms spherical particles, which can be set in the average size with the help of a classifier at the discharge. In many cases, the mean diameter d_{50} lies in the range of 0.4–1.5 mm.

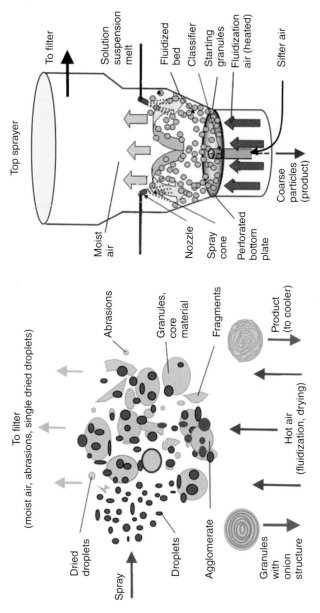

Figure 8.23 Principle of the fluidized bed spray granulation.

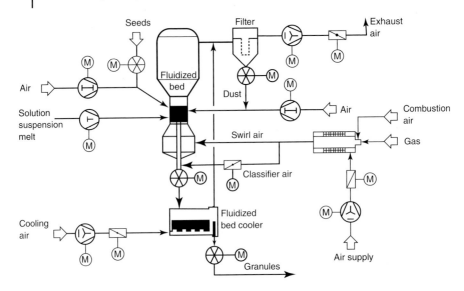

Figure 8.24 Process flow diagram of a fluidized bed granulation with integrated classifier and dust recycling (Courtesy of Glatt).

The process flow diagram depicts Figure 8.24. It contains the recycling of dust and the discharge of product after an integrated sifting and cooling in the small fluidized bed. In production plant follows usually a sieving of the cooled granules. The oversize material is recycled into the fluidized bed after grinding, for example, in a rolling mill.

8.9
Crystallization/Precipitation

As a unit operation of process technology, crystallization represents an interesting design technology and allows many ways of particle formation. The materials precipitate controllable out of the solution by an adjustable degree of supersaturation, preferably after addition of seed crystals. Crystallization enables a recovery of valuable products, concentration or purification of materials in the chemical and pharmaceutical industry as well as in food technology. Table 8.7 shows various methods. Crystallizations are possible from solutions, melting, and amorphous solids or out of gas phases as well as by recrystallization of redissolved crystals.

Crystallization from the solution, consisting of nucleation and crystal growth, can be controlled, depending on the material, in direction of a suppression of crystal growth. Therefore, the engineer sets the operating and supersaturation conditions accordingly, possibly assisted by addition of suitable substances. This results in very fine powder with a low degree of crystallinity. The opposite is also possible. By promoting crystal growth, large shaped crystals can develop also great monocrystals.

Table 8.7 Methods of crystallization from the solution.

Crystallization type	Technical process
Cooling crystallization	Cooling the solution
Evaporative crystallization	Evaporating the solvent
Vacuum crystallization	Evaporation of the solvent by depressurization
Displacement crystallization	Addition of a substance that reduces the solubility
Precipitation crystallization	Adding of a precipitant, reaction product is in supersaturated state
Freeze crystallization	Freezing out the solvent

The best example to demonstrate the variation of crystallization processes represents sugar. In Central Europe, sugar comes out of the sugar beets by extraction with water. After cleaning and boiling down the solution, in vacuo crystallizes the sucrose. Nature of the solutions and operation of crystallization generate different crystallite sizes, shapes, and colors (Figure 8.25). Specialty products such as candy move around slowly in a crystallizing cradle for long time. The brown color originates from the caramelization. Fine dried sugar serves as starting material for powdered sugar, produced in a grinding process, and for sugar cubes, arises in a compaction step. There exist in each case about 25 distinct products for home and industry.

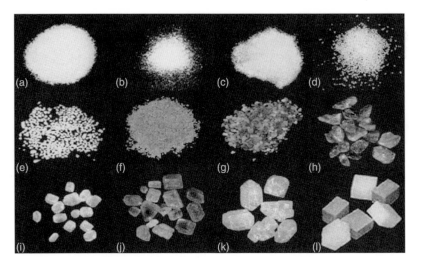

Figure 8.25 Sugar in various supply forms (a) finest sugar (refined); (b) powdered sugar; (c) jell sugar; (d) conserve sugar; (e) coarse sugar; (f) brown sugar; (g) Grümmel-candy; (h) crust candy; (i) white candy; (j) brown sugar candy; (k) Kluntje-candy; and (l) sugar cubes (Courtesy of Pfeifer and Langen).

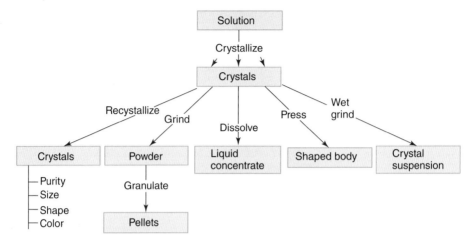

Figure 8.26 Possibilities for shaping crystals.

In sense of product design, the obtained crystals, separated from the mother liquor, enable a redesign in several ways. For this purpose, the crystals are either dried or solved again. Recrystallizations represent purification procedures, which go hand in hand by an adjustment of size and shape, color, and flavor. Further, crystals allow the manufacturing of concentrated solutions, powders, granules, or molded body (Figure 8.26).

8.10
Learnings

√ Spray drying is a world established method for the production of free-flowing powders from solutions/suspensions. Adapted procedures allow particles with average sizes from 25 µm (twin-fluid nozzle) up to 600 µm (single-fluid nozzle).

√ Big production towers show diameters up to 12 m and heights up to 50 m. The water evaporation may be in the range of $20-25\,\mathrm{t\,h^{-1}}$.

√ Several small nozzles are more effective than a few large and allow with increasing throughput a reduction in the specific energy consumption.

√ During drying, the structure formation in particle depends strongly on the formulation of solution and determines the product design (bulk density, voidage, flowability, and stickiness).

√ The developing particles are larger or smaller than the sprayed droplets. Due to the agglomeration in the course of drying, usually larger particles emerge in the dry product.

√ The drying with superheated steam allows an inert, emission-free drying with energy recovery.

√ Many organic substances require an inert drying with nitrogen in the gas circuit.

√ The spray agglomeration offers more design options as the normal spray drying, especially for the particle size, shape, and density, but is more expensive.

√ Sticky formulations dry in the spray agglomeration.

√ Different crystallization processes allow the preparation and purification of crystallizable substances.

References

1. Filková, I., Huang, L.X., and Mujumdar, A.S. (2007) in *Industrial Spray Drying Systems*, 3rd edn, Chapter 10 (ed. A.S. Mujumdar), CRC Press, pp. 215–256.
2. Rähse, W. and Dicoi, O. (2009) Produktdesign disperser Stoffe: Industrielle Sprühtrocknung. *Chem. Ing. Tech.*, **81** (6), 699–716.
3. Rähse, W. (2002) in *Laundry Detergents* (ed E. Smulders), Chapter 6, Wiley-VCH Verlag GmbH, Weinheim, pp. 122–144.
4. Walzel, P. (2009) Spraying and atomizing of liquids, in *Ullmann's Encyclopedia of Industrial Chemistry*, Wiley-VCH Verlag GmbH, Weinheim. doi: 10.1002/14356007.b02_06.pub2 (accessed 15 January 2010).
5. Walzel, P. (1990) Zerstäuben von Flüssigkeiten. *Chem. Ing. Tech.*, **62** (12), 983–994.
6. Blei, S. and Sommerfeld, M. (2005) Sprühtrocknung – Einfluss der Elementarprozesse auf Produkteigenschaften. *Chem. Ing. Tech.*, **77** (3), 278–282.
7. Masters, K. (1991) *Spray Drying Handbook*, 5th edn, Longman, London.
8. Bauckhage, K. (2003) in *Handbuch der Mechanischen Verfahrenstechnik* (ed H. Schubert), Chapter 5.2, Wiley-VCH Verlag GmbH, Weinheim, pp. 383–431.
9. Uhlemann, H. and Mörl, L. (2000) *Wirbelschicht- Sprühgranulation*, Springer-Verlag, Berlin.
10. Stiess, M. and Ripperger, S. (2008) *Mechanische Verfahrenstechnik – Partikeltechnologie 1*, Springer-Verlag, Berlin.
11. Chamberlain, H.R. (1990) The Procter & Gamble Co., DE 690 30 395 T2, Jan. 22, 1990 (EP: 90 30 0648.4).
12. Kröger, B. and Schulte, G. (2000) GVC Fachausschuss Trocknungstechnik, Würzburg, p. 16, Preprints.
13. Kind, M. and Stein, J. (2007) in *Product Design and Engineering*, vol. 1, Chapter 7 (eds U. Brökel, W. Meier, and U. Wagner), Wiley-VCH Verlag GmbH, Weinheim, p. 133.
14. Huang, L.-X., Zhou, R.-J., and Mujumdar, A.S. (2008) Review on the latest development of spray drying. *Drying Technol. Equip.*, **6** (1), 3–8.
15. Herbener, R. (1987) Staubarme Pulver durch Sprühtrocknen – Erfahrungen mit neueren Techniken. *Chem. Ing. Tech.*, **59**, 112–117.
16. Meinhard, B.-R. and Mortensen, S. *Stand und neue Varianten der Sprühtrocknung*, GEA/Niro.
17. Lauritzen, C. (1993) in *Proceedings of the World Conference on Oilseed Technology and Utilization* (ed. T.H. Applewhite), AOCS Press, p. 299 ff.
18. Kumar, P. and Mujumdar, A.S. (1990) Superheated steam drying: a bibliography. *Drying Technol. Int. J.*, **8**, 195–205.
19. Shi, Y., Li, J., Li, X., Wu, M., Zhao, G., and Wu, M. (2012) Research progress of the superheated steam drying technology. *Drying Technol. Equip.*, **10** (1), 3–9.
20. van Deventer, H.C. (2004) Industrial Superheated Steam Drying, TNO - Report R 2004/ 239 (via Internet); Printed version: Forschungsvereinigung für Luft- und Trocknungstechnik, Frankfurt/Main, Heft L 202.
21. Paatz, K. and Rähse, W. (1999) Henkel AG & Co. KGaA, DE 19900247A1, Jan. 7, 1999.
22. Rähse, W., Dicoi, O., and Walzel, P. (2001) Henkel AG & Co. KGaA, EP 1280591, April 24, 2001.
23. Monse, K., Linnepe, T., Groom, S., and Walzel, P. (2002) Einfluss der Düsengeometrie auf die Partikelbildung

bei der Entspannung überhitzter Suspensionen. *Chem. Ing. Tech.*, **74** (7), 963–966.

24. Rähse, W. (2009) Produktdesign disperser Stoffe: Industrielle Granulation. *Chem. Ing. Tech.*, **81** (3), 241–253.

25. Mörl, L., Heinrich, S., and Peglow, M. (2007) in *Granulation* (eds A.D. Salman, M. Hounslow, and J.P.K. Seville), Chapter 2, Elsevier, Amsterdam, pp. 21–188.

26. Rähse, W. (2009) Produktdesign disperser Stoffe: Industrielles Partikelcoating. *Chem. Ing. Tech.*, **81** (3), 225–240.

27. Gamse, T., Schwinghammer, S., and Marr, R. (2007) Erzeugung feinster Partikel durch Einsatz von überkritischen Fluiden. *Chem. Ing. Tech.*, **77** (6), 669–680.

28. Weidner, E., Petermann, M., Blatter, K., and Rekowski, V. (2001) Manufacture of powder coatings by spraying of gas-enriched melts. *Chem. Eng. Technol.*, **24** (5), 529–533.

29. van Ginneken, L. and Weyten, H. (2003) in *Carbon Dioxide Recovery and Utilization* (ed. M. Aresta), Chapter 3, Kluwer Academic Publishers, Dordrecht, pp. 123–135.

30. US: Clean Air Act, Environmental Protection Agency (EPA) (2002) *Technische Anleitung zur Reinhaltung der Luft, TA-Luft*, Carl Heymanns Verlag KG, Köln 1963, last accessed 1990.

9
Manufacturing of Application-Related Designed Plastic Products

Summary

No other field offers more possibilities for tailor-made production of very different products, as does macromolecular chemistry. There are a number of options for controlling the properties during polymerization (Polymer Engineering), by the addition of excipients (Polymer Design), and by variable shaping (Polymer Shaping). This chapter deals with the design of polymers for the detergent industry. First, suitable water-soluble polymers represent a part of the formulations of detergents to improve the washing result. Second, there are foils for the packaging of liquid detergent portions (PVOH, polyvinyl alcohol pouches). Third, insoluble thermoplastic polymers (PE, polyethylene; PP, and polypropylene; PET, polyethylene terephthalate) are needed as packaging materials for solids, pastes, suspensions, and liquids in the form of laminated cardboard, boxes, bags, bottles, buckets, and tubes. In all cases, the polymers can be tailored to suit the application. Adjustments to meet the required physical, technical, and optical properties, such as color, durability, elasticity or strength, and stiffness, are made during the reaction and through additives. In an extrusion process, the plasticized or melted material gets the product shape; any form (film, foil, hollow, or solid body) can be produced. The packaging should conform not only in form and functions, but also in aesthetics, relating to the specific brand and the designed color combinations, as well as pictures and logos. The product design is the sum of these parameters.

9.1
Polymers

No other chemistry opens up so many possibilities for targeted production of one-, two-, and three-dimensional products (see Figure 9.1), that is, fibers, films, and molded articles, as does macromolecular chemistry.

In a chemical reaction (polymerization, polycondensation, and polyaddition), monomeric organic molecules combine both with linear chains and with branched and/or cross-linked structures. The individual polymer molecules consist of several thousand monomeric units up to more than 1 million, depending on the reaction.

Industrial Product Design of Solids and Liquids: A Practical Guide, First Edition. Wilfried Rähse.
© 2014 Wiley-VCH Verlag GmbH & Co. KGaA. Published 2014 by Wiley-VCH Verlag GmbH & Co. KGaA.

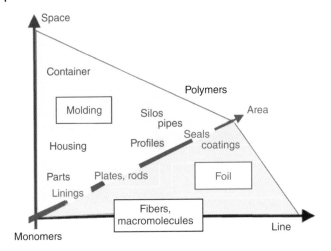

Figure 9.1 One-, two-, and three-dimensional plastic products.

The fundamental understanding of polymer chemistry goes back to Hermann Staudinger [1–3], who published pioneering ideas about the structure of polymers in the 1920s. In the following period, development of new polymers and novel synthesis methods has made rapid progress. In Germany, the industry produces and processes more than 20 polymers, based on different monomers, in quantities of 100 000 tons to 1 800 000 tons/annum (2007).

Selection of monomers, the production process, the catalyst, and further the operation conditions, allow a variation of material within wide limits. Physical properties of plastics, which include malleability, hardness, elasticity, breakage and tensile strength, temperature and heat distortion, creepage, and surface resistance as well as the chemical stability to solvents and oxidants, and against UV light further improve depending on the composition of different molecules. However, the chemistry and reaction engineering help adjust the physical and chemical properties and the downstream modifications of polymers by admixtures (compounding) offer other opportunities for the optimization of technical key figures and the aesthetics of plastic products.

The shape of a product arises through various methods, depending on the plastic and the intended use. After polymerization and compounding, the widely deployed thermoplastics occur usually as granules by extrusion. After melting of these granules at the industrial customer's end, the product shape originates at the extruder head through holes, slots, or flanged tools with cooling.

For thermoplastics, three stages exist to the final product (Figure 9.2): the production of the polymer (synthesis), its modification with additives, and the transfer into the application shape; all three steps occur in parallel. The coordinated processes allow targeted manufacturing of performance-oriented products and are responsible for the complete product design. The processes comprise diverse syntheses (polymer engineering) and polymer design by the addition of excipients as well as by polymer shaping, possibly supported by a blowing agent. These processes

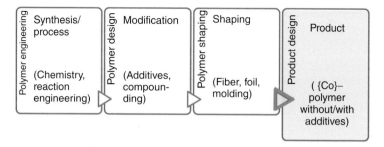

Figure 9.2 Three stages in the plastics manufacture up to the desired product design of thermoplastic products.

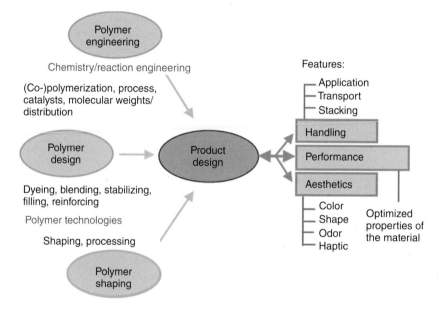

Figure 9.3 Composition and characteristics of the product design in plastics processing.

lead to the three characteristic features (see Figure 9.3): aesthetics, handling, and performance [4]. Polymer engineering at the molecular level determines the product's performance.

9.2
Importance of Plastics

The consumer encounters plastics in daily life in almost all products. Computers and television, cars, tires, telephones, and hobby and household appliances contain tailored polymers, as also liquids for the care of skin and hair, for cleaning the dishes, kitchen, and bathroom as well as the drains. Macromolecules operate as active agents and viscosity regulators. Consumers buy all cleansers as well as liquid and solid foods (water, yogurt, cookies, and chocolates) in plastic containers. The plastic bag is established as a carrying bag for shopping. In technical fields, many objects in the household consist of plastics, such as window frames, insulations, and gaskets, pipes, carpets, paints, and adhesives.

Many plastics are made in a custom shape and quality. They reach the consumer directly. A typical example is the glass-like water cup from high impact polystyrene (PS), which does not break and cause any scratches or spillings when dropped. The variety of polymers as well as the opportunities for targeted adjustments to specific parameters makes these materials universally useful. After the amount of steel produced (over 1 billion tons/annum [5]), which often goes in the capital goods sector, plastics represent the most frequently used material.

Many well-known plastics are made from molten thermoplastic polymers (Table 9.1). The melt presents the particular advantage that it runs via extrusion into any shape. In this way, it is possible to realize almost every requirement through

Table 9.1 Known thermoplastics, elastomers, and duromers: (highlighted: in the detergents industry used for packaging, adhesives not included).

Thermoplastics (fusible, soluble, and non-crosslinked)		Elastomer (infusible, insoluble, swellable, and widely crosslinked)	Duromer (infusible, insoluble, closely crosslinked, and thermally stable)
Polyethylene (PE)	Polyurethane (TPU)	Natural rubber (NR)	Polyester (resin) (UP)
Polypropylene (PP)	Polymethyl-methacrylate (PMMA)	Styrene-butadiene rubber (SBR)	Epoxy resins (EPs)
Polystyrene (PS)	Polytetra-fluoroethylene (PTFE)	Chloroprene rubber (CR)	Formaldehyde resins (different types)
Polyvinyl chloride (PVC)	Polyvinylidene fluoride (PVDF)	Acrylonitrile-butadiene rubber (NBR)	Polyurethanes (PU)
Polyacrylonitrile (PAN)	Polysulfone (PSU)	Fluoro rubber (FKM and FPM)	
Polyethylene terephthalate (PET)	Polyamide (PA)	Ethylene-propylene-diene rubber (EPDM and EPM)	
Polycarbonate (PC)	*Polyvinyl alcohol (PVAl/PVOH)*	(Chlorine) butadiene rubber (CR and BR)	

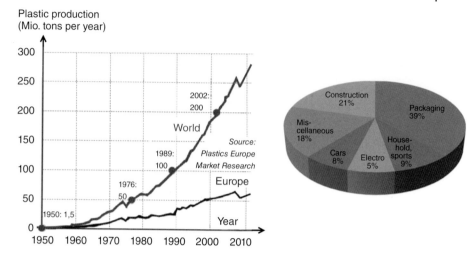

Figure 9.4 Production level over time and the breakup of plastics according to the applications in Europe in 2011.

design. The manufacturing of duroplastics takes place by irreversible cross-linking reactions. The injected monomers polymerize in the shaping tool. In some cases, for foaming, such as for the polyurethanes, the addition of carbon dioxide is necessary. Duroplastics are infusible, insoluble, and closely cross-linked. Particularly, hard and durable parts require a mechanical finishing. Flexible cross-linked rubbers represent the largest group of elastomers. The cross-linking reactions ensue at elevated temperatures over vulcanization with sulfur, peroxides, or metal oxides, as well as by UV-initiation or by addition of low molecular weight polyfunctional amines and alcohols.

In 2011, the plastic industry produced about 280 million tons worldwide, out of which almost 50% reached the consumer directly. About a quarter of the total number was produced in Europe (21%) and the CIS States (3%), 39% in Asia (thereof 23% in China and 5% in Japan), 7% in the Middle East and Africa, 5% in Latin America, and 20% in the NAFTA countries. Breakup of these figures shows that in Europe, 39% goes into packaging [6]. The other parts are depicted in Figure 9.4. The various types of plastics (Europe) are distributed as follows: polyethylene (PE) 29%, polypropylene (PP) 19%, polyvinyl chloride (PVC) 11%, polystyrene (PS) 7.5%, polyethylene terephthalate (PET) 6.5%, polyurethane (PUR) 7%, and others 20% [7]. It follows that well over 70% of the plastics are thermoplastics.

9.3
Task

First, a general explanation of the terms "polymer engineering, design, and shaping" that constitute product design for plastics is provided. Thereafter, a description of plastics in the detergent industry ensues. Different water-soluble macromolecules act in the washing process while water-insoluble plastics are used for packaging. The composition and structure of macromolecules are optimized in the polymer engineering step. The functions determine the product design.

Modern detergent formulations contain different water-soluble polymers, which take on various tasks in the washing process. These include the suppression of growth of calcium carbonate crystals (builders), the prevention of graying by dispersion of dirt, limiting of color transfer, and rejection of dirt by means of polymer coating on the textile fiber.

For washing and cleaning agents, the packaging material consists partially or completely of plastic. A special case for packaging represents the water-soluble films of polyvinyl alcohol (PVOH/PVAl) in the form of a pouch. These sachets contain low-water liquid detergents for one-time use. Solids are usually packed in laminated and unlaminated cardboards, some with an inserted plastic bag. Liquids, sometimes also solids, are packed in 100 ml to 10 l large plastic bottles or containers of different materials. The plastic packaging, optimized for application with additives, serves for transportation, protection, storage, and as dosing units. The packaging shape meets the requirements of problem-free handling and aesthetic demands. The transport safety of goods, stacked on pallets, is ensured through a shrink or stretch film. In general, packaging materials consist of different polymers. The question arises as to how an application-optimal formulation appears. It is essential to choose the right raw material mixtures for processing aesthetic and stable containers.

This contribution discusses various possibilities for the design of polymers, especially for detergents. In focus are the thermoplastics that play an important role in packaging. The manufacturing chain comprises the performance profiles and optimal shapes for handling and the aesthetics (*Plastics Engineered Product Design* [8, 9]).

9.4
Product Design for Plastics

The product design of plastics consists of the elements "polymer engineering, polymer design, and shaping."

9.4.1
Polymer Engineering

Polymer engineering is the selective synthesis of monomers into a desired polymer in terms of chemical and physical structure. Besides the variation of monomers in type, number, and proportions, some possibilities exist to control the reaction, further to choosing the polymerization process (such as mass or emulsion), the reactor type, and operating conditions (see Figure 9.5). This affects the design of reactors, the solvent, addition strategies of monomers, pressure, and temperature profiles, as also the residence time distributions and the achievable molecular

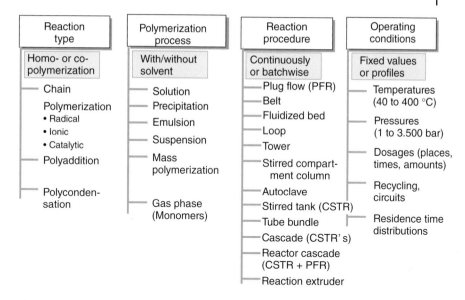

Reaction type	Polymerization process	Reaction procedure	Operating conditions
Homo- or co-polymerization	With/without solvent	Continuously or batchwise	Fixed values or profiles
Chain	Solution	Plug flow (PFR)	Temperatures (40 to 400 °C)
Polymerization	Precipitation	Belt	
• Radical	Emulsion	Fluidized bed	Pressures (1 to 3.500 bar)
• Ionic	Suspension	Loop	
• Catalytic		Tower	Dosages (places, times, amounts)
Polyaddition	Mass polymerization	Stirred compartment column	
		Autoclave	Recycling, circuits
Polycondensation	Gas phase (Monomers)	Stirred tank (CSTR)	
		Tube bundle	Residence time distributions
		Cascade (CSTR's)	
		Reactor cascade (CSTR + PFR)	
		Reaction extruder	

Figure 9.5 Variants and influences in production of polymers as well as the possibilities to control the product design by polymer engineering.

weights with their distributions. Other parameters for adjusting the polymerization represent the catalysts, initiators, and special regulators (inhibitors, promotors).

In addition to the manufacturing of monomers and polymer engineering (targeted polymerization), the chemical industry is responsible largely for polymer design (compounding/coloring). Thermoplastics can be synthesized as homopolymers or copolymers, and transferred via the plastic melt into powder or granular form. These include, in particular, PE, PP, polyamide (PA), polycarbonate (PC), PVC, PS, and thermoplastic polyurethane (TPU). A light material with high thermal insulation (e.g., plates) arises from PS by foaming with carbon dioxide or with alkanes.

The synthesis of PE, the most important thermoplastic, offers all possible settings, which are illustrated and discussed. Reactor and operating conditions, such as pressure, temperature, and catalysts, decide the molecular weight distributions and the degrees of branching (Table 9.2). The parameters allow the adjustment of molecular weight distribution over a wide range from mono- to bimodal. A proportional linear relationship exists between crystallinity (30–80%) and density (0.90–0.97 g cm^{-3}). The choice of monomers and co-monomers determines the structure of the polymer [10]. Common co-monomers are propene, 1-butene, 1-hexene, and 1-octene, and also vinyl acetate and acrylics (for polymer films). The following seven polymeric PE types are commonly found [5]:

- *VLDPE* ("very low density polyethylene"): linear PE with many short branches of co-monomers (alpha-olefins, such as 1-butene, 1-hexene, and 1-octene); very low density of 0.880–0.915 g cm^{-3}, produced with metallocene catalysts (polyolefin pastomere).

Table 9.2 Processes for the preparation of different polyethylenes [10, 11].

Process	Reactor	Pressure (bar)	Temperature (°C)	Catalyst	PE-type
High pressure	Tubular reactor; stirred autoclave	1000–3500	80–300	Oxygen, peroxide; regulator: propene, acetaldehyde	LD
Ziegler (suspension)	Stirred tank (paraffin oil)	2–8	40–140	Ziegler-Natta-cat.; $MgCl_2$, $TiCl_4$, and triethylaluminum; metallocene-comp.	LLD, MD, HD, HMW, and UMW
Phillips (suspension)	Tubular loops and loop-reactors (isobutane)	40	90–105	Fine silicon substrate with chromium compounds	LLD, MD, HD, and HMW
Gas phase	Fluidized bed	20–25	90–120	Ziegler-, Phillips-, and metallocene-catalysts	LLD, MD, and HD
Solution polymerization	Stirred tank; (1-octene; hexane; cyclohexane; and isopar)	50–150	150–300	Ziegler- or metallocene catalysts	Ethylene-propylene-elastomers

- *LLDPE* ("linear low-density polyethylene"): linear PE with short branches of co-monomers (alpha-olefins); narrow molecular weight distribution and low density, between 0.915 and 0.925 $g\,cm^{-3}$.
- *LDPE* ("low density polyethylene"): highly branched polymer chains achieved through radical high-pressure reaction with 1000–3500 bar; broad molecular weight distribution and low density of 0.910–0.940 $g\,cm^{-3}$.
- *MDPE* ("medium density polyethylene"): medium density between 0.925 and 0.935 $g\,cm^{-3}$.
- *HDPE* ("high density polyethylene"): slightly branched polymer chains; high density of 0.94–0.97 $g\,cm^{-3}$.
- *HMWPE* ("high molecular weight polyethylene"): molecular weights predominantly in the range of 500–1000 $kg\,mol^{-1}$ and density 0.942–0.954 $g\,cm^{-3}$.
- *UHMWPE* ("ultra high molecular weight polyethylene"): molecular weights from about 3 100–5 700 $kg\,mol^{-1}$ and density 0.93–0.94 $g\,cm^{-3}$.

Another very interesting polymer group, the ethylene vinyl acetate (EVA) results with vinyl acetate (10–40%) as copolymer. This thermoplast shows elastomeric properties. Depending on the amount of copolymer and polymerization, granules, powders and dispersions, or films originate in the polymer engineering and subsequent procedures.

In addition, some types cross-link using high-energy radiation (PEX or XLPE). The variations in polymer engineering lead to about 20–30 basic types of copolymers, distinguishable by the physical parameters (density, crystallinity, melting point, and modulus of elasticity). The number rises above the amounts of co-monomer as well as the adjustable molecular weights and distributions easily to more than a hundred (without special additives/blends).

In the high-pressure process, ethylene polymerizes in the reactor [10, 12] after compression in two stages of the reaction. The excess flows back after separation in a high- and low-pressure circulation. In another method, ethylene recirculation occurs after product separation in the flash tank. With gaseous ethylene, a fluidized-bed process works, where the powdered product arises directly in the vortex chamber. On using a solvent in the reaction, the liquid runs back into circulation after passing through separation and purification. The thermoplastic polymers formed, which may contain the catalyst, usually accrue as melt that flows into an extruder. In this machine, homogenizing takes place, well established as the second part of product design called *polymer design*.

9.4.2
Polymer Design

The property profile of synthesized polymers is adjusted according to the application during "polymer design." Single or twin-screw extruders, installed downstream of the polymerization in the house of the manufacturer, incorporate the different additives homogeneously. The flanged underwater pelletizing generates polymer granules (Figure 9.6), which are sold to industrial customers for further processing. Uniform melting during further processing in the extruder requires granules of predominantly uniform size as starting material. Finally, the granules are suitable for transport, storage, conveying, and dosing.

Compounding represents the homogeneous incorporation of additives into the melt. For optimization of polymers, there are two possibilities:

1) Incorporation of additives and/or
2) Mixing ("blending") of different polymers.

Both operations happen in specially adapted extruders, wherein the geometry and the combination of various screw elements vary according to the homogenizing task. The majority of plastic is mixed with additives. Polymer design helps in the replacement of glass and metal by plastics and in the development of new applications.

The improvement of material properties by additives (Table 9.3) leads to a reduction in the amount of polymer required. Furthermore, other additives support

Figure 9.6 Plastic granules (sizes 1–5 mm) by extrusion: (a) PE colored (from wet granulation); (b) PC; (c) recycled PE; and (d) PVOH (Courtesy of Bayer AG (b) and Kuraray (d)).

the shaping, which thereby runs smoothly and without problems. The products are always more specialized and stay increasingly longer. In particular, the chemical, electrical, and technical characteristics can be tailored to the application. Addition of UV inhibitors and antioxidants prevents the discoloration of products.

The stability of plastics increases by incorporation of particles, fibers (fiber-reinforced plastics), or platelets. Fabric mats and glass fiber mats offer even better stability. Many of the polymer melts are extremely hydrophobic. For the incorporation of reinforcing materials, the melts need some hydrophilic properties. This happens, for example, by the grafting of the melt with maleic anhydride, acrylic acid, or itaconic acid [11], or with any bonding agent. Flame protection represents an essential aspect. By adding suitable substances, it is possible to reduce the formation of hazardous gases in case of fire. Often, aesthetics (color, transparency) or economy (cheaper by mixing fillers) claims the focus. A well-known example is PP, reinforced with glass fibers or talc, sold in trade since 30 years. Glass-fiber-reinforced PAs are widely used in Europe (see Chapter 16).

By incorporation of a second polymer in the molten basis polymer, special blends emerge. These blends (see Table 9.4) combine the characteristics of two very different polymers. Therefore, novel properties arise, which otherwise would be unachievable. A targeted admixing of polymers leads usually to an improvement in several key figures, which continue to improve by addition of typical additives. In this connection, the highly complex combination of thermoplastic polymers (PS and PVC) with elastomers is highlighted for improving impact strength. In this way, unbreakable and even transparent objects arise from plastic – dyed or clear. In addition, mixing of two thermoplastics may improve the hardness/stiffness/resistance.

Table 9.3 Additives for the polymer design.

Additives	Tasks	Material examples
Softener	Reducing brittleness, hardness, and decreasing the glass transition temperature	Esters of phthalic acid (such as dioctyl phthalate), glycerol
Extender	Improve the processing	Mineral oils and paraffins
Antistatics	Prevention of electric charges	Fatty acid esters, amides, ethoxylated amines and alcohols, sulfonates
Antioxidants	Scavenging radicals	Phenols and amines
Deactivators	Preventing of free radicals	Amine
Sun protection	Reduction of the influence of UV rays	Carbon black (absorption), hydroxybenzo-phenone (UV transfer in IR radiation), and dialkyldithiocarbamates (scavenger)
Heat stabilizers (especially PVC)	Decomposition mechanism at double bonds; use of complexing	Heavy metals (Zn, Sn, Ba, Cd, and Pb) compounds
Flameproofing	Reduction of toxic gases in the thermal decomposition	Polybrominated diphenyl ethers and aluminum hydroxide
Coloring agent	Coloring of the total material	Dyes (dissolved), pigments (insoluble), and effect pigments
Fillers	Extenders to cheapen and reduction of fire hazard	Chalk, sand, diatomaceous earth, sawdust, and starch
Reinforcing materials	Improvement of the mechanical properties	*Fibers*: glass, carbon, plastic (aramid), natural products (jute) *Particles*: glass, graphite, and aluminum hydroxide *Platelets*: layer silicates, talc, and mica
Adhesion promoter	Combine polymers with reinforcing materials (molecular bridges at the interfaces)	Vinyl chloride copolymers, various resins, and so on

For example, blending of PP with ethylene–propylene elastomer in different amounts modifies the hardness in steps, from very firm to elastic. After addition of glass fibers, the polymer mixture shows not only increased elasticity but also a higher stability (polymer blend + additives = multiple polymer design). Besides the typical characteristics of PC, blends have an increased heat and chemical resistance. Therefore, they are widely used in automobiles, for electrotechnical products, and for devices in the household.

Table 9.4 Commercially available polyblends.

Main polymer (the continuous phase forms the matrix)	Blended polymer (disperse phase)	Improved properties
Polypropylene (PP)	Ethylene-propylene elastomer (EPM)	Impact strength
Polypropylene (PP)	Ethylene-propylene-diene rubber (EPDM)	Impact strength
Polystyrene (PS)	Elastomer	Impact strength
Acrylonitrile-butadiene-styrene copolymer (ABS)	Polyamide (PA)	Impact strength at low temperatures
Polysulfone (PSU)	Unknown	Electrical indicators
Polycarbonate (PC)	Polybutylene terephthalate (PBT)	Impact strength, heat, and chemical resistance
Polycarbonate (PC)	Polyethylene terephthalate (PET)	Impact strength, heat, and chemical resistance
Polycarbonate (PC)	Acrylonitrile-butadiene-styrene copolymer (ABS)	Impact and notched impact strength, stiffness
Acrylonitrile-styrene-acrylate (ASA)	Polyamide (PA)	UV-stability, haptic, resistance to chemicals

The blended macromolecules either dissolve homogeneously in the melt or spread or disperse heterogeneously. The properties of heterogeneous polyblends depend on the morphology (size, distribution, and spatial structure) and interfacial compatibility of materials involved [10]. These blends show two glass transition temperatures, unlike dissolved macromolecules.

9.4.3
Polymer Shaping

The chemical industry supplies different polymers to the plastics processing industry, including homogeneously incorporated additives and other polymers (blends). Thermoplastics are mostly available as granules. During remelting in an extruder, the plastic processor adds lubricants, and, if necessary, further excipients for supply and removal from the mold. Alternatively, the melt flows at the extruder head through bores, nozzles, or slots to form pre- or end-products (Figure 9.7). Other opportunities in processing are the grinding or dissolving of granules and powders for coatings and adhesives.

For the shaping of thermoplastics ("original form" and "reshaping"), many processes exist, all based on extruder technology. Important industrial procedures for the manufacture of fibers, moldings, hollow bodies, and films are demonstrated in Table 9.5.

Table 9.5 Technologies for molding of thermoplastic materials.

Technology	Products	Shaping
Extrusion	Fibers, rods, plates, pipes, profiles, and moldings	*Extruding* through nozzles, holes, and slots
Injection molding	Moldings, full, and hollow body	*Extruder* head with flanged tools; injection of melt into the tool shape under high pressure ($< 2\,500$ bar)
Blow molding	Bottles, barrels, buckets, containers, bellows, and fuel tanks	*Extrusion* of a tube, which receives in a downstream mold the product form by blowing up the pinched tube with pressurized air; through plurality of extruders (materials), it is possible to produce distinct layers
Blow-fill process	Filled hollow body	Hollow bodies, manufactured through blow molding, are filled with product before demolding
Injection blow	High pressure (seamless) bottles with a constant wall thickness	Injection molded *preforms* (seamless) are blown in a mold, after heating to the softening temperature (squeezing not required)
Stretch blow molding	High pressure (seamless) bottles with a constant wall thickness	Injection molded *preforms* with finished thread are heated in multiple temperature zones, and then stretched longitudinally before blown with pressurized air in two pressure stages
Coextrusion	Multilayer foils (up to seven) for packaging, building materials, and landfill covers	*Extrude* of different polymers through several extruders for the production of multilayer films, sheets, or tubes, usually with the use of recycled materials for the inner layers
Thermoforming	(Food) packaging, blisters, trays, boxes, and industrial parts	Shaping of *tempered foils* by applying a vacuum below the molds in the tools
Casting (mold and belt castings, precipitations)	Moldings and foils	Workpieces and foils
Calender foils	Foils and coatings	*Rolling out* of the melt by a tempered roll systems with narrow gaps
Blown foils	Foils and multilayer films (different thickness)	*Extrusion* through annular die and shaping to a thin film by air injection, up to seven extruders (layers) possible
Flat films	Foils, multilayer films (different thickness)	Extrusion through slot dies (chill roll process). For flat films, several layers are one above the other

(continued overleaf)

Table 9.5 *(Continued)*

Technology	Products	Shaping
Foam cast/gas counter pressure	Lightweight plastic parts for the office with a smooth surface (housing for calculators, printers)	Polymer granules with *gas-seceding additives;* by counter pressure arise smooth surfaces
Gas/water injection	Massive, light body (thick walls with cavity)	Injection of parts, which contain inside a cavity, received by blowing nitrogen or water (weight loss by hollow space)
Film insert molding	Engine covers (parts of high aesthetics)	A perfect fitted, stamped foil is placed in the cavity of an injection mold, followed by pre- and main injections, possibly with foam
Foaming	Very light body, high thermal insulation	Plastic expansion using chemical and physical blowing agents
Rotational molding (-casting)	Large hollow body	Plastic melt rotates in a hollow tool

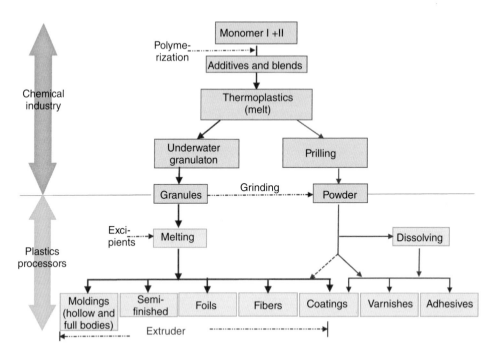

Figure 9.7 Shaping of thermoplastics by the plastics processors via remelting, grinding, and dissolving.

9.5
Polymers in Detergent Formulations

Suppliers of raw materials synthesize and improve the functions of polymers for the washing or cleaning process (Polymer Engineering). This results in a product range from which the industrial customer can choose. These chemicals are sold as liquid or melting, aqueous solution, or solid (powder, granules, or flakes), and as compounds in large containers, silos, or tank trucks. In the detergent industry, product developers and application engineers incorporate these nature-based or mostly synthetic polymers in different formulations. They study the effect and stability in conjunction with other components of the formulation, depending on the concentration and applied quantity, the washing temperature, and textile materials. This optimization takes place under practice conditions in different washing machines.

Some examples of polymers, found in solid and liquid detergents, in cleaners (hand and dishwasher) as well as in starches, are discussed in the following. Their tasks include water softening, dispersion of dust, dirt rejection on the fiber, preventing color transfers, and smoothing (starch) of textiles [13]. The polymers represent an integral part of the recipe (see Table 9.6) and, therefore, exert influence on the design of the final product, most visible in the gel-like liquid detergents.

9.5.1
Cobuilder

Two sodium salts of the polycarboxylates, used in many solid and liquid detergents and cleaning products, have gained great importance as multifunctional builders. The modern phosphate-free solid detergents include relatively large amounts of polycarboxylates. Polymeric builders are polyacrylic acids with an average molecular weight (MW) of about 4500 g mol^{-1}, and copolymers consisting of acrylic and maleic

Table 9.6 Polymers in solid detergent formulations.

Task in the washing process	Substance	Amount
Water softening Cobuilder	*Polycarboxylate (Na-salts)*	2–8%
Stain removal Enzyme[a] *Inhibitors*	Protease, amylase, cellulase	<1.5%
Greying	*Cellulose ethers*	0–1%
Discoloration	*Polyvinylpyrrolidone*	<0.5%
Anti-foam	Polydimethylsiloxane[b]	0.1–2.5%
Pollution	*Polyethylene glycol terephthalate*	<1%

[a]Bioengineered, see Chapter 10.
[b]Silicone oil, insoluble in water.

acid with $20\,000{-}70\,000\,\mathrm{g\,mol^{-1}}$, added in amounts of 2 to $>8\%$. Depending on the region, the water utilized for washing contains poorly soluble calcium and magnesium salts in low concentrations. The polymers prevent the precipitation of carbonate and phosphate salts of these cations (threshold effect). Polycarboxylates are negatively charged polyelectrolytes. The number of charges is based on the mixing ratio of different monomers. These polyelectrolytes can bind multiply charged cations instead of sodium. Furthermore, in this way, they hold solid dirt particles in suspension.

For water softening, high-performance detergents use a new builder system. These solid systems are formulated approximately since 10 years. With sodium carbonate, a targeted precipitation of calcium and magnesium ions takes place. In the washing process, added polymers not only delay crystal formation but also inhibit further growth of the formed calcium carbonate (calcite) by adsorption on the crystal nuclei. Further, they disperse the precipitates and prevent deposits on the laundry. Only the development of these specialized polymers allows the application of this new builder system based on soluble sodium carbonate. It realizes a superior performance compared to the previously used ion exchangers (zeolites) and complexing agents. The dirt removed remains in water through dispersion. It is pumped out at the end of the washing process with the wash liquor. Polycarboxylates are not readily biodegradable, but more than 90% drop out in the clarifier or adsorb on the sewage sludge. Therefore, this class of substances is not critical to the environment.

9.5.2
Inhibitors (Graying, Color Transfer, Foam, and Dirt)

During washing, graying inhibitors keep the removed dirt dispersed and prevent redeposition on the fiber of the textile. Various polymers based on starch and cellulose are appropriate chemicals for this purpose. Particularly proven are cellulose ethers, such as carboxymethyl cellulose, methylcellulose, and mixed ethers and admixtures thereof, in use between 0.1 and 1%. In addition, polyvinylpyrrolidone and polycarboxylates keep the dirt in suspension and prevent graying. Appropriate and the best substances are derived through experiments because the effect depends on other components of the formulation and the textiles.

Color transfer inhibitors [14] (polyvinylpyrrolidone, poly-N-vinylpyrrolidone, polydiallyldimethylammonium, ammonium chloride = poly-DADMAC; MW $10\,000{-}70\,000\,\mathrm{g\,mol^{-1}}$) avoid discoloration of colored laundry during washing. Therefore, color detergents contain these specialists. To reduce the formation of stains and to facilitate the removal of dirt in the next wash, formulations include the so-called soil repellents. There are a number of copolymers in active detergents that cover the fiber, anchor themselves, and repel the dirt. A compound known to act in this manner is polyethylene glycol terephthalate (MW $10\,000{-}50\,000\,\mathrm{g\,mol^{-1}}$). Even PVOH (MW $15\,000{-}100\,000\,\mathrm{g\,mol^{-1}}$) provides such an effect.

Polydimethylsiloxane shows low viscosity, is insoluble in water, and spreads quickly at foamy surfaces. Developing air bubbles disturb the washing process by hindering the wetting of textiles and by reducing the mechanics. Surfactants stabilize gas bubbles at the liquid/gas boundary surface. Special silicon oils, with an affinity to the interfaces, are able to displace the surfactants and to destabilize the foam. The bubbles collapse and disappear. When added in sufficient quantities, suitable silicone oil prevents bubble formation. For detergents, the defoamer exists in the form of granules or powders, produced through separate procedures.

9.5.3
Excipients and Starches

In the manufacture of granules and/or coating layers, polyethylene glycol (PEG; MW 2000–12 000 g mol^{-1}) constitutes an important auxiliary substance, processed in solution, as solid, or as melt. An example illustrates the coating layer of enzyme granules, in which PEG provides the elasticity of coating. Sprayed dispersions may improve the smoothness of laundry or laundry items (collar). Predominantly suitable are natural starches, as also PVOH, PEG, and others.

9.6
Plastics in Detergent Industry

In the production plants of the detergent industry, plastics represent widely used materials for apparatuses and machinery. Examples are storage tanks, some of them fiberglass reinforced, along with tank linings, containers, guide roller for belts/bucket elevators, pipes, and pumps, filter materials and conveyor belts as well as anti-stick coatings. Different plastic types are suitable for this purpose. Moreover, many packaging tasks are solved with thermoplastics, as illustrated in the following. Packaging, optimized according to the wishes of customers in function and aesthetics, represents an interesting example of product design.

9.6.1
Packaging

Packaging of solids and liquids plays an important role in the consumer industry. The container is necessary for transport, storage, and protection of contents against mechanical damages as well as against chemical changes. The used plastic may not release any substances and must be chemically inert to the content in the intended temperature range. In principle, the filled product can react with the plastic material and with components of formulation, in particular by the action of UV radiation and by the diffusion of oxygen and water vapor through the container wall, supported by strong temperature fluctuations. Mostly, this concerns chemical oxidation reactions. Clumps and water movements (adsorbed/bound) occur in solids, which cause a reduction in the rate of dissolution. Water and oxygen diffusion through the container wall supports the growth of microorganisms in fluids. The selected packaging material must ensure that over a period of at least 3 months no significant changes of plastic material, and, in particular, of the content, occur under varying climatic conditions. An appropriate choice of polymer for the container, for lining, or for coating of cardboards meets the requirements.

Solids are packed in bags, boxes, or laminated cartons with/without foil liner and in rigid plastic containers. The lamination occurs usually through a single- or double-sided mechanical/thermal coating or bonding of PE film. A high standard of blocking effect (gas, water vapor, and light tight) ensures a five-layer laminate, for example, from outside to inside: PE/carton/PE/Al/PE with a carton weight of 75%. Plastic films, used for solids as packaging and auxiliary material, are of great importance. For example, several foils serve as composites for refill bags, or in the form of a single layer as inserted bag. Furthermore, foils secure the products stacked on pallets as stretch or shrink wrap.

The refill normally is filled in flexible plastic bags for transport. In households, the refill or the solid product comes in a rigid box or in a tin can. The polymer film consists of two or more layers in which the layer structure ensures high stability. The preparation occurs by fitting together several films with a pulse or thermal welding. Other methods represent the bonding of foils as well as the extrusion lamination [15], in which a hot melt is inserted through a slot die between two uncoiled films. The composites run thereafter over cooled rolls. The foil surfaces change by a possible coating (paint), vaporization, or metallization. Thus, gas permeability comes close to zero and the sealability improves. The treatment of the surface allows optimizing the aesthetics.

For wall materials, optimal plastics exist, according to the requirements of minimal diffusion of oxygen and water vapor through the wall. In widespread PE, gas permeability depends on the manufacturing method. The denser HDPE has significantly lower values than LDPE (a factor of about 3–5 for oxygen, temperature dependent [15]). The packaging of liquid and pasty products, partly also of solid products, consists generally of thermoplastics PE, PP, and PET (see Table 9.7). Different plastics for bottle, label, and closure are possible. For manufacture of containers, the use of homopolymers and copolymers with addition of excipients is customary.

Typical additives provide protection against UV rays and cause coloring. For aesthetic reasons, some copolymerized ethylenes enhance the transparency in the production of PP (PP Raco, random copolymer). At multilayer walls, the processing of plastics takes place with two or three extruders, wherein the outer layer consists of a new polymer, and the interior of recycled material. The processing of different materials is possible in this way.

A tube arises from melted PE and PP granules through extrusion. The emerging tube runs out vertically. Then, it is disconnected in a mold and inflated with compressed air (7–10 bars) into the desired hollow body shape (see Figure 9.8). This simple process is called *blow molding* and is used worldwide for bottles and other objects.

The processing of PET starts with injection-molded preforms (Figure 9.9). The hot preform, heated above the glass transition temperature and inflated in a tool with air, yields bottles with seamless walls. Heating takes place in different

Table 9.7 Plastics in the packaging of detergents (partially copolymerized or recycled).

Packaging	Technology	Plastics
Foils		
Refill bags	Multiple composites	LDPE and LLDPE
Labels	Monolayered foils	LDPE
Bag	Monolayered foils	LDPE, PP, and PET
Blister	Multilayer films	PET
Stretching and shrinking	Monolayered foils	LDPE and LLDPE
Carton		
Folding box	Lamination foils	LDPE
Carrying package	Lamination foils	LDPE
Hollow body		
Bottles	Blow molding and stretch blow molding	HDPE, PP, and PET
Canister	Blow molding and stretch blow molding	HDPE, PP, and PET
Cans	Blow molding and stretch blow molding	HDPE, PP, and PET
Tubes	Blow molding and stretch blow molding	LDPE
Closure	Injection molding	HDPE, PP, and PET
Measuring cups	Injection molding	PP

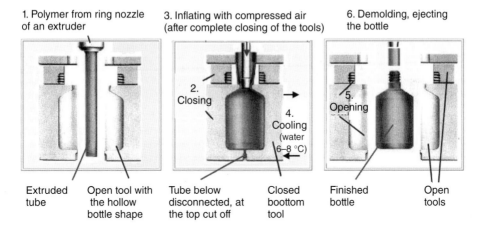

1. Polymer from ring nozzle of an extruder
3. Inflating with compressed air (after complete closing of the tools)
6. Demolding, ejecting the bottle

2. Closing
4. Cooling (water 6–8 °C)
5. Opening

Extruded tube — Open tool with the hollow bottle shape — Tube below disconnected, at the top cut off — Closed boottom tool — Finished bottle — Open tools

Figure 9.8 Principle of the blow molding to hollow bodies from an extruded tube.

temperature zones, so that the upper part with the thread remains dimensionally stable, and the lower portion is heated in a form- and location-dependent manner. In the "stretch blow molding" process, elongation of preform occurs not only in circumference, but also mechanically in the longitudinal direction [16, 17]. First, the preform is stretched with a mandrel, accompanied from simultaneous pre-blowing (5–15 bars). Thereby, the polymer molecules align preferably parallel to the stretching direction and bring about an improvement in mechanical properties.

1.	2.	3.	4.
Insert of the preheated, injection molded preform	Longitudinal stretching of the preform under preblowing	Finish blowing and cooling	Demolding

Figure 9.9 Principle of stretch blow molding of PET hollow bodies from an injection molded preform.

In this procedure, the temperature shows a big impact [18]. Then with final blowing (~ 28–40 bar), the final shape emerges. For energy optimization, a portion of the blown air is recycled with about 10 bars and used in the next stage for pre-blowing.

Blow molding and stretch blow molding are the most widely used methods for manufacturing of bottles. The transparent, pressure-resistant bottles for mineral water consist of PET as do the clear containers for detergents and cleaning products. The filling of the bottles and tubes produced takes place in separate filling lines.

Detergent producers utilize clear and colored plastic containers, sometimes with a glossy finish, for filling liquids, gels, and paste-like formulations. In some cases, containers are equipped with dosers, atomizers, and applicators (brush). Some of the liquid products, wrapped in special plastics for the application, such as toilet cleaners and fragrance beads, dissolve gradually in water. These products are often sold in thermoformed blisters (PET). The packaging enables easy storage, removal, and application. The form, flap, and colors, as also the surface design and quality of materials support the brand, which allows positive differentiation from the competition. In addition to the physical and chemical functions of the performance, convenience, and aesthetics, the brand represents a relatively large proportion of product design (Figure 9.10).

The bottles should not only be easy to transport and handle but also look attractive. A high product performance, caused by the content, is the essential quality aspect. Materials, colors, and shapes of containers, color combinations as well as images and lettering on the label yield the brand design. Figure 9.11 depicts the illustration of some packaging in various materials.

For most of the liquid products, the shipment takes place in large cardboard boxes, which contain usually 6 or 12 units per box. The closing of cartons is done by bonding with an adhesive or a sticky tape. Henkel, a famous manufacturer of detergents, launched its own production of adhesives in 1922, primarily to bond the detergent boxes. This later became a business by itself, which, today after expansions, represents the world's largest adhesives manufacturer.

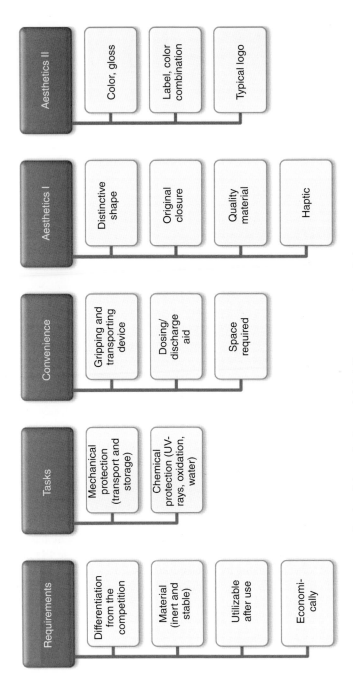

Figure 9.10 Requirements and tasks of the packaging and their influences on product design.

Materials

Bottle	PP	PET		PET, PET foil	PET	LDPE
Closure	HDPE, PP	PET, PP		PP	HDPE, PP	PP
Refill bag			LLDPE/PP			

Figure 9.11 Plastic packaging in the detergents segment with/without application aid. (Courtesy of Henkel AG & Co. KGaA).

9.6.2
Polyvinyl Alcohol as Packaging Material

For the detergent industry, PVOH/PVAl is an interesting thermoplastic polymer. It dissolves well in water, but not in many organic solvents. There are different types in the market, such as powders and granules, as well as thin films. Since PVOH is not synthesizable through the direct route, it is obtained from the alkaline hydrolysis of polyvinyl acetate (PVAc). First, PVOH quality depends on the polymerization of PVAs and of possibly used co-monomers, on the molecular weight distribution obtained, and on the proportion of head-to-tail linkages, controllable via the temperature. The molecular weight of industrially used atactic PVOH ranges from 9000 to 500 000 g mol^{-1}.

Secondly, the degree of hydrolysis, between 70% and almost 100%, as well as the distribution of OH groups (in blocks or random), determines the water solubility of the homopolymer. An 87% degree of hydrolysis and decreasing molecular weight yield the best values in cold water. Ash content (< 0.1–1%) represents another parameter.

For the 87% quality (residual acetyl content 10.8%), melting and decomposition regions are close together. Temperatures above about 180 °C lead to intra- and inter-molecular water cleavages that cause a change in the rate of dissolution. This temperature is of particular importance in the foil sealing at the edges; otherwise, the sealed parts remain unresolved in water. For lowering the melting range below 160 °C, it is advisable to incorporate additives or blend polymers. Advantageously, the addition occurs before reaching the melting temperature in the same extruder, which shapes the product. Refs. [22, 23] describe a number of substances that can be used for this purpose. The following substances are suitable as plasticizing additives – mono-, di-, and triglycerides, urea, glycerol, and/or nonionic surfactants in quantities of 2–10%. They not only lower the melting range but also affect positively the solubility. With solid PEGs, in particular, types of molecular weights ranging from 3000 to $15\,000\,\text{g}\,\text{mol}^{-1}$, blended PVOH is formed.

For packaging of a single washing portion, the detergent industry uses modified PVOH in the form of films and injection molded articles. Preferably, the foils emerge from blowing or casting methods [19]. Figure 9.12 shows conventional preparation processes for films, such as the extrusion of thermoplastics or casting/precipitations of polymer solutions. In the *blown film process*, the melt leaves the extruder through a flanged annular nozzle. The extruded tube forms a thin foil tube by blowing air. For production of multilayer films, an annular nozzle in the middle connects up to seven extruders. From the annularly extruded, blown film, vertical pulled up, arises after deflection, and lateral cutting indefinitely wound up foils.

For manufacturing of flat films [15] in a *chill roll process*, the melt flows through a wide slot nozzle onto a chill roll. On the way to the winder, a dancer roll regulates the film thickness. The *calendering process* is ideal for the production of PVC and thicker films. The added melt spreads out in the slot across the width of two opposing rollers before the passage begins, by heating to the foil temperature (see Figure 9.12 bottom/middle). By a plurality of directly successive calender rolls, specific shear stress over the entire film width arises to perform the stretching. The calendering technology serves preferably in the production of thicker films of constant thickness and further in coating or surface finishing.

The *casting process* is characterized by the use of a solvent with which the polymer dissolves in the tank. The solvent evaporates by heating the cast film on the belt.

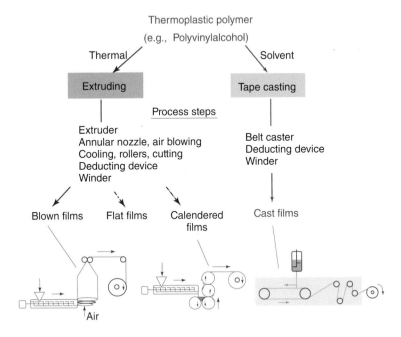

Figure 9.12 Manufacturing processes for films.

This part of the production line is encapsulated. The resulting film runs out of the enclosure to the winder. The microstructure of these foils differs from that of the thermally generated films, which become visible during welding.

In the detergent industry, PVOH films are used for wrapping of both solid and liquid products. One example is the toilet blocks, wrapped to prevent skin contact with the chemicals. Attached in the toilet water tank, the block supports cleaning during flushing. Another example is the transparent sachet for detergent powders and pastes. The third example may be the low-in-water (< 8% water) liquid detergents for single dosage wrapped in water-soluble films. Various methods are followed for producing the sachets, of which the packaging and thermoforming processes dominate, including the steps of forming, filling, and closing.

In *thermoforming for solids*, deformation of the film occurs in a horizontal continuous belt conveyor system by applying a vacuum on the underside. Product filling occurs in shaped tools on the belt, lined with foil, into the free spaces. Thereafter, another foil covers the filling on the upper side. Finally, the open edges are welded (Figure 9.13).

In the *thermoforming process for fluids*, two foils from the left and the right sides run between two rollers, which contain the shapes of two half-shells of the finished product. By applying a vacuum on the rollers, the films adhere and thus take the form of the container. During the welding of the edges, filling occurs centrally from above. Alternatively, foils from a roll run into special *packaging machines*. Folded in U-shape, the films are welded at the side and the bottom to the intended size. Then, the liquid flows into the open-top pouch through a centrally disposed pipe. After the upper edge is welded, the sachet (Figures 9.14 and 9.15) is ready.

Another method for the preparation of water-soluble shaped bodies involves the vertical extrusion of a polymer tube (as shown in Figure 9.8), which is disconnected by a two-piece, self-closing tool, and then inflated with compressed air. The cooling starts. Immediately, the hollow container is filled with the appropriate liquid through a tube that dips deeply into the hollow body and moves upward with the surface of the liquid. The completion requires the fastening of the closure or

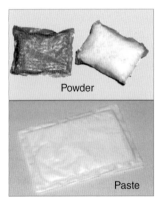

Figure 9.13 Detergent powders and pastes in water-soluble PVOH foils.

Figure 9.14 Design of liquid detergent with low water content, wrapped in PVOH foils (various market products).

Figure 9.15 Four pouches with liquid detergent in a blister pack (market product).

welding (*blow-fill-technology*). After further cooling, the tools open and the complete product is dropped. The "all in one" process is relatively simple, but process related; walls of different thickness and welds represent a disadvantage in the solubility.

Injection molding requires a two-part tool, made of polished stainless steel (Chapter 16). The cavities represent the shape of the manufactured container. The hollow body arises after the PVOH melt is supplied under pressure between 1000

(a)   (b)   (c)

Figure 9.16 Injection molded PVOH shapes for detergents and cleaning agents (a) sphere with solid and separate liquid filling; (b) three compartments for various solids (market product); and (c) liquid/liquid-filled ball with release films.

and 2000 bar into the extruder-flanged tool. The process is completed when the tool is cooled with water and the two parts are opened. The molding falls out. Demolding requires angles > 90° to the outer walls. For containers with partitions, the use of movable pistons facilitates the demolding. By injection molding, shapes with multiple compartments may originate. These moldings allow the separation of different liquids, solids with liquids, pastes, or melting (Figure 9.16). This extents several possibilities in product design.

In contrast to the PVOH films, which have a constant thickness in the range of about 55–75 μm, the wall thickness of injection-molded containers is about 450–700 μm. Appropriate wall thicknesses must be adapted to optimize the stability with respect to the solubility of these bodies. Foils allow the covering and separation. These shaped bodies, subdivided into compartments, may include different detergent formulations, without these coming into mutual contact [14, 20].

9.7
Shape and Function

The formulations of detergents contain water-soluble polymer molecules, which consist of linear, one-dimensional chains. Similar molecules are used as viscosity modifiers for several applications, also in liquid detergents. In contrast, foils have two dimensions. One form, which is water-soluble, can be used for portion packs [21], and as coating for enzymes. The other form, water-insoluble polymer films, can be processed as refill bags, and as stretch material and inside coating for tubes. The packaging materials consist of insoluble, three-dimensional objects (bottles, buckets, and tubes) so that in the detergent sector all kinds of plastic designs exist (Figure 9.17). PVOH is a possible material for bottles, filled with a low-water liquid. Therefore, in all dimensions water-soluble and insoluble tailor-made plastics are utilized, which contain additives and/or blended polymers. For each application, there are several problem-solving plastics, which differ not only in price, but also in aesthetics, for example, in the surface gloss. Interested customers judge the

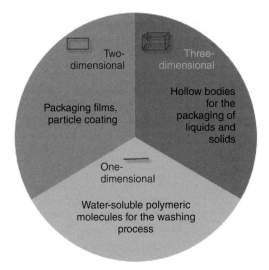

Figure 9.17 Plastics in the detergent area.

shape, function, and aesthetics of new products in focus groups, household use, and concept tests [4] before the product goes into the market.

9.8
Learnings

√ Macromolecular chemistry offers a large number of possibilities for customized manufacturing of application-optimized products.

√ Plastics (thermoplastics, elastomers, and duroplastics) differ with respect to their composition, solubility, cross-linking, and melt behavior.

√ The market share of thermoplastics is more than 70% from about 280 Mio tons/annum (2011) worldwide.

√ Product design occurs in three steps – polymer engineering, polymer design, and shaping.

√ Polymer engineering comprises the reaction type, polymerization process, reaction procedure (reactor), and operation conditions.

√ For optimizing of application properties in polymer design, excipients (additives) and other polymers (blends) are added.

√ For shaping of thermoplastics, many procedures exist. The moldings arise in two-part tools out of a melt, which originates from the plastic granules in an extruder.

√ In the household today, many plastic articles are made of transparent thermoplastics with blended elastomers for adjusting elastic properties (do not break).

√ In the detergent area, different water-soluble macromolecules are used in the formulation for softening the water, anti-graying, anti-foaming, soil repellency, and for avoiding discoloration. Further polymers (such as poly-acrylates) regulate the viscosity of liquid detergents.

√ Water-soluble PVOH foils and injection-molded shaped bodies exist for packaging of detergents. Water-insoluble materials (PE, PP, and PET) are widely used for bottles and tubes, obtained from blow molding and stretch blow molding processes.

References

1. Staudinger, H. (1920) Über Polymerisation. *Ber. Dtsch. Chem. Ges.*, **53**, 1073.
2. Staudinger, H. (1926) Die Chemie der hochmolekularen organischen Stoffe im Sinne der Kekuleschen Strukturlehre. *Ber. Dtsch. Chem. Ges.*, **59**, 3019–3043.
3. Staudinger, H. (1936) Über die makromolekulare Chemie. *Angew. Chem.*, **49** (45), 801–813.
4. Rähse, W. (2007) *Produktdesign in der chemischen Industrie*, Springer-Verlag, Berlin.
5. Baur, E., Brinkmann, S., Osswald, T.A., and Schmachtenberg, E. (2007) *Saechtling Kunststoff- Taschenbuch*, 30th edn, Carl Hanser Verlag, München.
6. PlasticsEurope (2009) *Communiqué de presse*, PlasticsEurope, Paris le 15 Octobre.
7. PlasticsEurope (2011) *Plastics – The Facts 2012: An Analysis of European Plastics Production, Demand and Waste Data for 2011*, Plastics Europe.
8. Rosato, D.V. (2003) *Plastics Engineered Product Design*, Elsevier Advanced Technology, Oxford.
9. Miller, E. (1981, 1983) *Plastics Products Design Handbook, Part A and B*, Marcel Dekker, Inc., New York.
10. Keim, W. (ed.) (2006, 2012) *Kunststoffe: Synthese, Herstellungsverfahren, Apparaturen*, Wiley-VCH Verlag GmbH, Weinheim.
11. Domininghaus, H. (2005) in *Die Kunststoffe und ihre Eigenschaften*, 6th edn (eds P. Eyerer, P. Elsner, and T. Hirth), Springer-Verlag, Berlin.
12. Luft, G. (2000) High-pressure polymerization of ethylene. *Chem. Unserer Zeit*, **34** (3), 190–199.
13. Smulders, E. (ed.) (2002) *Laundry Detergents*, Wiley-VCH Verlag GmbH, Weinheim.
14. Weber, H. *et al.* (2000) Henkel AG & Co. KGaA, DE 100 66 036 A1, Jul. 14, 2000.
15. Nentwig, J. (2006) *Kunststoff- Folien: Herstellung- Eigenschaften- Anwendung*, 3rd edn, Carl Hanser Verlag, München.
16. M. Häberlein, (2010) HTML-Vorlesungsskript, Polymere-Kunststoffe, Download der KUT-Vorlesung, FH Frankfurt am Main.
17. Michaeli, W., Brinkmann, T., and Lessenich-Henkys, V. (1995) *Kunststoff- Bauteile Werkstoffgerecht Konstruieren*, Carl Hanser Verlag, München.
18. Menges, G., Esser, K., Hüsgen, U., and Kunze, B. (2003) Process optimization in stretch blow molding. *Adv. Polym. Technol.*, **6** (3), 389–397.
19. Sanefuji, T., Fujita, S., and Kawai, T. (2006) Kuraray Co., LTD., DE 601 15 139 T2, Aug. 03, 2006.
20. Weber, H., Hoffmann, S., Rähse, W., and Jung, D. (2000) Henkel AG & Co. KGaA, DE 100 58 647 A1, Nov. 26, 2000.
21. Peters, S. (2011) *Material Revolution: Sustainable and Multi-Purpose Materials for Design and Architecture*, Birghäuser-Verlag (Springer), Berlin.
22. Rähse, W. and Hoffmann, S. (2002) Henkel AG & Co. KGaA, DE 102 41 466 A1, Sep, 6 2002.
23. Kawai, T. *et al.* (1997) Kuraray Co. Ltd., EP 0 794 215 A1, March 7, 1997.

10
Production of Tailor-Made Enzymes for Detergents

Summary

Incorporated in detergents, special enzymes are able to remove natural stains from laundry at temperatures of 60 °C and below. The various enzymes form complexes with the substrates (stains) and hydrolyze the fixed proteins and polysaccharides as well as fats. Surfactants rinse the resulting fragments into the wash liquor. The genes for the synthesis of these enzymes, which predominantly originate from *Bacillus* strains, are optimized for the desired performance of the enzymes. They work nonspecifically and are resistant to the detergent ingredients in the liquor. Industrial production of the enzymes takes place in fermenters 40–125 m^3 in size. During the fermentation, microorganisms secrete the enzyme into the medium to cleave proteins with high molecular weight that swim in the broth. The fragments penetrate into the cell and serve as energy sources. After the end of fermentation, the enzymes are separated from the microorganisms, concentrated and converted into the final product. Granulated products, containing the enzymes in the matrix, require a coating to ensure product safety. Product design is determined by protein engineering and, additionally, by design of the final product.

10.1
Product Design in Biotechnology

According to the rules of product design, the satisfaction of customer needs is at the center of developing a product [1, 2]. In the first step, through customer contact, the problem is identified. Then, in the second step, a development team is constituted. The team members create solutions, produce some samples in the laboratory and in a pilot plant, and realize the product idea in a production plant. This desirable approach is, however, the exception rather than the rule in the field of chemical products. Instead, development often actually begins with the available and known raw materials, with existing expertise in the company, and/or with the installed manufacturing processes. The question asked is what products could be achieved easily with the company's resources. Only much later are potential customers asked for their opinion of samples taken from the pilot plant. Subsequently, marketing

Industrial Product Design of Solids and Liquids: A Practical Guide, First Edition. Wilfried Rähse.
© 2014 Wiley-VCH Verlag GmbH & Co. KGaA. Published 2014 by Wiley-VCH Verlag GmbH & Co. KGaA.

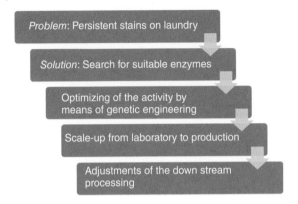

Figure 10.1 Customer-oriented development in biotechnology, explained by the example of detergent enzymes.

people look for a market. In this old type of development, the customer is usually not involved.

In contrast, development in biotechnology starts always from customer needs, as demanded by product design (Figure 10.1). After identifying a customer's problem, a customized solution follows. It often takes years, sometimes even decades, before the development generates a commercial product because the process is very difficult. An example from the field of detergents should explain the method. In a washing process, the customer's problem consisted of an insufficient cleaning of clothes at 60 °C and below. Some stains could not be completely washed out. Many of these spots are of biological origin. In biology, usually for such substances there exist enzymes, which break down the organic macromolecules. The research approach consisted in finding suitable enzymes that remove the difficult soils under conditions of a typical washing process. Translated into the language of chemistry, this means a nonspecific hydrolysis of proteins and polysaccharides in the presence of anionic surfactants in a strong alkaline solution at 40–60 °C.

Because today's customers want to save energy, a new challenge arises in developing modified enzymes with the peak of activity at 15–30 °C. The trend to lower wash temperatures can be clearly seen, partly because the colors of popular, colorful fabrics can then undergo frequent washing.

10.2
History

O. Röhm, a German visionary scientist and founder of the company Rohm and Haas (1907), was the first chemist to isolate enzymes for industrial usage. Consequently, he filed a patent in 1913 [3], which described the use of enzymes in detergents. The breakthrough in their actual use, however, took about half a century. From the mid-1960s onwards, proteases have been a component of many detergents. Such proteases cleave insoluble protein molecules adhering on the fibers. Since 1971, the enzymes in detergents are no longer present as a powder but only as granules, encapsulated in the core and coated. The coating of particles became necessary because enzyme dust can trigger allergies in exposed humans.

In 1972, a novel heavy duty detergent first reached the consumer. In addition to the proteases, they contained the starch-splitting amylases. These new enzymes allow selective degradation of polymeric carbohydrates with a starch structure. The combination of proteases and amylases enabled complete removal of sauce stains in the washing process at 60 °C for the first time. In addition to the coated granules for solid detergents, the biotechnology industry launched liquid enzyme preparations for liquid detergents. This was followed in 1986/1987 by cellulases, which eliminate pilling and protruding microfibrils on cotton and cotton blend fabric. In high-quality detergents, since 1988 lipases have been used to remove fatty soils at 20 °C [3]. The surfactants need higher wash temperatures for the same effect.

The first generation of proteases and amylases already showed the desired polymer degradation under washing conditions. In the second generation, more powerful enzymes emerged through genetically modified microorganisms, which brought higher enzyme yields. Assessed by enzyme activity and washing performance over the last 10–15 years, the use of enzymes in liquid and solid detergents has increased significantly from year to year. Worldwide, proteases are utilized in increasing quantities as stain specialist in detergents. In the predetermined time of washing programs, the optimum action of almost all enzymes is near 60 °C, only with α-amylase it is significantly higher. In an alkaline wash solution (pH 9–11), most enzymes are already active even at wash temperatures below 30 °C. Depending on the formulation, simultaneous use of oxidizing bleaches may inactivate the enzymes. The formulator must consider this possible effect during development, especially in terms of the release kinetics.

All enzyme names relate to the substrate or to the function and end with "ase." Detergents require hydrolases, incorporated in high-performance products. These enzymes perform hydrolytic cleavage, for example, of esters. Esterases are ester-splitting enzymes. Lipases hydrolyze natural oils (lipids), including the esters of triglycerides. However, all three names describe chemical processes, occurring in fat splitting. The term "lipase" is the most accurate description and the term "hydrolase" is superordinate.

Solid and liquid detergents contain an enzyme formulation (granule or solution) in amounts of 0.2 to <2%, within which the pure enzyme represents a small part (between 0.3% (cellulases) and 5% (proteases)). Because of high prices, the added enzymes affect the market price of the final product significantly. Cheap detergents often contain only 0.2–0.3% protease formulation, while expensive laundry detergents contain ~0.3–0.6% of each enzyme. The precise amounts depend on the activity of enzyme formulation per gram. As a rough estimate, the following figures result if we take 100% as representing the cost of an enzyme-free product. With the addition of 0.2% enzyme, the raw material costs rise to about 102%. With 2% enzyme the price increases to 120%.

10.3
Enzymes

This section describes the classifications within biotechnology, the mode of action, and the structure of enzymes.

10.3.1
Enzymes as Part of White Biotechnology

Biotechnology uses biological reactions in analytical and technical methods and in productions. It uses the findings of biochemists and (micro- and molecular) biologists, supported by technical chemists and process engineers. As an inter-disciplinary science, biotechnology deals with the use of living microorganisms, plant and animal cells, tissues, and cellular enzymes. This includes the specific modification of genes in microorganisms, and the protein structure of enzymes, known as *protein engineering*, another segment of biotechnology.

Biotechnological applications cover many areas, such as food technology and chemistry, medicine and pharmacy, through to agriculture and energy and waste management. It has become common to indicate the segments with colors (Table 10.1). "White" biotechnology utilizes biotechnological processes in indus-trial production [4] to carry out reactions and/or synthesize products. Therefore, all products that are produced in large quantities on a regular basis and pass through a biotechnological reaction, at least in one stage, belong to white biotechnology. Their influence will grow significantly in the future [5] and gain in importance in Germany [6].

Global sales of white biotechnology in 2004 yielded €55 billion [7]. This was achieved with various organic acids (citric acid and lactic acid) and amino acids (L-glutamate and L-lysine) and further with solvents (ethanol), antibiotics (penicillin and cephalosporin), and vitamins (ascorbic acid-vitamin C and riboflavin-vitamin B_2) as well as biopolymers (dextran, xanthan, and polylactide) and carbohydrates (fructose syrup and glucose). From 2006, sales increased to reach €125 billion in 2010 [8] and by 2015 are projected to reach US$ 300 billion (about €220 billion at 1.35$/€) [9].

White biotechnology includes enzymes. Special microorganisms in fermenters produce them, followed by work up of the fermentation broths and isolation of the enzymes and transfer into solid or liquid formulations. Enzymes are needed worldwide in increasing quantities for different tasks. They typically work as essential excipients, usually unnoticed by consumers. Examples are the production of bakery products, fruit juices, wines, and cheeses, as well as jeans, on which cellulases generate the "stonewashed" effect.

The enzyme business is very dynamic and is growing currently at about 10% per year. The value of world's enzyme production in 2005 was €1.7 billion [10] and reached about €2.55 billion in 2010 [11]. The market leader, with a market share of 47%, is the Danish company Novozymes, which covers all major fields

Table 10.1 Applications of biotechnology.

Description of biotechnology	Application	Examples (tags)
White	Industrial biotechnology Enzymes Fine chemicals Agricultural and pharmaceutical (pre-) products Food Replacing fossil fuels (biomass)	Enzymes, acids, antibiotics, vitamins, insulin, amino acids, biopolymers, bioplastics, bio-pesticides, starch, hyaluronic acid, cheese, starch, wine, animal feed, carbohydrates, bioethanol, biogas, and biohydrogen
Red	Medical and pharmaceutical field, products for Therapy Diagnosis Vaccines Cosmetics	Biopharmaceuticals, interferon, growth hormones, monoclonal antibodies, clotting factors, regenerative medicine (tissue engineering), gene and stem cell therapy, and special proteins
Green	Plants in agriculture Higher yields Pest resistance	Molecular pharming, gene transfer, gene modification, and biopesticides (R&D)
Blue	Microorganisms (bacteria) from the sea	Obtaining temperature-resistant bacteria
Gray	Waste management Wastewater Waste recycling and waste Contaminated soils Exhaust and emission control	Cleaning of industrial and municipal effluents, treatment of waste, biological treatment of soil and sea (petroleum), and biological waste gas treatment

of application. After them comes the DuPont (formerly Danisco Genencor) with 21% and the DSM (6%). Manufacturers of detergents and cleaning products are the main customers. They need about 34% of the enzymes produced, representing a value of almost €0.9 billion. The technical enzymes (ethanol from cellulose and starch, leather, textiles, and paper) follow with 34%, followed by food (baked goods, beverages, and syrups) with 23%, and the feed enzymes (phytases) with 9% [11]. Furthermore, diagnostic and therapeutic procedures require a small portion of the enzymes produced.

Red, green, and blue biotechnologies [12] develop special products and innovative applications. Additionally, in red biotechnology, small amounts of specialties, especially of genetically engineered drugs, are produced. In this context, "gray" waste management represents a special case. It is part of environmental technology and complements the chemical process industry. On industrial sites, large plants clean the wastewater and the exhaust air as well as contaminated soils sometimes. In addition, most of the waste is recycled. Normally, there are no (quality) products, in contrast to industrial white biotechnology.

10.3.2
Enzymes as Catalysts of Metabolism in Living Cells

Essentially, the metabolism in living cells consists of numerous individual reactions, catalyzed by different enzymes. These cause life support as well as the proliferation of cells and cell substances. Among the important cellular substances are proteins, nucleic acids, cell wall components, and reserve materials [13]. The building blocks and necessary energy for the biochemical reactions originate from nutrients, preferably from glucose. In addition to nutrients, the synthesis requires some mineral salts, trace elements, and vitamins. Many nutrients are first broken down into small units (catabolism) and then synthesized to the desired polymers (anabolism). All processes run by enzyme catalysis. Figure 10.2 illustrates the differences in terminology of biotechnology compared to chemistry and chemical engineering.

Higher developed cells with a nucleus contain thousands of different enzymes. All biochemical reactions need a specific enzyme for catalysis, including for the copying of genetic information. The specific enzyme is a monomeric or oligomeric, globular protein with/without nonprotein constituents (prosthetic groups). In the first step, metabolites adhere specifically (substrate specificity) at the catalytic center of an enzyme. This active site is formed from parts of polypeptide chain and/or cofactors and prosthetic groups. The metabolite is an intermediate product of a biochemical cellular process, which subsequently reacts in the direction intended by nature, determined by the enzyme responsible by lowering the activation energy (effect specificity).

Enzymes accelerate all reactions in cells and make them possible, without changing themselves. They work very efficiently in neutral medium, at mild temperatures, and at very low concentrations. In summary, synthesized in living

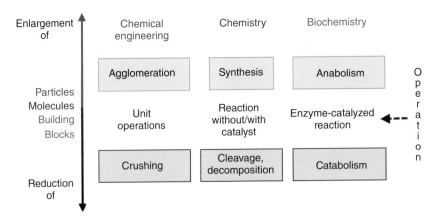

Figure 10.2 Operations for enlargement and reduction of particles, molecules, and cell blocks.

cells, enzymes are proteins with an active site [14], which catalyzes highly specific biochemical reactions.

10.3.3
Structure of Enzymes

Proteins are macromolecules having molecular weights of 10 000 to several hundred thousand daltons [15]. About 20 different amino acids form the building blocks of proteins; a few hundred to several thousand can be linked together by peptide bonds. The sequence of amino acids constitutes the primary structure, while the spatial convolution via hydrogen bonds, partly also by disulfide bonds, in the β-sheet and α-helix form [16] indicates the secondary structure. Depending on the environmental conditions (pH, salt, temperature, and concentrations), there arises in aqueous solution a spatial, globular (= substantially spherical) protein structure, also referred as *tertiary structure*. Here, the nonpolar protein chains organize themselves more interiorly, and the polar side chains more exteriorly (quaternary structure). This arrangement causes good water solubility. An exact folding of the protein molecule forms the active site. The spatially appropriate substrate fits exactly in this opening (key–lock principle, partly under spatial adaptation [15]). A substrate-specific binding ensues and, subsequently, catalytic conversion in exactly the direction of the enzyme type (hydrolyzation in the case of hydrolases). The products (such as debris) then leave the active site (Figure 10.3).

The active sites for a particular application are optimized with the help of genetic engineering by varying the outer, terminal amino acids, as done, for example, with protease. In this way, the substrate specificity is decreased, broadening the protease applications. In addition, a significant improvement of activity gives an increased reaction rate on the same substrate and/or an additional hydrolysis of other metabolites. Optimization of the active site is a significant component of product design (product performance).

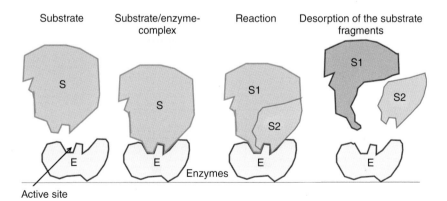

Figure 10.3 Model representation of enzyme catalysis.

10.4
Enzymes in Detergents

As a result of several product and process optimizations, the cost of manufacturing enzymes is reduced significantly. Such progress enables global use of enzymes in detergents. Effectiveness at low concentrations and at low temperatures, and complete biodegradability, are further advantages of these substances used in solid and liquid detergents.

10.4.1
Significance

Without the use of specific enzymes and specially tailored enzyme mixtures, the high performance of modern detergents would be not feasible [17, 18]. These catalysts show high hydrolytic activity in combination with low substrate specificity. They work in alkaline, surfactant media up to about 60 °C, even in the presence of bleaching agents. Newly developed in the recent years, some modified enzymes have high activity at relatively low temperatures (20−40 °C). This is the reason for a possible reduction of washing temperature without significant lowering of washing performance.

Washing at lower temperatures (30 °C) allows considerable energy savings compared to washing at 60 °C. In the process chain "production of detergents and transportation, washing the laundry," a heating of the washing liquor in the machine requires most energy. In addition, enzymes provide a significant contribution to reducing the amount of detergent needed. In 1980, 5 kg of dirty laundry needed about 275 g of detergent powder. Today, by using solid compact detergents, about 65−80 g is needed. Additionally, enzyme-containing detergents have significant ecological advantages [19]. Enzyme-free detergents are no longer competitive in terms of performance and ecology.

In the wash liquor, soil is removed partly by surfactants and by enzymes. To wash white laundry, solid heavy-duty detergents include a bleaching system and materials for whitening the fabrics. The enzymes in these detergents act primarily as soil and stain remover for natural dirt, particularly the protease, amylase, and lipase. They are suitable for many types of fabrics (except wool and silk) and colors, and for all washing programs up to 60 °C. The washing conditions (pH, T, c) are crucial to the visible product performance. In addition, the synergetic effect of one or more enzymes in combination with the other detergent ingredients as well as the chemistry of pollution and substrate (stains) determines the success of washing.

Owing to their molecule structure, cellulases exhibit different functions, which are important in all detergents. Therefore, most detergent manufacturers use mixtures of different cellulase types for biofinishing. This means that, firstly, one type removes protruding fibers and fuzz from cotton or other cellulose-based fibers and smoothes the fabrics. Other types remove solid dirt and prevent the reattachment of pigments, dust, soot, and cosmetics on the fabrics. These cellulases enhance the whiteness and improve the color impression. Color and wool detergent, containing such cellulase mixtures, succeed in refreshing the colors. After a few washes, the clothes appear new. Table 10.2 displays the enzymes used in powerful detergents (without division into subgroups).

Despite the relatively high production costs, the great success of enzymes in detergents [20] is based on four pillars:

1) *Successful adaptation to detergents and to washing liquor*: suitable enzymes work in alkaline, surfactant-containing, and oxidizing environments. The effect is detectable already at room temperature, and optimal in the temperature range 50–60 °C.
2) *Product performance*: visual and measurable success after washing. For removal of difficult soils, no suitable chemicals exist for the temperature range 20–40 °C, as alternative to the newly developed enzymes.
3) *Economics*: economic productions is improved firstly by extreme increases in product performance using genetic engineering, and secondly, by optimizing the manufacturing process.
4) *Safety*: potential hazards associated with enzyme and detergent manufacturing (work safety) and with the application (product safety) are known and controlled.

10.4.2
Optimizations of Production Strain

As an example for enzymes, we shall use protease to demonstrate the search, identification, and optimization, and the industrial manufacture of appropriate microorganisms (which synthesize protease). Subtilisins from the strain *Bacillus licheniformis*, known since 1959, are effective, nonspecific serine proteases for detergents. They show a high degree of compatibility with respect to alkalinity and temperature in the application. Complexing agents and even bleaches and, further, most surfactants do not diminish significantly the protease activity at the usual temperatures of up to 60 °C.

In a fermentation process, first the microorganisms multiply rapidly. Subsequently, they produce more serine proteases. The protease penetrates through the cell wall and is then present extracellularly in the fermentation broth. In this solution, they break down large protein molecules present in the culture medium [13, 21]. The bacilli act in this way because in the cell they only can utilize small peptide molecules and amino acids for building blocks; for energy generation, both are essential for proliferation. Accordingly, the nutrient solution is adjusted to be rich in protein. The optimized process continues until the growth phase of the microorganisms and the production phase are completed. At this stage, as much serine protease as possible is present in the fermenter broth, and the harvest can follow. By a combination of cooling and switching off the air supply the fermentation is terminated, before the downstream process for the broth starts.

The search for a high-performance variant of *Bacillus licheniformis* was aimed initially at an increased yield in the fermenter. A second optimization task was the development of enzyme variants with improved performance and stability in the

Table 10.2 Biocatalysts (hydrolases) in high-performance detergents.

Enzyme	Molecular weight (Da)	Effect	Optimum pH (T (°C))	Functions in detergent; removal of
Alkaline proteases	24 300 (about 274 amino acids)	Hydrolysis of peptide bonds, removal of proteins from the fiber	10–11 (60)	grass, mucus, excrements, blood, sauce, spinach, cosmetic formulations, and food (egg yolk)
α-Amylases	40 000–50 000	Hydrolysis of α-1,4 glycosidic bond of polysaccharides; liquefaction of starch	8 (> 60)	pasta, sauces, meat juice, pudding, cocoa (chocolate), baby food, and potatoes
Cellulases	52 000	Hydrolysis of β-1,4-D glycosidic bond in amorphous cellulose, anti-pilling; decomposition of microfibrils, removal of solid dirt	7–10 (<60)	protruding fine fibers and pills; intensifies the color impression, reduces lime and pigment deposits; and prevents graying
Lipases (esterases)	50 000–67 000	Hydrolysis of glycerides	10 (60)	oils, butter, margarine, fats, and cosmetics (make up, sunscreen, lipstick)

washing process. The continuing review of performance characteristics secured practical relevance during optimization. Previously, the search for powerful strains followed some selection methods, on the "search-and-throw-away principle." Further, a selection can occur after the triggering of an artificial mutation, which is a modification in the genome via high-energy radiation or via chemical mutagenesis. The genome represents the entire genetic information of a cell. After complete sequencing of the bacterial genome (*Bacillus subtilis* 1997 and *Bacillus licheniformis* 2004), it became possible to influence the genome in terms of improved productivity and quality by elimination or duplication of certain genes.

Since around the mid-1980s, genetic modification of the protease molecule (protein engineering) has taken place to optimize the washing performance of protease. This technique brought about a large jump in enzyme performance. In addition to helical DNA (deoxyribonucleic acid), some small circular DNA molecules exist in the cells; they are called *plasmids* and were initially used as a gene "taxicab" into the cells. Specific enzymes, known as *restriction enzymes*, were then used to cut open plasmids, suitable for the enzyme production. Subsequently, there followed the addition of genes for protease production as a gene cassette (manipulable fragments of DNA). Ligases connect the cut genes at the interfaces [13]. From the resulting pool, appropriate search methods find effective plasmids. After identification and isolation, they fit in the production strain. The specifically modified microorganisms produce the desired enzyme in larger quantities and/or increased activity. For economic reasons, all production strains used today are optimized via "genetic engineering." The use of plasmids has decreased. Today it is common to integrate the gene directly into the genome.

A look into the future [22] suggests three interesting new approaches for the development of superior enzymes. First, a search ensues through metagenome screening, in which all organisms are involved from the ground and not only those that are culturable in the laboratory. Second is the method of "gene shuffling." Here, the genes, which contain an instruction manual for the enzymes, would be cut into small sections and randomly reassembled ("protein engineering"). The new genes, inserted into the bacillus, produce other enzymes, better and worse. Then, selection according to the requirements occurs. Third, a method that will be increasingly used employs synthetic gene libraries, as they arise, to vary or set certain amino acid positions.

10.5
Industrial Manufacture of Proteases

Fermentation processes have special requirements for materials and types of machines and apparatus for sterility and cleanability. The process consists of fermentation and downstream processing. In the product, enzymes are available, highly diluted, in granular form, and as liquid.

10.5.1
Materials for the Plant

The process consists of two parts. The first, fermentation, is sterile according to pharmaceutical quality. Before starting fermentation, the system is sterilized with steam. Downstream processing, the second part, is non-sterile, but is disinfectable such as in the cosmetic industry. To ensure high-quality cleaning, the whole plant consists of stainless steel, especially of polished quality for all product contact surfaces. The microorganisms cannot adhere to such a surface and therefore do not penetrate into the material. Steels with the material numbers 1.4301, 1.4307 (AISI 304, V2A), 1.4401, 1.4404 (AISI 316L), and 1.4435 are preferably processed, usually surface treated. For plant hygiene, the installation must be free of dead spaces, including in the pumps, valves, pipes, seals, and all measuring devices. This allows complete emptying of the system and easy cleaning. All machinery as well as other parts of the plant should be constructed in accordance with the regulations of "good manufacturing practice" (GMP) and the "European Hygienic Engineering and Design Group" (EHEDG); in the United States, the American 3-A Sanitary Standards, and those of NSF International apply.

As is common in the pharmaceutical industry, the welds are formed in special machines and are virtually seamless. A gradient in piping ensures complete emptying. The plant is usually cleaned with help of a fully automatic system (for details see Section 13.10), called "cleaning in place (CIP)" [23]. Alkaline and acid solutions, which may contain oxidizing agents, are available in tanks for cleaning at higher temperature. First, water rinses out the residues. Subsequently, an acid and then an alkaline cleaning follow over pumping, usually at temperatures between 60 and 80 °C. Finally, the plant is rinsed with water and blown empty and dry.

In comparison to chemical reactors, several additional features characterize fermenters, owing to the aeration during sterile operation. Sterilization with steam and several fine filters prevents an infection. To avoid the settling of microorganisms on rough steel surfaces inside the facilities, the surfaces are polished. First, fine grinding with a paste of grade 180/240 is carried out, and then the final polishing is usually with suitable pastes (grain 320/400), polished to a mirror finish.

In the preferred steel qualities, 1.4435 or 1.4404 crafted, electropolished fittings and components show roughnesses of less than 0.8 μm. The sterility and easy cleanability require partially specialized, complex construction of equipment. The seals, in particular the double mechanical seals, must meet high demands in terms of material and manufacturing accuracy. Measurements verify the integrity, which is documented. For the steam sterilization of the fermenter plant and of the connected pipings and fittings, usually an automatic "sterilization in place (SIP)" system is available.

10.5.2
Three-Step Process

The entire production process consists of three stages (Figure 10.4). Stage 1 represents the fermentation, in which the synthesis of enzymes by microorganisms

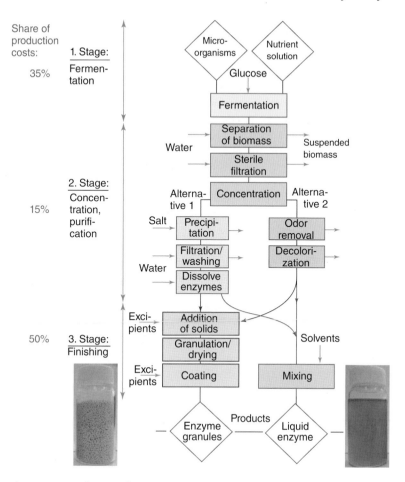

Figure 10.4 Production of enzymes (protease) depicted as block flow diagram.

ensues. The subsequent stages 2 and 3 are summarized under the common term "downstream processing." Stage 2 is the working up of the fermentation broth. This includes the separation and deactivation of microorganisms and, the required concentration and purification of enzymes. In the third stage, product finishing is applied, in either the direction of solid-coated granules or to an adjusted solution.

Manufacturing costs, which consist of the raw materials and the production costs, have been analyzed. For enzyme granules, alternative 2 (Figure 10.4) has the following breakdown: fermentation about 30–40%, treatment (concentration, isolation) 15–20%, and granulation/coating around 40–50% of the total cost. These values depend on the process stages and enzyme activity in the final product. The higher the activity, the higher the weighting of the fermentation costs. For liquid enzymes, the finishing is simple and inexpensive. Here, the cost shares of the fermentation exceed 50% in depending on the enzyme activity.

10.5.3
Fermentation

Microorganisms of the selected strain multiply from the agar culture in the Petri dish via the shake flasks and then in different sized fermenters (25 l, 250 l, 2.5 m³, …). Subsequently, the microorganisms are available in sufficient quantities, and are sterile in the prefermenter under growth conditions. The raw materials for the culture broth of fermentation are in tanks and barrels, silos, and containers. According to the particular recipe, these substances converge in a separate stirred tank for suspension in water. Sterilization of this nutrient solution (20 min, 121 °C, 2 bar) happens preferably in the main fermenter. In this process, the solution turns to brown because of the Maillard reaction, which gives compounds of amino acids and reducing sugars. In addition, the glucose solution is caramelized by the temperature effect over the oxidation with the oxygen of air, so that the fermentation broth is a dark brown color.

The enzyme preparation differs from other fermentation processes owing to the high oxygen demand in the growth phase. Therefore, they use batchwise production fermenters of high performance with an overall volume of 40–125 m³. The containers often show a H/D ratio of 3, and are equipped with three to five disk stirrers ($d/D = 0.25$–0.33). A slim tank increases the residence time of the gas bubbles for the same total volume. The higher level of liquid makes necessary a higher pressure at the aeration ring or aeration nozzles. This alone is not sufficient for the oxygen supply of *Bacillus licheniformis*. The oxygen transfer to the liquid phase needs an unusually high power input of 10 kW m^{-3}. Thereby the dissipated energy generates undesirable temperature increases in the suspension. A constant operating temperature in the range 35–40 °C is ensured with cooling water by either an increase in heat exchange surfaces or lowering of cooling water temperature. The stirred vessel has a heating and cooling jacket outside. In addition, to increase the exchange surfaces in fermenters, internal tubes are used, which are designed as single or double spiral or as meander, realizing the necessary additional surfaces (Figure 10.5). The exchange surface increases by this measure to a factor of 2–5. This allows a useful shortening of periods for heating and cooling during sterilization, which depend on the ratio of the volume of the fermenter broth to the exchange surface. The sterilization normally lasts for 3–5 h.

After sterilization and cooling to a fermentation temperature of 35–40 °C, inoculation occurs and aeration starts (Figure 10.6). The fermentation runs for 40–65 h under powerful stirring and introduction of large volumes of sterile air (about 500 m³ m^{-2} h^{-1}). Portion-wise addition of a sterile glucose solution provides the essential nutrient for energy. To control the process [21], various measuring and control devices are installed for temperature, gas pressure, and oxygen saturation in combination with gas flow rate, and, further, for power input, pH, and foam as well as weight and exhaust gas composition. Optical measurement of turbidity indicates the cell density and allows the growth of microorganisms to be followed. The fermentation ends when the number of microorganisms begins to fall and the enzyme concentration shows a maximum.

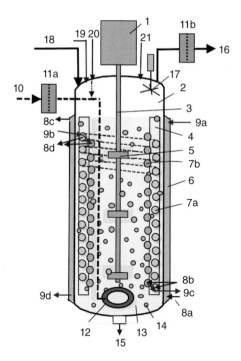

Figure 10.5 High-performance batch fermenter (without measuring and control technology): 1: drive of agitator shaft; 2: fermenter with diameter *D* and height *H*; 3: stirrer shaft; 4: baffles; 5: agitator blades (rotating disc) with diameter *d*; 6: heating-cooling jacket; 7a, 7b: internal heat exchangers as double spiral; 8a, 8b: cooling water inlet; 8c, 8d: cooling water outlet; 9a, 9b: steam inlet; 9c, 9d: steam/condensate outlet; 10: air supply; 11a, 11b: sterile filters for supply and exhaust air; 12: aeration; 13: nutrient solution, fermentation broth; 14: gas bubbles; 15: outlet for the broth; 16: exhaust; 17: mechanical foam breaker; 18: addition of nutrient medium, sterile strain, and sterile glucose solution; 19: pH adjustment with acid; 20: steam inlet; and 21: anti-foaming agent.

10.5.4
Downstream Processing of Fermentation Broths

In enzyme workup, the microorganisms are usually separated from the fermentation broth by centrifugal forces. Because of small differences in density between the germs and solution, successful use of separators requires significant dilution with water to reduce the density and viscosity. Water for the dilution of fermentation broth originates mainly from ultrafiltration (UF). The UF-permeate flows directly and via microfiltration (MF) as MF-concentrate to the stirred dilution vessel (Figure 10.7).

First, the diluted fermentation broth runs into the separators for removal of biomass. Subsequent to separation, performed in several stages, the microorganisms are present in a concentrated suspension. By pumping, this liquid flows for sterilization continuously through a heat exchanger at about 150 °C (Figure 10.8).

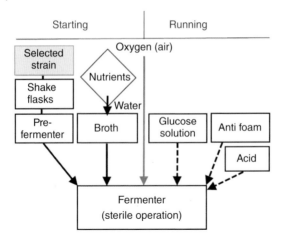

Figure 10.6 Block flow diagram of fermentation.

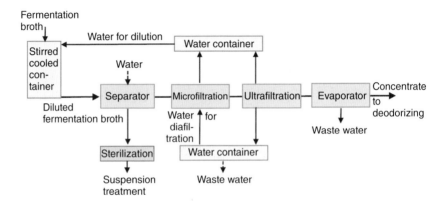

Figure 10.7 Block flow diagram of enzyme workup.

For larger amounts of fluids, the process expires usually continuously. After partial mechanical removal of water from the suspension and drying of wet material, an organic fertilizer with high performance, but bad odor, is obtained.

After separation of microorganisms via centrifugal forces in a centrifuge, a sterile filtration of the enzyme-containing solution takes place in a MF plant with 0.14 μm inorganic membranes, supported by diafiltration [24]. Membrane filtration is necessary to ensure a strain-free product, which regulations demand. Permeates of MF flow successively into the UF units (Figure 10.9), to concentrate the diluted solution by a factor of 20–40, and thereafter in a thin-film evaporator by factor 2. While enzymes in the MF diffuse through the membrane pores, the cells are held back, as well as other solids. In the UF (cut-off: 10–20 kDa), the opposite happens. Water and dissolved salts penetrate through the membrane, whereas the enzymes are restrained and concentrate in the solution.

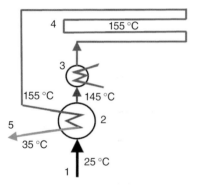

Figure 10.8 Sterilization process: killing of microorganisms in aqueous solution by means of continuous flow through a heat exchanger, heated with 5 bar steam: 1: feed of suspension; 2: heat exchanger with heat recovery; 3: steam superheater; 4: residence time line; and 5: outlet of sterile suspension, containing dead microorganisms.

Figure 10.9 Industrial centrifuge for the separation of biomass and membrane systems for sterile filtration (small image) and concentrations (Courtesy of Biozym).

In path 1 of the workup (as shown in Figure 10.4), the protein precipitates by the addition of salts (ammonium or sodium sulfate). After filtration and washing, the enzyme dissolves in water and forms a bright solution. In the alternative path 2, first the concentrated, brown enzyme solution is pumped to a newly developed deodorizing plant. Here under vacuum, superheated steam (Figure 10.10) releases the odorous substances such as dimethyldisulfide and dimethyltrisulfide [25, 26], which are present in very low concentrations, from the fermentation broth. The resultant brown solution then flows into the finishing step; alternatively, a bright solution is first obtained by means of a discoloration step on adsorption resins [27]. The liquid products arise only from the decolorized enzyme solutions, which are adjustable to the desired activity with diols and water.

Each step of the production chain, from harvest to finished product, causes appreciable yield losses that lie in total up to 25–35% for a variety of reasons. For efficient production, these losses must be minimized at all stages of production. In this connection, careful workup is of decisive importance. Recycling in the separation steps as well as an optimization of the diafiltration reduces losses in the individual stages. There, large amounts of liquids flow almost completely back for increasing the yields.

10.5.5
Manufacturing of Enzyme Granules

A block flow diagram demonstrates the finishing process and the conversion of a liquid in particles (Figure 10.11). Figure 10.12 illustrates this subprocess for the enzymes. After spraying on the dry components in a mixer, the liquid enzyme concentrate penetrates into the powder. Subsequently, in a mixer or extruder, after

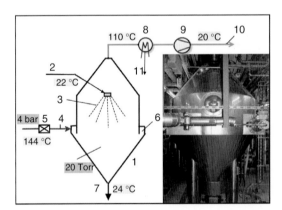

Figure 10.10 Deodorizing to remove odors from unpurified enzyme solutions, interior cleaning with CIP-solutions via spray heads: 1: insulated container; 2: feeding of solution through a nozzle; 3: spray; 4: supply of superheated steam; 5: reducing the pressure; 6: ring channel for uniform distribution of steam; 7: product, finished after several cycles; 8: heat exchanger; 9: vacuum pump; 10: exhaust gas for combustion; and 11: smelly water to the treatment plant or combustion.

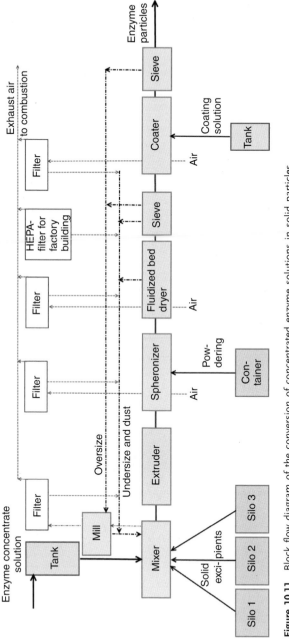

Figure 10.11 Block flow diagram of the conversion of concentrated enzyme solutions in solid particles.

Flow of the particles:
Extruder,
spheronizer,
fluidized bed dryer

Figure 10.12 Concentrated enzyme solution in a tank, which is sprayed on excipients in the mixer (top right), and subsequently shaped and dried to enzyme particles (Courtesy of Biozym).

the addition of binders the granules in desired size and form arise from the wet mixture; shaping is then performed in a spheronizer. Suitable excipients consist predominantly of natural polymers, such as various starch qualities. After drying in the fluidized bed and sieving, a coater is used to apply the protective layers (Figure 10.13), employing aqueous solutions or suspensions [28, 29]. The coating suspension includes, preferably, highly water-soluble salts (sodium sulfate) and/or

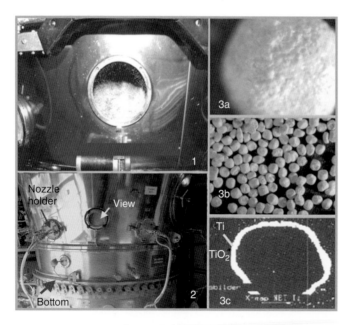

Figure 10.13 Coating of enzyme granules with a protective layer: 1: view into a fluidized bed coater for enzymes with top spraying; 2: lower part of coater, bottom plate, and fluidized bed; 3: coated extrudates (colors are changed in the micrographs); 3a: close-up of particle surface; 3b: extrudates, oversight; and 3c: enzyme extrudate, in resin (Epon)-embedded energy dispersive X-ray microanalysis (EDX).

polymers (e.g., poly(ethylene glycol)). The granules should be readily soluble in cold water or suspendible. Applied layers improve significantly the mechanical properties, especially the stability of particles. Enzyme particles tend to separate in the detergent powder. Segregation is reduced by adjustment of density as well as particle size distribution and shape. References [2, 30, 31] describe these problems in the granulation and coating in detail.

After coating, the enzymes are stored in a silo until the quality is approved. A mixture of different enzyme activities allows precise setting of the specified quality for proteases. Filling in "big bags" ensues either as pure enzyme (protease particle) or mixtures (protease, amylase, cellulase). The individual components are taken from the silos, controlled by weight, and then placed in a mobile container (Figure 10.14). From there, the enzyme particles flow into a gently running mixer. Immediately after mixing, filling into big bags (about 750 kg) or cardboard drums (20 kg) takes place (Figure 10.15).

Enzymes have sensitizing properties. This means that after first contact, via the respiratory tract, sensitization may occur and after renewed contact an allergic reaction may develop. Here, each person reacts differently. As with hay fever, only a minority will develop allergic reactions. To exclude all risks as well as for reasons of health and work safety, the inhalation of even tiny amounts of enzyme dust during the manufacturing and processing must be avoided. Therefore, the

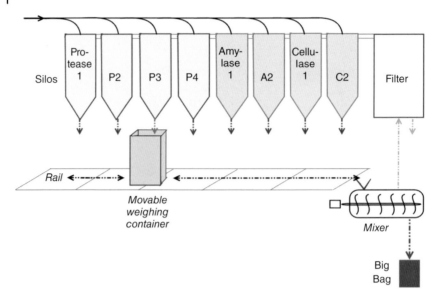

Figure 10.14 Mixing plant for adjusting the activity and for the production of enzyme mixtures.

manufacturer coats the granules with a protective layer. The coating meets two essential requirements for product safety. First, the operation must avoid respirable abrasion (<10 μm). Second, the coating layer ensures that the granules do not release enzyme dust – in fact, not only during operation and in application but also on heavy use. The coated enzymes satisfy consumer protection. The readily marketable enzyme granulates meet all requirements in the normal application in detergents [32].

Further, a layer on the granules allows improvements in both color, by the addition of white pigments, and storage stability. Therefore, the covering prevents direct contact between enzymes and other ingredients of detergent formulations. A good coating shows a high uniformity of layer, which should both adhere firmly and, have certain elasticity, so that they do not flake off under mechanical stress. Furthermore, the particles must be sufficiently stable and not break, so that no exposed surfaces arise. Ductile or elastic formulations for the matrix and coating layer thus have advantages over hard, brittle granules.

In production, control samples exist for each batch of coating. An important quality parameter is the abrasion behavior. To quantify the values in the laboratory, parallel tests in fluidized beds yield the amount of dust and its enzyme activity, executed worldwide under standardized conditions. To measure the fracture behavior with a standardized method, major detergent and enzyme manufacturers designed a new measuring apparatus. This allows determination of the developing enzyme dusts from granules after impact and shearing. The applied particle stress in measuring systems is far beyond the usual applications ("worst case").

Figure 10.15 Bottling of enzyme particles, and transport containers for solid particles and enzyme solutions (Courtesy of Biozym).

The known methods of processing technology enable us to determine quantitatively the formation of abrasion and breakage. The granules strength testing system (from ETEWE) measures the tensile strength and compressive strength, and provides information about the fracture behavior of granules in the compressive stress. Other possibilities are to shoot the pellets pneumatically on an inclined wall and furthermore to determine the wear after a friction stress in a 360° loop. Measurements by laser diffraction spectrometer display the changing size distribution of particles. All methods are of particular interest for comparative measurements under identical conditions.

A thick coating layer requires large amounts of materials up to the same weight as for the core. Therefore, the coating process causes high-level material costs,

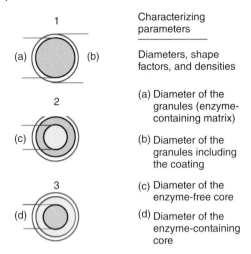

Characterizing
parameters
_____

Diameters, shape
factors, and densities

(a) Diameter of the
granules (enzyme-
containing matrix)

(b) Diameter of the
granules including
the coating

(c) Diameter of the
enzyme-free core

(d) Diameter of the
enzyme-containing
core

Figure 10.16 Possibilities for the structure of coated enzyme granules: 1: granules with inert coating layer; 2: granules with inert core, an enzyme-containing coating layer in the middle, and an outside inert layer; and 3: enzyme-containing core, inert coating layer 1 in the middle, and outward inert layer 2.

in addition to relatively high production costs. A uniformly applied film with a minimal thickness of 15 µm frequently suffices to perform the desired protection. An additional layer is usually not necessary. According to Figure 10.16, there are three common variants of granule structure for coated particles, depending on the used granulation process.

10.6
Workplace Safety

Detergent enzymes are sensibilizing agents, even at concentrations below 1% [33]. Therefore, the handling of enzymes and enzyme-containing intermediates, and their formulations, must be safe. This refers to both in the laboratory and in the pilot plant as well as to the manufacturing of enzymes and detergents. The safety concept includes intensive training of employees before starting employment and thereafter every 6–12 months as well as the use of personal safety equipment. This usually involves a suitable respirator and protective goggles, and also gloves and work shoes; furthermore, in the factory a helmet may be needed, depending on location. The first preventive medical examination of an employee occurs prior to starting the job, the second after 6–12 months [32]. All other tests ensue every 2 years, unless the physician specifies shorter periods.

In the enzyme factory, potential dangers are present to all people. Buildings and facilities must always be in clean condition. Workers are to remove immediately spilled solids, such as from sampling, with a special vacuum cleaner. Furthermore, regularly, the floors are thoroughly wet cleaned. In the workup, all enzyme solutions are handled in closed facilities. Any detected leak (e.g., at pumps) is to be quickly eliminated. In this part of the factory, there are to be no permanent jobs; monitoring is fully automated from the control room. Sampling takes place only when wearing personal protective equipment, and repairs and maintenance only after CIP-cleaning and flushing.

In contrast, the area for processing the solids requires that employees constantly use handgrips, for example, in the regular flushing of nozzles and nozzle holders of the coater, as well as when exchanging the plate at the extruder head. To minimize the potential risk of inhalation of dust, in addition to the point suction at the process stages (as well as at the screens, the mill, the mixers), some suction distributed in the plant must exist. Complete encapsulation of equipment ensures free movement of all persons in the production building. In addition, at several points in the production plant, the air is monitored for safety. The enzyme activity in the ambient air must be below strict limits, measured in micrograms.

In the detergent factory, a comprehensive safety concept is more difficult to implement because less hazardous substances in large quantities make up the product. In relation to other components, enzymes represent a very small portion. Awareness of the potential danger and the desire for cleanliness is less pronounced. Encapsulation of affected plant parts can protect employees. In the plant, the route from supplying the enzyme and subsequent pathways including the filling lines (Figure 10.17) should be fully enclosed. The enzyme-containing exhausts must be kept independent from other exhausts. Recirculation of enzyme-containing filter dust as well as oversize from the screen occurs by dissolution and oxidative and/or thermal deactivation in the slurry container. A blending of recycled products, which are outside of specification limits, ensues in amounts of 0.5–5% in the post-addition step – only as small amounts as possible, to maintain the specified quality. At the end of the conveyor belt, addition of these enzyme-containing materials is possible (point 6 in Figure 10.17).

In the production of liquid enzymes careful attention must also be paid to cleanliness. The focus is on avoiding aerosol formation and on removing immediately leaked or spilled liquids. Dried product on a floor may be dangerous by contact or, especially, after the small solid particles swirl in the air. In this process, encapsulation up to the filled bottles makes the manufacturing secure.

10.7
Product Design of Enzymes

An economic production of detergent enzymes has two main objectives. One is the development of application-optimized enzymes by "genetic engineering" (= design of enzyme performance), using a powerful production strain. The second is that, by intensive optimization, the enzyme-containing granules are obtained in the right size, shape, density, and solubility with almost perfect coating (= design of final product). The example of enzymes demonstrates that two very different but complementary design types are essential for needs-based products. Here,

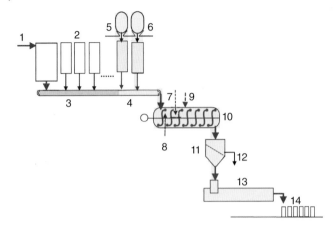

Figure 10.17 Encapsulation (grey; from 4 to 13) of routes for enzyme-containing materials in detergent processing: 1: tower powder or basis granules or extrudates; 2: supply hoppers with dosing screws for various components of post addition; 3: collecting belt; 4: encapsulated part with separate suction for enzyme-containing dusts; 5: big bag and silo for enzyme granules; 6: big bag and silo for recirculated final product; 7: perfume dosage; 8: nonionic surfactant dosage; 9: material for powdering; 10: preparation mixer; 11: sieve; 12: oversize; 13: filling line; and 14: detergents in the package.

this requires strong cooperation and understanding between biotechnology and chemical engineering for the successful development of a new product for the market.

10.8
Learnings

√ The development of a biotechnology-based product is difficult, lengthy, and expensive. Therefore, it must solve big customer problems.

√ Enzymes are part of the white industrial biotechnology.

√ They represent proteins of high molecular weights and catalyze the metabolism in living cells, causing the life support as well as the proliferation of cells and cell substances.

√ The catalysis of enzymes occurs specific to the substrate and effect.

√ High-performance detergents contain optimized hydrolases, such as proteases, amylases, cellulases, and lipases for lower temperature applications.

√ In high yields, these specialized enzymes arise through the genetically modified production strain "*Bacillus licheniformis*" (genetic engineering).

√ Production of these enzymes takes place in aerated fermenters with high-energy input.

√ After separation of biomass, the enzymes are concentrated and transferred into an adjusted liquid or into granules.

√ Enzyme dust is hazardous. Sensitization after first contact via the respiratory tract, and allergic reaction after renewed contact may occur to sensitive persons.

√ For safe handling, all granules obtain an effective coating layer.

√ Product design of detergent enzymes stand for high performance by genetic engineering and for optimization of physical properties by chemical engineering.

References

1. (a) Rähse, W. and Hoffmann, S. (2002) Produktdesign-Zusammenwirken von Chemie, Technik und Marketing im Dienste des Kunden. *Chem. Ing. Tech.*, **74** (9), 1220–1229. (b) Rähse, W. and Hoffmann, S. (2003) Product design – the interaction between chemistry, technology and marketing to meet customer needs. *Chem. Eng. Technol.*, **26** (9), 931–940.

2. Rähse, W. (2007) *Produktdesign in der chemischen Industrie*, Springer-Verlag, Berlin.

3. Smulders, E., Rähse, W., von Rybinski, W., Steber, J., Sung, E., and Wiebel, F. (2003) *Laundry Detergents*, Wiley-VCH Verlag GmbH.

4. Bundesministerium für Bildung und Forschung (BMBF) (2008) *Weiße Biotechnologie – Chancen für neue Produkte und umweltschonende Prozesse*, Bonn, Berlin, *www.bmbf.de/pub/weisse biotechnologie.pdf* (accessed 27 November 2013).

5. Festel, G., Knöll, J., Götz, H., and Zinke, H. (2004) Der Einfluss der Biotechnologie auf Produktionsverfahren in der Chemieindustrie. *Chem. Ing. Tech.*, **76** (3), 307–312.

6. Nusser, M. (2008) Industrielle Biotechnologie: Wirtschaftspolitische Bedeutung, Wettbewerbsfähigkeit und Standortattraktivität Deutschlands. *Chem. Ing. Tech.*, **80** (6), 713–724.

7. Bundesministerium für Bildung und Forschung (2004) Positionspapier der Dechema e. V. Weiße Biotechnologie – Chancen für Deutschland, Stand November 2004, *www.dechema.de/dechema media/ Downloads/ … /wbt04.pdf* (accessed 27 November 2013).

8. Garthoff, B. (2006) in *Dokumentation der vom Bundesministerium für Umwelt, Naturschutz und Reaktorsicherheit, dem Umweltbundesamt und der Deutschen Industrievereinigung Biotechnologie gemeinsam veranstalteten Fachtagung "Weiße Biotechnologie – Ökonomische und ökologische Chancen" am 18. Oktober 2006 im Bundespresseamt*, Berlin, (eds Hrg. W. Dubbert and T. Heine), S. 17, *www.umweltdaten.de/publikationen/fpdf-l/3260.pdf* (accessed 27 November 2013).

9. Schnee, M. and Heine, T. (2008) *Weiße Biotechnologie am Kapitalmarkt*, DVFA-Fachpublikation, Dreieich, *www.dib.org/Publikationen* (accessed 27 November 2013).

10. Braun, M., Teichert, O., and Zweck, A. (2006) *Übersichtsstudie: Biokatalyse in der industriellen Produktion*, VDI Zukünftige Technologien Nr. 57, Zukünftige Technologien Consulting der VDI Technologiezentrum GmbH, Düsseldorf, *www.vditz.de/publikation/biokatalyse-in-der-industriellen-produktion/* (accessed 27 November 2013).

11. Novozymes (2010) Novozymes Report 2010, Sales and Markets – Enzyme Business, *http://report2010.novozymes.com /Menu/Novozymes+Report+2010/Report /Sales+and+markets/Enzyme+Business* (accessed 27 November 2013).

12. Wink, M. (2011) *Molekulare Biotechnologie*, Wiley-VCH Verlag GmbH, Weinheim, S. 508 ff.

13. Präwe, P., Faust, U., Sittig, W., and Sukasch, D.A. (1994) *Handbuch der Biotechnologie*, 4 Aufl., R. Oldenbourg Verlag, München.

14. Kriegel, T. and Schellenberger, W. (2007) in *Biochemie und Pathobiochemie*, 8. Aufl., (Hrg. G. Löffler, P.E. Petrides, and P.C. Heinrich), Springer Medizin Verlag, Heidelberg, S. 107 ff.

15. Karlson, P., Doenecke, D., Koolman, J., Fuchs, G., and Gerok, W. (2005) *Karlsons Biochemie und Pathobiochemie*, 5 Aufl., Kapitel 3.5, Georg Thieme Verlag, S. 55 ff.

16. Kalbitzer, H.R. and Petrides, P.E. (2007) in *Biochemie und Pathobiochemie*, 8. Aufl. (Hrg. G. Löffler, P.E. Petrides, and P.C. Heinrich), Springer Medizin Verlag, Heidelberg, S. 69 ff.

17. van Ee, J.H., Misset, O., and Baas, E.J. (eds) (1997) *Enzymes in Detergency*, Surfactant Science Series, vol. 69, Marcel Dekker, Inc..

18. Maurer, K.-H. (1999) in *Industrielle Nutzung von Biokatalysatoren: ein Beitrag zur Nachhaltigkeit* (eds S. Heiden, A.-K. Bock, and G. Antranikian), Bd. **14**, Erich Schmidt-Verlag, Berlin, S. 173–185.

19. Roth, U., Hoppenheidt, K., Hottenroth, S., and Peche, R. (2004) *Entlastungseffekte für die Umwelt durch Enzymeinsatz in Vollwaschmitteln*, Studie, Bayerisches Institut für Angewandte Umweltforschung und –technik, Augsburg, *www.bifa.de/download/waschmittel kurz .pdf* (accessed 27 November 2013).

20. Aehle, W. (ed.) (2008) *Enzymes in Industry: Production and Applications*, Wiley-VCH Verlag GmbH, Weinheim.

21. Schügerl, K. (1997) *Alkalische Proteaseproduktion mit Bacillus Licheniformis*, Kapitel 3.3, Birkhäuser-Verlag, Basel, S. 166 ff.

22. Maurer, K.-H. and Wieland, S. (2005) *Saubere Arbeit: Innovative Waschmittelenzyme, Forschung und Entwicklung bei Henkel: heute für morgen*, Henkel AG & Co. KGaA, S. 22–25, *www.henkel.de/forschungsmagazine-27678.htm* (accessed 27 November 2013).

23. Tamime, A.Y. (ed.) (2008) *Cleaning-in-Place: Dairy, Food and Beverage Operations*, 3rd edn, Wiley-Blackwell.

24. Rähse, W. and Carduck, F.-J. (1985) Mikrofiltration von Fermenterbrühen. *Chem. Ing. Tech.*, **57** (9), 747–753.

25. Rähse, W., Paatz, K., Pichler, W., and Upadek, H. (1995) Verfahren zur Desodorierung und Stabilisierung biotechnologisch gewonnener Wertstoffe und ihrer wässrigen Zubereitungen. Henkel AG & Co. KGaA, PCT/EP95/02142, DE 595 03 532.9, Jun. 06, 1995.

26. Paatz, K., Rähse, W., and Dicoi, O. (1994) Vorrichtung zum destillativen Reinigen von Einsatzgut. Henkel AG & Co. KGaA, DE Patent 44 42 318.7, Nov. 29, 1994.

27. Baur, D., Pichler, W., Rähse, W., and van Holt, J. (2003) Verfahren zur Veredelung konzentrierter Enzymlösungen. Henkel AG & Co. KGaA, DE 103 04 066.8, Jan. 31, 2003.

28. Pawelczyk, H., Rähse, W., Carduck, F.-J., Kühne, N., Runge, V., and Upadek, H. (1991) Enzymzubereitung für Wasch- und Reinigungsmittel. Henkel AG & Co. KGaA, DE 591 05 338.1, Dec. 06, 1991.

29. Paatz, K., Rähse, W., Pichler, W., and Kottwitz, B. (1997) Coated enzyme preparation with an improved solubility. Henkel AG & Co. KGaA. PCT/EP97/06744 (DE 597 09 528.0), Dec. 02, 1997.

30. Rähse, W. (2009) Produktdesign disperser Stoffe: Industrielle Granulation. *Chem. Ing. Tech.*, **81** (3), 231–253.

31. Rähse, W. (2009) Produktdesign disperser Stoffe: Industrielles Partikelcoating. *Chem. Ing. Tech.*, **81** (3), 225–240.

32. Schneider, W. (1972) Der Einfluß enzymhaltiger Waschmittel auf den Eiweiß- und Lipidmantel der Haut, Fette, Seifen. *Anstrichmittel*, **74**, 420–423.

33. BGI (2005) *Auswahlkriterien für die spezielle arbeitsmedizinische Vorsorge nach dem Berufsgenossenschaftlichen Grundsatz G 23 "Obstruktive Atemwegserkrankungen": Enzymhaltige Stäube*, 504-23e (ZH 1/600.23e), Kooperation des HVBG mit dem Carl Heymanns Verlag.

11
Design of Solid Laundry Detergents According to Consumer Requirements

Summary

In the past 50 years, powdery and granular detergents have improved significantly in performance and ecology: more concentrated (less chemicals), increased washing effect, biodegradable surfactants, almost no phosphates, reduced water, and energy consumption in the washing process. Use of optimized enzymes and bleaching boosters enables lowered wash temperatures and removal of stains. The production of solid detergents expires in spray towers either without or with a compaction step or, alternatively, in "non-tower" agglomeration plants. Latest generation of granulation with an integrated mill for simultaneous drying and grinding allows the production of spherical particles, which can be adjusted small or large, heavy or light. Future formulations will permit lower washing temperatures (15–30 °C).

11.1
Market Products in Germany

According to latest test results, powdery and granular laundry detergents of high quality provide the best wash performances [1]. They contain alkalis, water softeners, surfactants, and enzymes, as well as, in the case of heavy-duty detergents, several bleaching agents, and optical brighteners. The superiority of bleaching appears on colorfast laundry, but especially on white wash. White is not only clean but lights bright white owing to the lightening with optical brighteners. Currently in Europe, washing usually takes place in machines in the range between 40 and 60 °C. Bleaching substances contribute significantly to reduction of existing germs [2, 3]. For reasons of hygiene, machine washing of bed linen and towels, as well as body linen occurs at least at about 60 °C with heavy-duty detergents. Even if liquid detergents show some advantages in handling and therefore in Germany gain market shares [4], solid detergents of high quality remain (still) indispensable. Only solid forms contain enzymes in the presence of bleaching agents without loss of activity, as the enzyme particles are provided with a coating. This layer prevents a reaction with adjacent particles. In the absence of

Industrial Product Design of Solids and Liquids: A Practical Guide, First Edition. Wilfried Rähse.
© 2014 Wiley-VCH Verlag GmbH & Co. KGaA. Published 2014 by Wiley-VCH Verlag GmbH & Co. KGaA.

UV light, dry solids cannot react in the package, neither with each other nor with air. There is, in contrast to liquid products, no risk of contamination with microorganisms.

Color detergents contain a color transfer inhibitor [5], but no bleach and whitening agents. Typical wash temperatures are in the range of 20–40 °C. Ink transfers represent a big problem in the washing of colored textiles. New chemicals are able to prevent the transfer from one garment to the next. The steady increase in the use of colored clothing necessitates this development. Owing to inhibitors, detached colors remain in the washing liquor. At the end, the dissolved colors flow via the liquor pump into the sewer. Otherwise, many color detergents correspond largely to the composition of heavy-duty detergents.

A small proportion of the population shows allergic reactions after using detergents. Critical components may be perfume oils, optical brighteners, dyes, and preservatives (for liquid detergents), possibly inadequately coated or broken enzyme granules (enzyme dust). It was for this reason that the development and production of skin-friendly formulations commenced. These include detergents, which pass tests on dermatological compatibility and contain approved ingredients.

Conceptually, the previously mentioned products belong to the universally usable laundry detergents, which means they are suitable for all textiles except specialties. For these, there are a variety of special detergents for curtains, sportswear, fine synthetics, wool, and silk. Other products provide color boosters, remove stains, or act as a wash enhancer. The agents for wool and silk must be formulated without bleach and without enzymes (proteins), because both damage (protein) fibers. Except for the curtain detergents, liquid products outweigh.

According to the Body Care and Detergents Industry Association (IKW in Germany), the expenditure in the German market for universal detergents amounted to €1176 million a^{-1}. In the past 12 years, this sum remained almost constant, with small fluctuations. In addition, sales of specialty detergents, fabric softeners, and additives amounted to €823 million (2012[1]). Detergents are available in the market worldwide as powder, in packets of different sizes and as refills. The detergent industry sells liquid formulations and gels in bottles. Usual package sizes lie between 0.6 and 10 kg. Besides customary powders and liquids, volume and/or formulation concentrates (via the recipe) arise, in fact, as granules or as gels. Owing to smaller space requirements, detergent concentrates are handy. They dominate the market in Japan [6]. In Germany, liquid concentrates increase their market share, while solid concentrates decrease. The customary powdery commodity remains in demand. As predosed concentrates, detergent tablets and sachets make the dosing easier in application. In the German market, they play only a minor role.

1) Google: IKW – Reinigungsmittel – Marktdaten.

11.2
Identification and Consideration of Customer Needs

Customers are interested in the look of their clothing and the cleanliness of their laundry in the household. For this, detergents represent an indispensable tool, but get little attention ("low-interest product"). They have to work and spread a pleasant smell. At the center stands product performance, depending on the textiles, washing machines, and operation conditions (Figure 11.1). Performance arises through a concerted chemistry and biotechnology in the formulation. Convenience (handling) and aesthetics (particle sizes and shapes, colors, fragrances) come from the process technology, adjustable in wide ranges [7]. Here, a discussion takes place about the possibilities of influencing the product design via the production process and formulation. In focus are the customary powdery and granular detergents.

For successful introduction of new products, brand companies want to know the opinion of consumers in an early development phase. Searching for and finding of the real customer's needs are difficult and costly. Only big brand companies have the necessary resources. They instruct agencies to question the target audience. This occurs in group discussions (focus groups with 10–15 participants), where participants air their views and wishes [8]. Questions refer to the performance

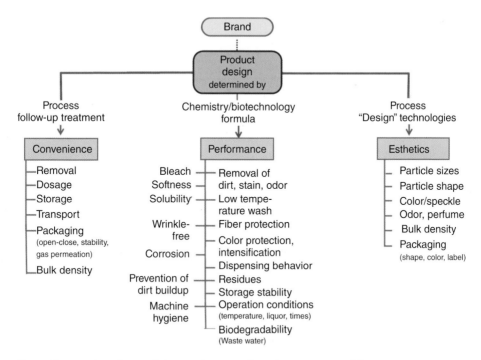

Figure 11.1 Product design of disperse detergents from the perspective of the manufacturer.

spectrum of new or changed products. Customers should define essential elements of the product and packaging to influence the finished appearance.

Specific questions often concern the washing of sensitive, colored textiles (mixed fabrics, fine synthetics, lingerie, and wool). Sometimes, new fragrances, optimized enzymes, or modified products are in the limelight. Usually, from direct questioning to an innovative proposal arise several insights for optimizing of product designs. New aspects are rare, but the weights of each parameter yield valuable information to development and marketing. Furthermore, from assessment of brand and willingness to buy, indications for the market success result [9]. Performed with samples from the pilot plant, tests in many households take place subsequently, each with 100–250 participants (concept and home use tests [10–12]). Tests occur in one or in several countries. The novelty of the tested detergent and the credibility of claims about the product, personal relevance, as well as buying intention are the focus of investigations.

In the market, brand- and no-name detergents differ in product design. Customers purchase either owing to the promise of brand manufacturers (consistent high quality and tested) or they go for the lowest price. The reasons for buying detergents are often the experienced quality of previous purchases or the demonstrated performance in independent tests. In Germany, current heavy-duty detergents in the market do not differ significantly in the main constituents. Differences arise from some ingredients of the formulation. They depend on the presence of expensive raw materials and their quantities. Examples represent enzymes, bleach boosters, and dye transfer inhibitors, apart from some surfactant mixtures.

Second, special components provide fiber protection (pH regulators and complexing agents), which is missing frequently in cheap detergents. Many ingredients and their formulation are known from patents, lectures, and publications of raw material and from brand manufacturers. In addition, commodity producers present their findings at conferences and trade shows. Furthermore, brochures offer significant information. Consequently, the market is largely transparent for professionals.

Essential customer requirements are easily written down (Table 11.1), but difficult to realize. Also, this is because new fashion fabrics/colors are always appearing on the market. Modern detergents have to meet many requirements concerning product and packaging, washing performance, hygiene, handling, and aesthetics.

Brand manufacturers offer their employees comprehensive social services. They differ from other manufacturers in that they develop customer-oriented new processes. Brand companies search for novel raw materials and enzymes, often in cooperation with raw material manufacturers. In addition, they support European bodies and organizations through their membership, participating through dedicated collaboration on regulations and laws. Development of analyses for safety and tolerability, as well as test methods, which can be used by all, brings further contribution to the community. Besides, environment, sustainability (resources), and social responsibility (global social engagement) represent important cornerstones of company policy in brand companies [13].

Table 11.1 Significant customer wishes for washing of laundry in machines.

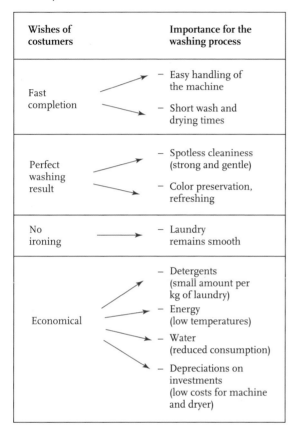

Wishes of costumers	Importance for the washing process
Fast completion	– Easy handling of the machine – Short wash and drying times
Perfect washing result	– Spotless cleaniness (strong and gentle) – Color preservation, refreshing
No ironing	– Laundry remains smooth
Economical	– Detergents (small amount per kg of laundry) – Energy (low temperatures) – Water (reduced consumption) – Depreciations on investments (low costs for machine and dryer)

Successful marketing of detergents requires strong brands [9, 14]. For their orientation, many customers seek known products that are available everywhere. In contrast, they buy at discount stores, where the name of the market gives orientation (private label products). Largely, brands influence purchase decisions and therefore constitute a superordinate part of product design.

In the field of detergents and cleaning agents, the umbrella brands outweigh. A superior strategy, known as *multi-brand strategy* of product families, is performed in the area of detergents (shown in Figure 11.2). Here, the same (product) brand represents all detergent types (heavy duty, color, and sensitive) in liquid and solid form (powder, granules) together in a brand family. If more than one umbrella brand is available for brand families, each is assigned different price segments. In this way, multi-brand families emerge. A matrix (Figure 11.2b) displays linking of comparable brand families with multiple brands.

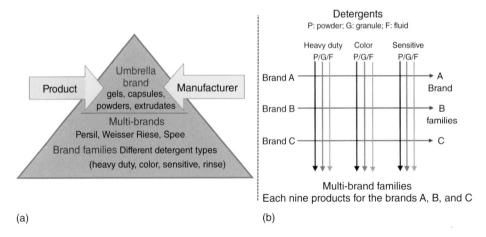

(a)

(b)

Figure 11.2 (a,b) Product umbrella brands (example: brands from Henkel), multi-brand, and brand family strategies for detergents.

11.3
History of Laundry Washing

Homer (about 850 BC) in his epic, the Odyssey, describes Nausicaa washing laundry along with other girls on the beach; the dresses are then placed in the sun to bleach. Usually, the release of contaminants from the fiber in clear water is supported by the strong mechanical action (rubbing, hitting, and kicking). The discovery of soap is attributed to the Sumerians around 2500 BC. They made soap from oil and vegetable ash (potash and potassium carbonate). Soap served both as a remedy for injuries and as a cosmetic cream. To manufacture soap, the ancient Egyptians overcooked animal fats with plant ash and additionally with trona, a natural sodium-containing mineral. The Roman scholar Pliny the Elder (23–79 AD), known from Naturalis Historia [15], described the Gauls' use of overcooked beech ash with tallow from buck and goat. Similar formulations offered "De simplicibus Mediaminibus" [16], attributed to the Greek physician Galen (129–199 AD). He first mentioned the cleansing action of soaps (oldest surfactants) and added that German soap was the purest and greasiest.

The Romans had washermen who cleaned the customers' laundry with rotten urine. Microbial degradation of ingredients (urea, uric acid, and metabolic products) generated ammonia. Developing alkalinity improved the washing results significantly. Emperor Vespasian (9–79 AD) is said to have raised taxes from the rich washers ("Pecunia non olet").

The Arabs knew about the production of caustic soda and caustic potash from their carbonates with slaked lime. In the seventh century, they brought the technology of improved soap fabrication to Spain. This knowledge spread rapidly in Europe. The Spaniards and French (Marseille) used fruits of the olive tree as the source of oil, and established the center of soap making (Savon de Marseille). In the fourteenth century, guilds of soap makers arose. Lumpy soap, sometimes scented and mainly used for cleaning the body, remained a luxury item. Finally, the technical production of soda succeeded after Leblanc (1790), and especially after Solvay (1865). Available in larger quantities, soda allowed the hydrolyzation of fats and oils at elevated temperatures.

During this time, laundry washing with wood ashes (in bags) and hot water developed. Around 1880, the first powdered detergent came on the market, consisting mainly of soap powder (surfactants), mixed with soda and water glass (sodium silicate) as alkali carriers. In 1907, Fritz Henkel invented heavy-duty detergents. In addition to the constituents previously mentioned, it contained sodium perborate tetrahydrate as the source of hydrogen peroxide. Therefore, it was no longer necessary to bleach clothes out on the grass in the sun. From perborate and silicate arose the brand name "Persil." In same year, Otto Röhm suggested the use of enzymes in laundry detergents for removing stains [17]. Both innovations fundamentally altered the detergents [5].

With the hydrogenation of natural fatty acids, and the reaction of the resulting fatty alcohols with sulfuric acid to anionic surfactants, the modern surfactant chemistry began in 1928. In the 1950s, soap was replaced by petrochemical-based, branched tetrapropylene benzene sulfonates (TPSs, also abbreviated as TBS). This group of substances shows poor biodegradation and therefore generated mountains of foam on the rivers. During this time, the change from hand to machine wash occurred. Readily biodegradable linear alkyl benzene sulfonates (LAS) replaced TPS in 1961 (German detergents regulation, 1964).

From 1982 onwards, insoluble ion exchangers replaced eutrophying tripolyphosphates in Europe as water softeners. Zeolites (Sasil) allowed the production of phosphate-free detergents by binding of water hardness (calcium and magnesium ions). In the new millennium, combinations of soda with special polyacrylates (soluble builder; Section 9.5) displaced the zeolites. In the 1970s and in the following 20 years, crucial breakthroughs ensued in the washing performance. The highlights were, first, enzymes [18] (protease, amylase, cellulase, and lipase; Section 10.4) and then the activator for bleaching at 40–60 °C in heavy-duty detergents, produced as customary powders and as superconcentrates. Third, color detergents got their importance by addition of color transfer inhibitors, available in liquid and solid formulations.

Figure 11.3 shows examples of detergent packages for solids during 1877–1977. In these times, solid detergents dominated the market. Currently, the offer is much broader. Besides the heavy-duty detergent, there are light-duty detergents and many specialties in solid and liquid forms (Figure 11.4). Modern packages are discussed in Section 9.6.

11.4
Washing Process

In addition to detergents, the washing process includes textiles, dirt, the washing machine [19], and water. In parts of Asia, the Middle East, and Europe, people use

Figure 11.3 Historical solid detergents and their packagings from 1877 up to 1977 (brands from Henkel).

Figure 11.4 Historical and current detergents and their packagings from 1980 up to 2013, solids (powder, Megaperls®, tabs), liquids, and gels (brands from Henkel).

front-loading washing machines that have horizontal axis and allow high washing temperatures. They differ significantly from the top-loading, vertical-axis machines in Japan, Australia, New Zealand, Canada, the United States, and Latin America (Figure 11.5). In these countries, the wash temperature is lower, and in most cases the washing water softer. Deviant eating habits lead to other stains. This explains the differences in detergent formulations, and the similarities in the types of surfactants and enzymes.

Textiles are completely listed within the Textile Labeling Act [20]. They consist of natural and synthetic fibers, but often of blended fabrics (Table 11.2). Depending on the care label, textiles may be washed (tub with temperature or hand), bleached (triangle), machine dried (circle in a square), and ironed (with iron points), as shown in Figure 11.6. The number of points indicates the permitted temperature, an underscore for gentle handling.

Figure 11.5 European frontloader and top loader for India (Courtesy of Samsung).

Table 11.2 Selected natural and/or synthetic fibers for textiles.

No.	Basis	Origin of fibers	Textile fiber
1	Cellulose	Plant fibers	Cotton and linen (flax)
2	Cellulose + proteins	Plant fibers + silkworms	Semi silk (a mixture of cotton and silk)
3	Proteins (insects)	Cocoon of the silkworm and various types of spinners	Silk, processed in chiffon, satin, and taffeta
4	Proteins (mammals)	Sheep, merino, angora, cashmere goat, llama, angora rabbits, camels, small camels, alpacas, and muskoxen	Sheep wool, merino wool, mohair, cashmere, llama wool, angora, camel hair, yak, alpaca wool, and qiviut
5	Cellulose + chemistry	Modified cellulose (regenerate fiber and natural-based synthetic fibers)	Viscose, Viscose silk (Rayon), Modal, Lyocell, Cupro silk, and Cellulose acetate (artificial silk)
6	Petroleum	Chemical fibers	Polyester, usually polyethylene terephthalate (Diolen, Trevira); polyamide (Nylon, Perlon); polyacrylonitrile (Nylon, Orlon); polytetrafluoroethylene (Gore-Tex); polypropylene (Polycolon); and polyurethane + poly(ethylene glycol) (elastane)

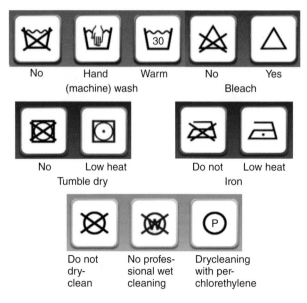

| No | Hand (machine) wash | Warm | No | Yes |

Bleach

| No | Low heat | | Do not | Low heat |

Tumble dry

Iron

| Do not dry-clean | No professional wet cleaning | Drycleaning with per-chlorethylene |

Figure 11.6 Some textile care labels.

Most stains arise during cooking, eating, and drinking and by leisure activities at home (DIY = do-it-yourself) or in nature (sports). Sometimes, there is blood or traces of cosmetic. The causes of contamination in the laundry (Table 11.3) are bodily excretions (sweat, feces, urine, and skin flakes). While most dirt can easily wash out in the machine, it is advisable to pretreat stains or, if possible, immediately wash and rinse out. Most of these contaminants do not dissolve in water.

In Europe, conventional drum washing machines with horizontal axis have a variety of programs. The right program can be selected to include textiles and degree of contamination. The program controls the addition of detergent powder and fabric softener as well as the wash cycle and temperatures. The existing load sensor adjusts the water level. The timer oversees washing, rinsing, and removal of washing liquor, besides spin cycle times, revolutions, and final spin. Vibrations, caused by imbalances, avoid a sensor that shuts down the spin speed and starts up. Often, a program sequence includes the following steps: prewash or soak, wash, rinse, rinse stop, pump, spin, and anti-crease. Special features of machines represent the short programs, besides addition of more water or an additional rinse cycle. Frequent washing of delicate fabrics and the attempt to save energy lead to more low-temperature applications (<30–40 °C) [21, 22]. Optimal detergents, weight of the wash load, and energy consumption depend on the textiles to be washed. Further, pollution levels and water hardness affect the amount of detergents used. The costs of the wash load are given in Table 11.4.

During the washing process, the wash liquor contains the detergent and dissolved/dispersed dirt. At the end of the process, the liquor runs out completely by being pumped into the sewer and then into the water treatment plant. Therefore, the use of avoidable substances is forbidden. Because of the effect of eutrophication,

Table 11.3 Dirt and stains on clothing

Solubility	Occurrence	Examples
Soluble in water	Food	Salt, sugar, and clear juices
	Skin secretions	Body sweat
	Body exudates	Urine
Insoluble in water, with surfactants rinseable	Cosmetics	Pigments, fats, oils, waxes, and skin fat
	Edible fats	Butter, lard, cooking fat, oil, and sausage
	Environment	Dust, dirt, and soot
Insoluble in water, after enzymatic degradation with surfactants rinseable	Food	Egg, milk, sauce, spinach, noodles, meat juice, pudding, cocoa, carrot, and baby food
	Body	Blood, feces, and mucus
	Cosmetics	Makeup and lipstick
	Environment	Grass
Residues, but bleachable	Food	Wine, coffee, tea, fruits, vegetables, and berries
Totally insoluble, not removable	Household	Varnishes (solvent based), marker and printer inks, pens, and candle wax
	Environment	Tar

Table 11.4 Estimated costs of washing at 60 °C for a wash load (~5.5 kg cotton) and laundry drying in modern machines (price basis: Düsseldorf 2012).

No.	Parameter	Specific costs (estimated values)	Costs	Sum of the costs
1	Detergent 80 g[a]	€19.95 for 48 loads	€0.42	—
2	Water 55 l	€1.95 m^{-3}	€0.11	—
3	Waste water 55 l	€1.52 m^{-3}	€0.08	—
4	Energy 0.85 kWh	€0.24 kWh^{-1}	€0.20	1–4: €0.81
5	Depreciation washing machine	€600 for 2000 loads	€0.30	1–5: €1.11
6	Energy dryer 2.6 kWh	€0.24 kWh^{-1}	€0.62	—
7	Depreciation Condensation dryer	€600 for 2000 loads	€0.30	6 + 7: €0.92 1–7: €2.03

[a] Market leader.

the formulations may contain phosphorus compounds only in prescribed limits [23]. These depend on the hardness of water and detergent type, and are in the order of 0.5 (to 1) g P/l detergent solution. Calculations result in maximum values of 2.5 (to 5) g of phosphorus per wash load in 5 l liquor. The German Washing and Cleaning Agent Law [24] allows only biodegradable surfactants, as well as minimal water and energy consumption.

The drinking water that is available in Germany is generally of high quality. The calcium and magnesium ions contained in it cause carbonate hardness (soft water <1.5, hard water >2.5 mmol l^{-1}). Therefore, dosages depend on the water hardness according to the label on the package. Insufficient binding of water hardness may cause precipitation of soaps and reduce the effect of surfactants. Furthermore, the carbonate crystals formed are visible on dark laundry. Possibly, some sharp crystal edges damage the textile fibers.

Heavy metal contents in the tap water are well below allowable values that are specified in the drinking water regulation [25]. However, these values increase significantly by transportation in old galvanized iron pipes, especially when chloride levels in the water are above 50 ppm. Damage to zinc coatings cause rust. After a long stagnation in pipes, the rust flows out with the first liters in the form of brown water. In addition, heavy metal content of some ingredients in the detergent requires the use of effective chelating agents.

11.5
Recipe

Recipes for detergents are tailored to textiles, the way of washing (machine, hand), and wash temperatures. The frame formulation [26] for heavy-duty and for color detergents with selected ingredients is given in Table 11.5. For several chemical ingredients, comparable alternatives exist; but they are not considered here.

Suitable ion exchangers or precipitants reduce or prevent the negative effects of hard water. However, insoluble ion exchangers (zeolites) are sometimes visible on dark colors (black) as a white precipitate. Therefore, in modern detergents water-soluble builders are preferred as water softeners. Novel water softeners, based on sodium carbonate/polycarboxylates, dissolve easily in all temperature ranges. Formulated complexing and sequestering agents delay the precipitation of calcium carbonate after exceeding the solubility (threshold effect). These special polymers prevent crystal growth of precipitated carbonates by adsorption on the crystallization nuclei. They keep water hardness in the wash liquor dissolved and/or finely dispersed, up to pump off. Effective complexing agents represent, in particular, copolymerized polycarboxylates [26], which bind calcium because of their many charges, and keep the dirt in the liquor.

The surfactants in detergents reduce interfacial tension and improve the wetting of fibers as well as of hydrophobic soils. In this way, surfactants remove much of the soil from the fibers. With the help of mechanics, dirt reaches the liquor. Best results are achieved by mixtures of anionic with nonionic surfactants in the

Table 11.5 Frame formulation for disperse detergent (heavy duty and color).

Tasks in the washing process	Substances	Amounts
Water softening		
(1) Builders	Zeolite and/or sodium carbonate, silicates	15–30%
(2) Cobuilders	Polycarboxylates (sodium salts)	2–8%
Washing alkalis		
(3) pH (1% solution) for optimal washing conditions	Sodium carbonate pH 11.4	See (1)
	Sodium bicarbonate pH 8.4	0–8%
	Sodium silicate pH 12.5	0–7%
Removal of fats and dirt		
(4) Anionic surfactants	Alkyl benzene sulfonates (Na)	8–14%
	Fatty alcohol sulfates (Na)	0–4%
(5) Nonionic surfactants	Fatty alcohol ethoxylates	0–6%
	Alkylpolyglycosides	0–3%
(6) Amphoteric surfactants	Betaine	0–3%
Stain detachment		
(7) Enzymes	Protein, amylase, cellulase, and lipase	0.2–1.5%
Detergent boosters		
(8) Cationic surfactants	Esterquats	0–1.5%
Brightening (only heavy-duty detergents)		
(9) Bleach	Sodium percarbonate	8–18%
(10) Bleach activator	Tetraacetylethylenediamine	3–7%
(11) Optical brighteners	Stilbenes	0.1–0.3%
Inhibitors		
(12) Graying (antiredeposition agents)	Cellulose ethers	0–3.5%
(13) Discoloration	Polyvinylpyrrolidone	0–0.5%
(14) Foam	Polydimethylsiloxane	0.1–2.5%
(15) Pollution (soil repellent)	Polyethylene glycol terephthalate	<1.5%
(16) Heavy metal complexation	1-Hydroxyethane-1.1-diphosphonic	<1%
Excipients		
(17) Powder formation	Sodium sulfate pH 9	0–35%
(18) pH-value	Citric acid	0–1%
(19) Softener	Bentonite and esterquats	0–1%
(20) Disintegrants (for tabs only)	Cellulose	<5%
Fragrance		
(21) Perfume	Mixture of 5–30 synthetic fragrances	0.2–0.6%
Washing machine (inhibitors)		
(22) Corrosion	Sodium silicate	See (3)
(23) Lime deposits	Polycarboxylates	See (2)
	Citric acid	See (18)

wash liquor, in which either more anionics are combined with little nonionics or the other way round. Formulations of detergents that are sensitive on the skin should contain alkylpolyglycosides, but preferably betaines. Betaines reduce significantly the irritation potential of anionic surfactants on the skin, as is known from cosmetics that contain sodium laurylsulfate (shampoo). Foam regulators prevent the formation of foam. Air in the foam significantly impedes the washing process by reducing the mechanic effects.

Washing alkalis adjust the pH, usually to values between 10 and 11.5. With increasing pH, natural fibers swell up. The zeta potential increases significantly. This strengthens repulsive forces, which facilitate the detachment of dirt by surfactants and the incorporation into micelles. The negative charge of micelles and fibers prevents redeposition of dirt. Substances such as cellulose ethers and polyethylene glycol terephthalate support the antisoiling effect.

High performances of modern detergents would not be feasible without the use of specific enzymes and enzyme mixtures. Genetically modified bacilli produce biotechnologically detergent enzymes in fermenters (see Section 10.5). Many difficult stains are of biological origin (grass, egg, and potatoes). In nature, there exist usually suitable enzymes for biodegradation of these substances (see Section 10.4). The degradation takes place in the washing process by a nonspecific, enzymatic hydrolysis of proteins (proteases), polysaccharides (amylases), and lipids (lipase) in the presence of anionic surfactants, in a strongly alkaline solution [17, 27].

Peracetic from tetraacetylethylenediamine (TAED) at about 40 °C or perhydrox-ylions (ionic mechanism) from hydrogen peroxide at over 60 °C brighten bleachable spots. These spots are generally caused by wine, fruit, tea, and coffee. Bleaching results in the spots being no longer perceptible. Furthermore, optical brighteners (stilbenes) improve the visible whiteness of fabrics. Heavy metal ions (>5 ppm) catalyze radical bleaching. Complexing of cations prevents the formation of radicals. Strong bleaching is not preferred because it leads to fabric damage (holes). Optical brighteners improve whiteness, the fluorescing substances absorbing UV light and emitting visible light. Their molecular structure transforms yellow light into white. Especially in sunshine, the brighteners show a strong effect.

11.6
Design of Finished Products

The final product always consists of base powder/granules and an admixture of components as well as one or two small portions of liquids, which are not visible in the powder. In case of stickiness, additional powdering allows the manufacture of free flowing products (Figure 11.7). Powdery base materials, in combination with admixed components, in particular with color speckles, determine the design.

Starting from powders, different supply forms may arise (Figure 11.8). For example, in granulation or extrusion emerge particles in any sizes by enlarging the particle diameters. They show the shape of powders and spheres or rods (see Chapter 5). Grinding of powders leads to fine powders as well as micro- and

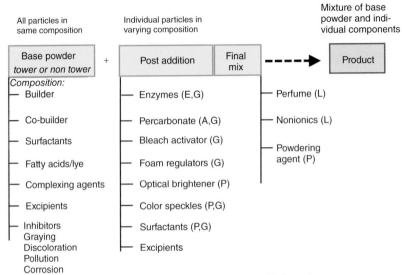

Figure 11.7 Simple presentation of detergent manufacturing.

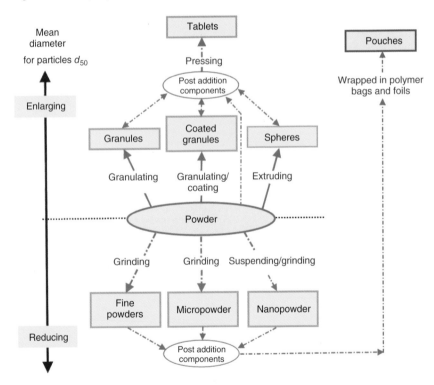

Figure 11.8 Possible forms, starting from powdery materials, as variants of product design.

Figure 11.9 Solid, portioned laundry detergents. Above: single-phase tablets, also with speckles, and a three-phase tablet; below: granules in soluble poly(vinyl alcohol) (PVOH) films, and powder in porous insoluble polymer bag.

nanopowders. Nanoparticles occur by grinding in stirred bead mills (see Section 7.5). Tablets and pouches emerge from these preproducts, after dosing of post-addition components (Figure 11.9). Wrapped in water-soluble or water-insoluble films, pouches can contain detergents of different particle size distributions.

What follows is a discussion about solid detergents. Over time, two forms of dispersed detergents could prevail (Figure 11.10). First, customary powders (d_{50} ~350–550 µm) with bulk densities of 500 ($\pm$150) g l^{-1} represents the dominant market standard. Powders originate by spray drying. Second, of importance are granules (d_{50} 600–800 µm) in powdery form with bulk densities of about 700 ($\pm$100) g l^{-1}. These products emerge from compacting/granulating of spray-dried

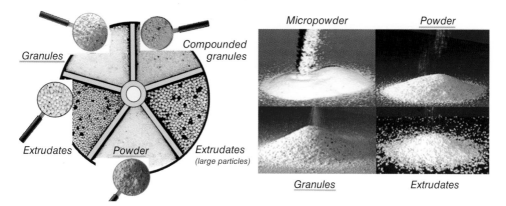

Figure 11.10 Final mixed heavy-duty detergents with color speckles (powders, granules, extrudates; underlined in blue: conventional products).

powders or from non-tower methods. The extrusion process is a special case of granulation.

Substances that are sensitive against temperature, dissolution, and alkalinity or reactive as bleach and enzymes must be added to the base powder in dry form, as spray dried powder, or as agglomerated granules. This happens in a final mix ("post addition"). To prevent segregation, all particles correspond in density and shape as well as in particle size. In the case of sodium percarbonate, an adaptation is relatively easy to realize. The production of this raw material takes place in a spray agglomeration process (see Section 8.8) by adjusting of desired particle size distribution. Other admixture components require more effort for setting the right size. For example, enzyme granules usually have particle sizes between 300 and 1100 μm. The lower range of these sizes is suitable for customary powders, the upper region for powdery granules, but not really for the extrudates (1.2–1.4 mm).

Assessed according to separation tendency, normal powders cause the least effort in processing. At the end, post addition of powdery granules is easy after screening of some components. However, the larger, spherically rounded cylinders from the extrusion process are attractive, but require an intensive adjustment of all components in size and shape to minimize segregation.

11.7
Manufacturing Processes

Three processes dominate the manufacturing of solid detergents: atomization in spray towers and atomization with a subsequent compaction, as well as the "non-tower" process (Figure 11.11). After these processes, the "base powders" arise, which are processed in the post-addition step and then packaged. One line allows the manufacturing of base powders up to $35 \, t \, h^{-1}$. Both the base powder and ready mixture are stored usually for some time in silos. This decouples the process stages, and the filling of the product into packages runs in multiple filling lines with higher capacities.

It is also possible to combine a tower with a non-tower procedure. Then, only small amounts of about 10–15% of the final product arises from the tower by drying a mixture of various solutions and components. The far greater proportion of "base powders" runs over non-tower steps. This split is usually the most economical solution.

An analysis of the cost structure shows that all raw materials together determine the production costs, with a proportion of 75–85%. The remaining 15–25% is the depreciation of equipment and buildings, apart from insurance and energy and labor costs. The division results in raw material costs having a strong influence on recipes and procedures. This is increasingly valid for solid as well as for liquid detergents.

Procedures for liquid detergents are comparatively simple and consist mainly of pipes, tanks, and containers. In the case of discontinuous production, the mixing of all components occurs in a stirred tank. Continuous productions are characterized

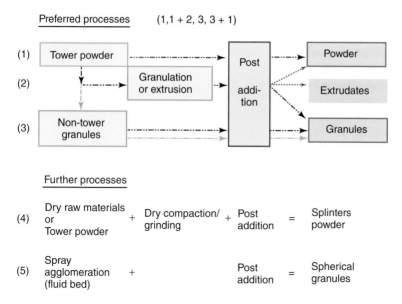

Figure 11.11   Manufacturing processes for detergents.

by a plurality of metering pumps with a common drive and a downstream static or dynamic mixer.

11.7.1
Drying in Spray Towers for Customary Powders

Worldwide, almost all detergent manufacturers use countercurrent towers for the generation of base powders [28]. Metered from containers, liquid and solid materials flow into the slurry tank. There, solids dissolve in water at 60–80 °C. In this vessel of the preparation plant run several chemical reactions with sodium hydroxide, namely, neutralizations of anionic surfactant acid and complexing acids. From there, pressure atomization of suspension with simultaneous drying in hot air ensues, producing free flowing powder. The drying in the spray tower exhibits many advantages, if the neutralization of acids takes place in the slurry tank, and if the quantity of dissolved and/or suspended raw materials is high.

1) Cost advantages in raw materials (depending on location);
2) Low costs by drying of concentrated slurries with high throughputs;
3) Simultaneous shaping (beads);
4) Preventing segregation of powders;
5) Flexibility concerning bulk densities and particle size distributions;
6) Simple return of enzyme-containing dusts and coarse fractions by dissolving in the slurry vessel.

For reasons of plant and personnel safety as well as storage stability, reactive or chemically sensitive substances are admixed. Therefore, the post addition takes

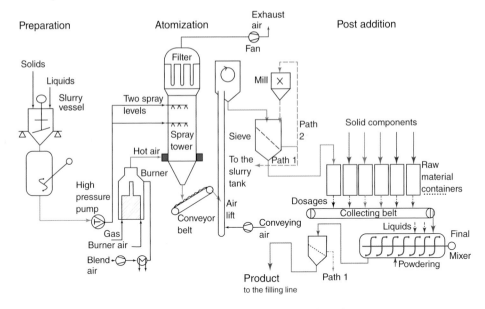

Figure 11.12 Production of powdery detergents with traditional countercurrent spray tower for base powders (variants: one or two spray levels; integrated or external filter; and gas or oil fired).

place by dosing problematic powders or granules (Figure 11.12). The mixing includes coated enzymes, coated bleaches (sodium percarbonate), and activators (TAED). Granulated antifoaming agents (silicone foam regulator concentrate, SIK) show an enhanced efficacy. This is the reason for a specific separate granulation. To simplify production, the manufacturing of colored speckles occurs separately before the actual production of detergents. However, processing of brighteners is possible both in the spray tower as well as in the workup. Drying in the tower requires subsequent cleaning of the tower in case of product changes to brightener-free base powders. Intensive cleaning eliminates even traces of optical brighteners. Washings are necessary because color detergents should not contain these substances.

Some liquids, namely, nonionic surfactants and perfumes, are sprayed very fine on the powder via twin-fluid nozzles. The possible stickiness eliminates subsequent powdering with finely ground powder. Good powdering are in particular zeolites, as well as ground base powder or ground salts. Enzyme-containing dusts and coarse particles from the final screening, in addition possibly oversized from tower screen, return on path 1 (Figure 11.10) into the slurry vessel. There, decomposition of enzymes and bleaching agents occurs under hydrothermal conditions.

Variants of widely used spray drying in countercurrent mode with hot air represent a drying of superheated slurries (flash drying) and drying with superheated steam. Another case is the drying of slurries for light-duty detergents. They need lower product temperatures, realized by a hot air drying in cocurrent (see Section 8.5).

11.7.2
Combined Spraying with Compacting Processes for Concentrates

In the mid to late 1980s, work began for the development of detergent concentrates. The obvious idea was that customers get a much smaller package with less weight for more easy transportation and space-saving storage. An open question reads whether customers accept less mass and volume for same or better performance. Different answers gave the market in various countries.

In Western Europe, several developments ran in process engineering and chemistry. A procedural approach consisted in the gradual increase of bulk density from about 430 to 480–550 g l^{-1} (1989) and later on to 700 ± 50 g l^{-1} (1993). As a first step, a dry compression of base powder occurs in a high-performance ring-layer mixer during continuous spraying of nonionic surfactants. The high-energy input leads to strong shearing of powders during the short residence times. Thereby, a slight reduction of the powder size happens, wherein intraparticle cavities disappear. A passage through the mixer leads to an increase in bulk density by 50–100 g l^{-1}. More expensive, but also more effective is dry compaction by roller compaction with grinding/screening (see Section 5.1). The process is fully encapsulated in a facility [29].

Alternative methods exist to increase the bulk density of powders, such as granulation or extrusion (Figure 11.13). These processes use the addition of fluids (see Section 5.2) and include either drying and/or cooling in a fluidized bed. Many manufacturers perform simple, inexpensive granulations. Thereby, bulk densities increase, along with a slight increase in particle sizes. Owing to short residence times, the product remains powdery, without significant improvement in design. Formerly protected by patents, the technologically sophisticated extrusion created aesthetically pleasing products. As agglomeration liquid, the detergent extrusion uses poly(ethylene glycol) (PEG)-melt, which solidifies on cooling and ensures particle hardness. Both processes guarantee an increase in bulk density by compaction of powdery raw materials. This happens by forming new particles with less porosity and voidage. In addition, the shape factor improves, depending on the process equipment.

There were three *chemical approaches* to concentrate products and to improve formulations:

1) Less water
   a. Reduction of residual moisture content from 9–11% to 4–6% brought a gain of 5% (absolute value).
   b. Switch to a bleach with little crystal water or to water-free compounds enables lowering the weight by 5–6% (perborate tetrahydrate to monohydrate and then to percarbonate).
2) Extensive renunciation of fillers/processing aids: the reduction of sodium sulfate from 25–50% to 0–25% means a decrease in the total amount of about 35%. Added processing aids may improve economically the quality of sprayed powder products, especially the free-flowing properties.

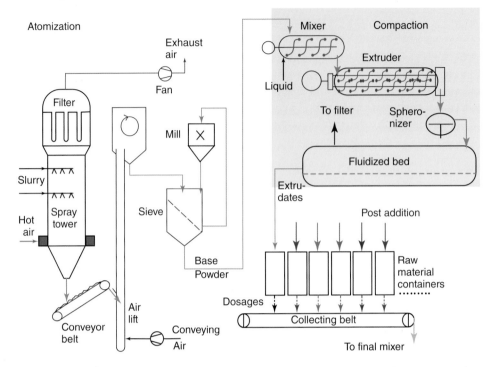

Figure 11.13 Process for preparing detergents with high bulk densities from tower powder (compaction step: extrusion; alternatives: granulation, pelletizing, and roller compaction).

3) Reducing the less relevant substances (sodium carbonate, silicates, and bleaching), and increasing the active compounds (surfactants, enzymes, and bleach activators) lead to better results at low temperatures.

Despite use of only half of the detergent per load, these measures cause an improved wash performance in the temperature range between 30 and 60 °C. Shortly before the changeover, the standard powder was dosed with 150 g (=333 ml) per 4.5 kg of laundry. Washing with the first generation of compacted detergents needs 105 g, and with second generation (extruded) 76 g. Present-day super concentrates work with 67 g = 100 ml for a 5.5 kg wash load, standard powder with 85 g = 165 ml. The extent of concentration is visible, in particular at the volume reduction over time (Figure 11.14).

11.7.3
Non-Tower Method

As a characterizing feature, a spray tower for drying and for forming the base powder is lacking in "non-tower" processes. Shaping of detergent powders by spray drying makes sense only if relatively high percentages of the materials

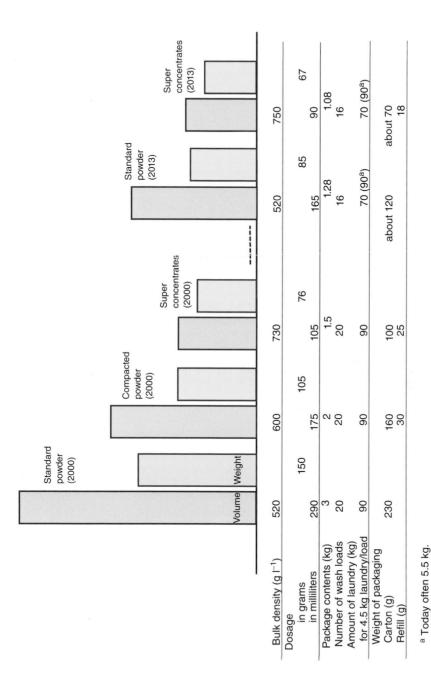

	Standard powder (2000)	Compacted powder (2000)	Super concentrates (2000)	Standard powder (2013)	Super concentrates (2013)
Bulk density (g l⁻¹)	520	600	730	520	750
Dosage					
in grams	150	105	76	85	67
in milliliters	290	175	105	165	90
Package contents (kg)	3	2	1.5	1.28	1.08
Number of wash loads	20	20	20	16	16
Amount of laundry (kg) for 4.5 kg laundry/load	90	90	90	70 (90ᵃ)	70 (90ᵃ)
Weight of packaging					
Carton (g)	230	160	100	about 120	about 70
Refill (g)		30	25		18

ᵃ Today often 5.5 kg.

Figure 11.14 Effects of the concentration of laundry detergents during the past 20 years in Germany.

are advantageously present in solution or in suspension. Furthermore, anionic surfactant acids neutralize economically in the slurry-mixing tank. Use of solid sodium carbonate or dried zeolite instead of a suspension significantly reduces the need for drying. Alkylbenzene sulfonate (LAS-S), used in quantities above 8%, is available as neutralized paste (65%), which eliminates the necessity of a neutralization process in the detergent plant. It would also be possible to place a small neutralization mixer in front of the granulation mixer. Water from solution and from neutralization facilitates the granulation. Hot air in the fluidized bed removes the excess moisture.

Modern detergent facilities, whether equipped with a spray tower or a granulating mixer, operate wastewater-free. Investment costs of these alternative processes are of the same magnitude, because core elements (tower vs mixer and fluidized bed) represent only a small part of the total investment with equal post-addition and filling lines. Depending on the costs of raw materials for the same formulation, non-tower procedures offer economic advantages. Furthermore, air pollution at the production site is lower. Here, a reduced amount of evaporated water permits advantages in economy and ecology. However, non-tower methods depict lower flexibility compared to spray towers. Low bulk densities for special detergents, preferred because of higher dissolution speed, are not yet achievable. Such specialties arise advantageously in spray towers under cocurrent flow of air.

In many cases, continuously working processes consist of two cascaded mixers with a subsequent fluidized bed dryer and downstream a grinding/screening unit [7]. First, all dry materials are metered into a continuous ring layer mixer. There, a part of the nonionic surfactants is sprayed on powders under high-energy premixing. This homogenized and rubbed off solid mixture then flows into a ploughshare mixer, in which the powder granulates after addition of liquids (Figure 11.15). Owing to short residence times, shaping ensues usually only up to the powder-like state, not perceptible as real granules. In the subsequent fluidized bed, the granules dry up to customary moisture contents of 4–8%. Sieves separate coarse particles from the dried granules. After passage through a mill, they flow back on the screen or into the dryer. The optimal way back leads into the high-speed mixer (along with filter dust). High shear forces in the granulation mixer and the relatively high amounts of liquid effect high bulk densities. The powdery granules replace tower powder. In the usual manner, subsequently the post addition of solid components takes place.

According to this method, parts of the detergent formulation in the shape of small granules can also be produced, and are referred to as *compounds*. Compounds consist of one main component together with several agglomerated substances. The main component may be an anionic surfactant, nonionic surfactant, or silicates. A mixture of several compounds represents the "base powder." Depending on the formulation, a division may allow more flexibility in composition. However, this mixture displays visually a nonuniform powder image. This method is known as *compound process*.

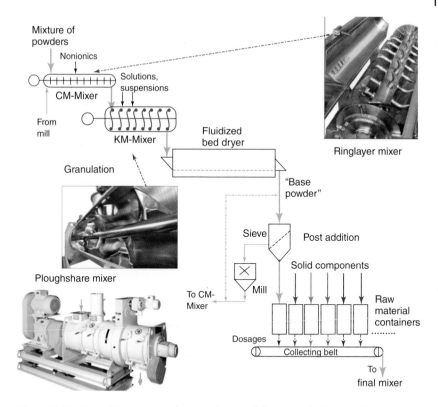

Figure 11.15 Granulation process for manufacture of "base powder" (or single compound), followed by post addition (Mixer courtesy of Lödige).

11.8
Novel Manufacturing Method for Granules

According to their needs, customers choose the best product, concerning performance, convenience, and aesthetics. As solid laundry detergents, some people prefer compact extrudates or granules; others find customary powders good or good enough. From several studies it is known that spherical particles are preferred in comparison to powder or powdery granules. A forward-looking, universal "non-tower" process would have to cover the following areas:

- Spherical particles in all sizes;
- Optionally small, medium, or large diameter (e.g., d_{50} about 350, 550, and 900 µm);
- Optionally medium or high bulk densities (e.g., $\gamma = 500$, 750, and 1000 g l^{-1}).

This would cover most areas of marketable powder and granules. A super concentrate with an even higher bulk density is added. In addition, new combinations (such as 350 µm and 750 g l^{-1} or 900 µm and 500 g l^{-1}) would be possible. The mentioned requirements fulfill a novel granulation [30]. Various products emerge

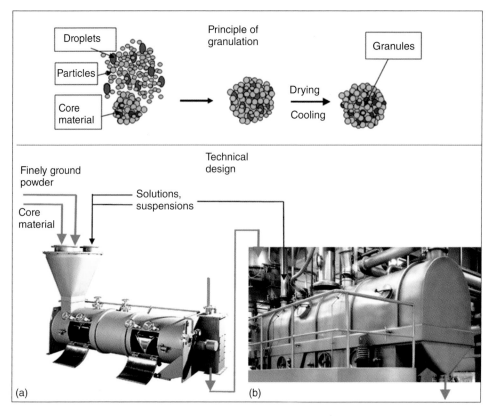

Figure 11.16 Granulation of finely ground particles on core granules of same composition: (a) KM mixer and (b) Vibro Fluidizer (Courtesy of Lödige (a) and GEA Niro (b)).

successively and economically in a large plant with only one production line. In this innovative process, particles in spherical shape arise with the help of liquids. The special feature of this method represents the granulation of finely ground powders on a core material of defined sizes (Figure 11.16). Extremely fine grinding of components ensures that each ball has exactly the same composition. A 500 μm particle contains about 10 000 ground particles, so that a 1% component still provides about 100 particles for each ball. The granulation expires fast because of the high adhesive forces of small particles (see Section 5.2). Therefore, the granulation mixer and equipment may be comparatively small.

In comparison with conventional granulations, this process shows some special features. Downstream of the fluidized bed dryer, a screening plant is arranged to recover core material. In continuous operations, the equipment allows an adjustment of the particle sizes and width of distribution by choosing the right mesh widths. Besides gaining the onsize, there is a sieving of core material as well as of coarse and fine fractions of granules. All three screenings preferably run simultaneously on a triplane (Figure 11.17). The most distinguishing element of

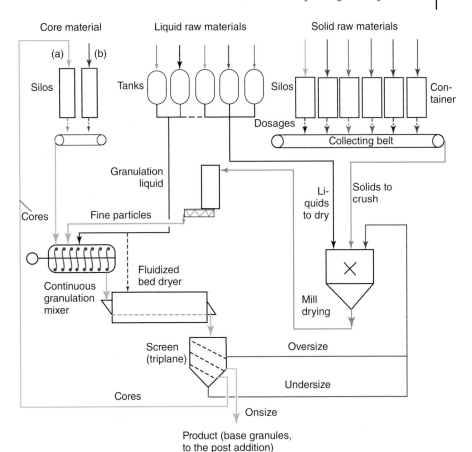

Figure 11.17 Innovative granulation with options for setting the mean particle size, particle size distribution, and bulk density. Core material: (a) granules from the steady state and (b) sieved solids only for first startup.

the innovative process is a powerful mill-drying unit, in which wet solids, solutions, and suspensions dry at the same time.

Bulk density depends on the densities of raw materials. On the other side, the distribution of liquids between mixers and the fluidized bed and mill dryer allows an adjustment of bulk densities within wide limits. The granulation mixer must be suitable for introduction of high shear forces. Preferably, grinding runs in impact mills with multi-levels and adjustable gaps. Other opportunities represent the settings of the rotational speeds, temperatures, and air velocities in mill drying. Transported with grinding air, the ground particles fly into a silo with integrated filters. After deposition, a metered addition into the granulation mixer ensues. Only for the first startup, the process required sieved raw material as core. In the stationary mode, sufficient quantities of core material accrue as undersize, which

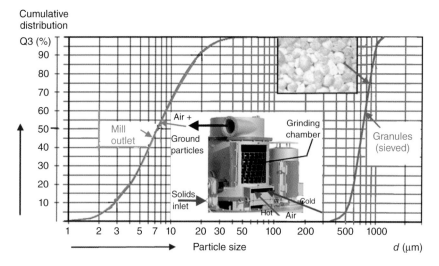

Figure 11.18 Cumulative frequency distributions of ground powder at mill outlet and of produced granules after sieving: the image in the upper part of diagram shows microscoped granules, in the lower part an air turbulence mill (Courtesy of Jäckering).

are slightly smaller than the onsize. All other solids run through the mill, apart from some solutions and suspensions on particles.

Figure 11.18 demonstrates the principle. With the used mill and set operating conditions, grinding provides reproducible particle sizes of $1-40\,\mu m$ ($d_{50} = 7\,\mu m$). In this case, the granules show after sieving sizes of $750\,\mu m$ (d_{50}) and bulk densities of about $730\,g\,l^{-1}$. During collection of core material, the mesh sizes of the sieve determine the mean particle sizes. Sizes may be set from very small to large. Width of the distribution arises from screening of fine and coarse particles. Narrow distributions require great circulations (=high costs). This special process always leads to spherical granules, wherein the shape factor depicts values above 0.82, usually as much as 0.85. A higher addition of aqueous suspensions and/or solutions into grind drying increases the bulk density, wherein grinding should be intensified. Reduction in bulk density is achievable by introducing fewer liquids into the slower running mill and partly adding liquids along the fluid bed dryer. In addition, the amount and type of granulation liquid influences the bulk density, especially directly in the granulation.

11.9
Economic Considerations

First, non-tower productions reduce the manufacturing costs in many cases. An additional way represents the optimizing of the "base powder" concept. The idea includes production of all detergent variants in a large-scale plant with only one line. In the process, product-characterizing differences are realized by operation conditions and the components in the post addition. It is possible to generate heavy-duty and color detergents in customary quality (approximately $\gamma = 550\,\mathrm{g\,l^{-1}}$) and as concentrates ($\gamma$ about $800\,\mathrm{g\,l^{-1}}$). Adjustable parameters are operating conditions and distribution of liquids. Dosed liquids flow separately or in mixture both into grind drying as well as into granulation and further into the fluidized bed, preferably in coordinated amounts. Particle sizes are calculable and adjustable by choosing the right screen mesh sizes for the core material. With same post-addition materials, the final products are in the range of $450\,\mu\mathrm{m}$ (d_{50}) up to about $750\,\mu\mathrm{m}$. Figure 11.19 shows a possible plant design of the post-addition step.

11.10
Outlook

Major steps of development in the past 50 years led to large-scale changes in the formulations of detergents. These are biodegradable surfactants and phosphate-free soluble builders as well as various enzymes, bleaches, and activators, apart from color transfer inhibitors. Introduction of concentrates provided a significant contribution to development. All activities exhibit ecological aspects. Currently, the washing process needs less chemistry, energy, and water, and wastewater is less polluted.

Also, in future developments, ecology stands at the center. In particular, by lowering the wash temperatures, power consumption should drop. The trend towards more sensitive textiles requires low mechanics and high water level in the machine for fiber and color protection, apart from low temperatures of $15-30\,^\circ\mathrm{C}$. For these reasons, the brand manufacturers focus developments on low-temperature washings with liquid and solid detergents. The key to success seems to lie in modified enzymes, with reasonable efficiency in desired temperature range. Different surfactants bring progress by optimizing the types and amounts. Further, an effective low-temperature bleach must be sought and found. During declining washing temperature below $20\,^\circ\mathrm{C}$, exchange from sodium salts to the better soluble potassium salts is possible. In Japanese detergents, these salts are components for the cold wash ($5-18\,^\circ\mathrm{C}$) for many years. Nevertheless, the wash method is not comparable to the European conditions.

The duality seems to remain, which means normal powder and concentrates are still customary forms. Perhaps a manufacturer might dare to introduce super concentrates (more enzymes, fewer builders, and $900\,\mathrm{g\,l^{-1}}$ bulk density). Esthetically pleasing spherical granules of a modern, real granulation would certainly be well received by customers. However, detergent manufacturers possess depreciated production facilities with low manufacturing costs. Therefore, a speedy conversion to modern granules is unlikely.

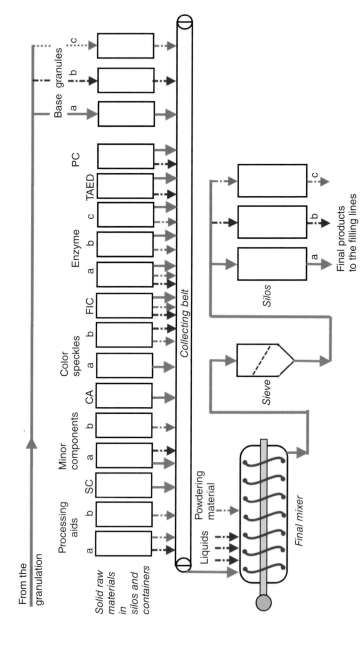

Figure 11.19 Processing plant (post addition) for three base granules, distinguishable in particle sizes, and bulk densities. Abbreviations: SC, silicates; CA, citric acid; FIC, foam inhibitor concentrate; TAED, tetraacetylethylenediamine; and PC, sodium percarbonate.

11.11
Learnings

√ Solid laundry detergents exist as powder and as granules, apart from spherical granules and extrudates.

√ Powdery and granular heavy-duty detergents of high quality provide best wash performances.

√ Bleaching substances in heavy-duty detergents contribute significantly to the reduction of germs.

√ Liquid detergents show some advantages in handling and gain market shares. Solid detergents remain indispensable because of the bleach.

√ Without preservation, liquid products show a risk of contamination with microorganisms.

√ Color detergents are free of bleach and whitening agents, but contain color transfer inhibitors.

√ Brand manufacturers question their customers in all stages of development in order to increase the market success.

√ Consumers want a fast completion, perfect washing results and no ironing; besides, all steps should be economical.

√ Washing processes include the textiles, dirt, washing machine, and water.

√ Our clothes consist of natural and synthetic fibers, often as blended fabrics.

√ Manufacturing of the base detergent powder follows atomization in spray towers without/with subsequent compaction and then in non-tower processes.

√ Afterwards, post addition follows, in which sensitive components are metered and mixed with the base powder.

√ A novel granulation enables the production of base granules in various sizes and wide ranges of bulk densities.

√ Large single-line manufacturing without a tower provides economic benefits.

References

1. Stiftung Warentest (2012) Weißes wird grau, p. 67–71.
2. Bockmühl, D.P. (2011) Wäschehygiene bei niedrigen Temperaturen – Erkenntnisse und Herausforderungen. *SOFW-J.*, **137** (10), 2–8.
3. Sajitz, M. and Grohmann, J. (2011) Hygiene effects of bleach systems in laundry detergents. *SOFW-J.*, **137** (10), 16–24.
4. Google: Sevenone Media (2010) Märkte im Visier, Waschmittel. *www.sevenonemedia.de/c/document_library/get'file?uuid=aa149b4f-d2e8-4add-8816-df54f9383334&groupId=10143*
5. Rähse, W., von Rybinski, W., Steber, J., Sung, E., and Wiebel, F. (2003) in *Laundry Detergents* (ed. E. Smulders), Wiley-VCH Verlag GmbH, Weinheim.
6. Tsumadori, M. (2003) The changing marketplace in Asia-developed markets in Japan, Korea and Taiwan, in *Proceedings of the 5th World Conference on Detergents (2002, Montreux)* (ed. A. Cahn), (2003) AOCS Press, p. 33–37.
7. Rähse, W. (2007) *Produktdesign in der chemischen Industrie*, Springer-Verlag, Berlin.
8. Stewart, D.W., Shamdasani, P.N., and Rook, D.W. (2007) *Focus Groups: Theory and Practice*, Applied Social Research Methods, vol. 20, Sage Publications, Thousand Oaks, CA.

9. Morwind, K., Koppenhöfer, J.P., and Nüßler, P. (2002) in *Markenmanagement. Grundfragen der identitätsorientierten Markenführung. "Mit Best Practice – Fallstudien"* (eds H. Meffert, C. Burmann, and M. Koers), Th. Gabler GmbH, Wiesbaden, pp. 477–506.

10. Kahn, K.B. (2011) *Product Planning Essentials*, 2nd edn, M.E. Sharpe, Inc., New York.

11. Bradley, N. (2007) *Marketing Research: Tools and Techniques*, Oxford University Press, Oxford.

12. Rähse, W. and Hoffmann, S. (2003) Product design – interaction between chemistry, technology and marketing to meet consumer needs. *Chem. Eng. Technol.*, **26** (9), 1–10.

13. Henkel (2012) Sustainability Report 2012, *http://sustainabilityreport.Henkel.com* (accessed 27 November 2013).

14. Ahlström, C. and Gesper, T. (2007) *Der Erfolg von Handelsmarken – Welche Strategien die Position der Marke stärken*, Diplomica Verlag GmbH, Hamburg.

15. (1840) *Cajus Plinius Secundus Naturgeschichte. Übersetzt und erläutert von Dr. Ph. H. Külb*, Vollständige Ausgabe in Google Books, Bd. **1-6**, Verlag der J. B. Meßler'schen Buchhandlung, Stuttgart.

16. Knapp, F. (1866) *Lehrbuch der chemischen Technologie: zum Unterricht und Selbststudium*, 3rd edn, vol. 1, Verlag Friedrich Vieweg & Sohn, Braunschweig, p. 619 ff.

17. Rähse, W. (2012) Enzyme für Waschmittel. *Chem. Ing. Tech.*, **84** (12), 2152–2163.

18. Aehle, W. (ed.) (2007) *Enzymes in Industry: Production and Applications*, Wiley-VCH Verlag GmbH, Weinheim.

19. Wagner, G. (2010) *Waschmittel, Chemie, Umwelt, Nachhaltigkeit*, 4th edn, Wiley-VCH Verlag GmbH, Weinheim.

20. Textilkennzeichnungsgesetz (2010) Textilkennzeichnungsgesetz vom 14.8.1986, zuletzt geändert am 26.8.2010, *www.gesetze-im-internet.de/bundesrecht/textilkennzg/gesamt.pdf* (accessed 27 November 2013).

21. Hauthal, H.G. (2012) Veranstaltungsbericht – Waschen bei niedrigen Temperaturen. *SOFW-J.*, **138** (1/2), 73–80.

22. Laitala, K. and Mollan Jensen, H. (2010) Cleaning effect of household laundry detergents at low temperatures. *Tenside Surfactants Deterg.*, **47** (6), 413–420.

23. Ein Service des Bundesministeriums der Justiz in Zusammenarbeit mit der juris GmbH (1980) Verordnung über Höchstmengen für Phosphate in Wasch- und Reinigungsmitteln (Phosphathöchstmengenverordnung – PHöchstMengV), vom 04.06.1980, *www.gesetze-im-internet.de/bundesrecht/ph˙chstmengv/gesamt.pdf*.

24. WRMG (2007) Gesetz über die Umweltverträglichkeit von Wasch- und Reinigungsmitteln (Wasch- und Reinigungsmittelgesetz – WRMG), 29.4.2007, letzte Änderung 2.11.2011, *www.gesetze-im-internet.de/bundesrecht/wrmg/gesamt.pdf* (accessed 27 November 2013).

25. Bunddeministerium für Gesundheit (2011) Erste Verordnung zur Änderung der Trinkwasserverordnung vom 3. Mai 2011, Anhänge, Bundesgesetzblatt Jahrgang 2011 Teil 1 Nr. 21, ausgegeben zu Bonn am 11. Mai 2011, *www.bmg.bund.de/ministerium/presse/-pressemitteilungen/2011-02/aenderung-der-trinkwasserverordnung.html* (accessed 27 November 2013).

26. Rähse, W. (2010) Produktdesign von Kunststoffen für die Waschmittelindustrie. *Chem. Ing. Tech.*, **82** (12), 2073–2088.

27. Hauthal, H.G. (1995) Gentechnisch hergestellte Waschmittel-Enzyme. *SOFW-J.*, **121** (11), 795–803.

28. Rähse, W. and Dicoi, O. (2009) Produktdesign disperser Stoffe: Industrielle Sprühtrocknung. *Chem. Ing. Tech.*, **81** (6), 699–716.

29. Rähse, W. (2003) in *Laundry Detergents* (ed. E. Smulders), Chapter 6, Wiley-VCH Verlag GmbH, Weinheim, pp. 122–145.

30. Rähse, W. and Larson, B. (2007) Henkel AG & Co KGaA, DE 10 2006 017 312 A1, Oct. 18, 2007.

12
Product Design of Liquids

Summary

The purity of liquids (technical grade, for analysis) and their composition (mixture, emulsion) as well as the presence of solids constitute a substantial part of product design. Furthermore, the performance and convenience in the application indicate the quality. Possibilities for varying the design from the same starting material are described for sulfuric acid, milk, and tomatoes, and subsequently for a hydrophobic liquid (perfume) in detail. The design of perfume oils emerges both from the composition of individual fragrances to an individual overall smell and from further processing to another form of application. Perfume oils consist of ~30–200 natural, nature identical, and synthetic fragrances. The task-oriented formulation of odor regarding the top, middle, and base notes as well as the required fixation represent an artistic performance of the perfumer. In the consumer sector, odorants are used for fragrancing of the body and for aromatherapy, and for scenting rooms and products. The product fragrance originates partly or wholly from syntheses. The control of room and product scenting with respect to time and intensity occurs through the concentration and the "physical state" of the oil, which is present in dilute solution, undiluted, encapsulated, and sprayed on, emulsified, foamed, and thickened, and in the form of a gel or solid bar. Any solid form of fragrances (beads, blocks, and gels) occurs with the help of carriers.

12.1
Introduction

The setting of the desired quality and product shape as well as converting into other phases and forms represents the key to product design. Some examples are used to describe the procedures for improving the liquids with respect to the application, namely, for hydrophilic and hydrophobic liquids and for emulsions. The main part of the discussion is centered over the perfume oils, showing the possibilities for changing the physical state of essential oils. Some examples demonstrate that many designs are already realized. Nevertheless, product design

Industrial Product Design of Solids and Liquids: A Practical Guide, First Edition. Wilfried Rähse.
© 2014 Wiley-VCH Verlag GmbH & Co. KGaA. Published 2014 by Wiley-VCH Verlag GmbH & Co. KGaA.

facilitates a possible transfer of learnings, ideas, and designs to other areas/fields of application.

The purity of liquid (technical grade, for analysis), assessed from the amount of impurities and from physicochemical measurements, constitutes a substantial part of product design. Furthermore, the quality of mixtures and their composition represent a base for particular effects in the application.

12.2
Water-Based Liquids

Several food and beverages are liquids or contain free water. Moreover, industrially produced water-based liquids, for example, different consumer goods, include dissolved or suspended solids, liquids, and gases. These originate from nature or from organic and inorganic chemistry. In the presence of water, a microbiological contamination can happen. To increase the durability, there are three possibilities, namely, short-term heating of the product, addition of preservatives, and microfiltration.

First, the heat kills harmful microorganisms during pasteurization in 15–30 s at 72–75 °C. More effective is the ultrapasteurization or ultra-high temperature (UHT) treatment, where the heat acts at >135 °C for about 4 s. Batch-wise real sterilization ensues under pressure at 121 °C for 20 min to get a sterile solution. Therefore, suitable equipment (EHEDG, European Hygienic Engineering and Design Group; see Section 13.10) and materials (Chapter 16) are needed. Alternatively, microfiltration holds back all microorganisms (see Section 10.5.3). Industrially manufactured water mixtures may contain alcohols, acids, alkalis, salts, macromolecules, and emulsions. Depending on the formulation, the water has to be preserved with a few parts per thousand of suitable chemicals. The addition of different preservatives, partly approved for food, is described in Section 13.6.2.

A pure liquid can evaporate by heating, whereby molecules diffuse into the gas phase. On the contrary, by cooling below the melting point, a liquid can solidify, and then become a solid product. The physical states of solutions, such as sodium hydroxide, depend on the concentration and additionally the temperature (see Section 2.3). The solid shape of lye at higher concentration is because of the process used.

Figure 12.1 displays the phases of well-known sulfuric acid at normal pressure. This chemical is available in different purities, up to an ultra-pure quality. Purity influences the melting point significantly, which shows values at one end with −15 °C for the highest quality and about +10 °C on the other end for technical grade (100% acid). The chemical industry sells sulfuric acid in different degrees of concentration. At 10% concentration, it serves as battery acid. Fuming sulfuric acid (oleum) is formed on dissolving gaseous sulfur trioxide in sulfuric acid. Acid mixtures represent additional possibilities to get optimal results in the application. Further diversity arises through chemical reactions, for example, for preparation of fertilizers or salts.

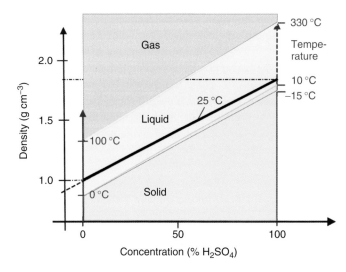

Figure 12.1 Phase diagram of sulfuric acid under normal pressure; the straight line shows the dependence of the density at 25 °C from concentration.

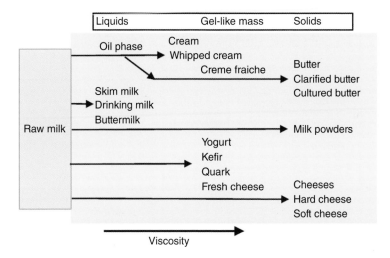

Figure 12.2 Some simplified opportunities in milk processing (characterizing the product design).

Milk, as a natural product from cow, sheep, or goat, represents a starting material for different products. It is used extensively in different processes (see Figure 12.2). The product design is discussed here along with some brief highlights. Technically, milk is a three-phase emulsion (aqueous and oil phases and solids). The oil phase (rahm) converts via crystallization into a solid (butter), into a gel-like mass (crème fraiche), or viscous liquid (cream). From raw milk, the preserved liquid milk for drinking and milk powder arise, as well as fresh cheese; from pasteurized milk not only quark but also different hard and sliced cheeses originate. Product design for milk is well established, remaining only in niches, especially in connection with mixed micro-/nanopowders.

Processing of tomatoes, an example for the water-based liquid/solid system, leads to different products. Firstly, the peeled, cored tomatoes pass through a sieve at elevated temperatures. The mass is concentrated in a multistage evaporator under vacuum. In this way, different products emerge, from the sauce to the paste. The tomato powders arise on heated rollers or in a spray-drying tower. The methods generate a 96% powder that is also suitable for coloring of noodles. Figure 12.3 explains some more details.

The juice originates from higher concentrates through dilution. The cooking of the mass with spices, sugar, and vinegar results in ketchup. This process allows the addition of more powdered spices or pieces of other spices and different vegetables. Therefore, the manufacturers generate a plurality of various flavors. An interesting point is the adjustment of viscosity, preventing segregation of the liquid from the solids. The pastes are usually thixotropic, so that they liquefy by shaking. Therefore, there is a need for thickeners to reduce this effect. The usual agents are starch- and cellulose derivatives and xanthan gum, as well as silica nanoparticles.

Design of suitable packaging materials represents a big issue for tomato paste, especially for ketchup. Rigid glass or squeezable plastic bottles are opposing alternatives for ketchup. Owing to the thixotropic behavior, the discharge openings

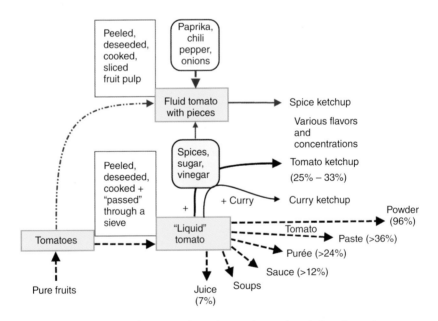

Figure 12.3 Processing of tomatoes for explaining the product design of two-phase systems (liquid/solid).

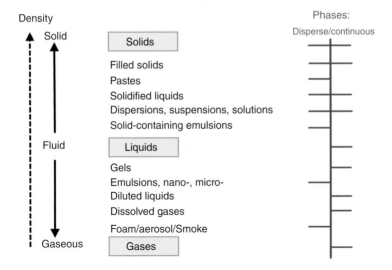

Figure 12.4 Design-variations of liquids by addition of solid, liquid, or gaseous substances.

must be adjusted by trials with the product. Food cans, tubes, or jars with screw caps are in use for tomato pastes, and tetra packs for the juice and sauces.

Both for hydrophilic and lipophilic liquids, further possibilities exist to develop optimal products. Many new product variants result by the addition of appropriate solid, liquid, and/or gaseous substances, which are soluble or insoluble (Figure 12.4), depending on the application. The additives differ in their composition, in phases (dispersed or continuous), in particle- or droplet sizes, and in many characteristic measured values (density, refractive index, and fixed points). Often, additional processing is necessary for producing the optimal consistency for use. Thus, it is possible to stir liquids in melting and pour the mixture into blocks or other shapes.

Granules may arise by dropletization of melt, and powders by spraying from a molten mixture under cooling conditions. Other possibilities represent both the application of a melt as coating layer, and the homogeneous incorporation of liquids into solids. An extruder, kneader, or mixer allows the mixing of liquids and melting into powders and excipients, forming downstream shapes such as granules, strands, or pieces (see Section 2.2).

Immiscible liquids and liquid mixtures are of interest because of the hydrophilic and hydrophobic molecular structure (characterized by the hydrophobic/lipophilic balance HLB), described in detail (Chapter 13). Emulsions arise in several ways through the addition of suitable emulsifiers. The two phases partially or completely dissolve into one another. After mixing, one phase emulsifies as droplets in the continuous phase. In addition, targeted segregations are possible. By a short shaking immediately before use, biphasic products form a short-term emulsion (Figure 12.5). Further possibilities for structural changes originate from unsolved/dissolved solids and consistency regulators.

Figure 12.5 Hydrophilic/hydrophobic liquids in colors, forming an emulsion after short shaking.

12.3
Water-Insoluble Liquids (Example: Perfume Oils)

The product design of water-insoluble, organic liquids is described using the example of perfume oils. Because of its complex composition, perfume oil allows particularly diverse designs. The targeted making requires a more detailed description of the interesting perfume oils. This section focuses on the origin, chemical composition, and isolation of fragrances. Thereafter, a description of solidification of organic liquids occurs.

12.3.1
History of Perfume Oils

The word perfume derives from the Latin word "per fumum," meaning, by smoke, and incense from "incendere," which means, to burn. It was believed that the smoke of incense burnt as part of religious ceremonies transported the prayers to heaven. In history [1–4], the earliest evidence for the making of perfumes dated approximately from 2000 up to 5000 BC. The priests in Mesopotamia used perfumes, scented ointments, and resins for honoring the gods and anointing the dead. The development and manufacturing of perfumes began in ancient

Mesopotamia and Egypt. Women known as the Assyrians manufactured perfumes by crushing of various plants. In hot salted water, the fragments stood overnight. Subsequently, the perfume arose on filtering the liquids and mixing them with hot oil. In India, people used perfumed ointments and oils as well as scented plant parts especially for medical purposes and for cleansing the body as well as for rituals.

In the second millennium BC in Mesopotamia, a female chemist (Tapputi) distilled various flowers with oils and other aromatics for the extraction of fragrances. Around this time, the upper class of Egyptians had already discovered the perfume for personal care and therapy. The world's oldest perfume factory, discovered in Pyrgos/Cyprus, dated from this time. They processed flowers, herbs, and spices. Still in use today in the perfume industry, the word "Chypre" indicates the importance of Cyprus in the scent culture.

In the ninth century, an Arab chemist (Al-Kindi) described in the book *Chemistry of Perfume and Distillations* more than 100 recipes and manufacturing equipment, such as the alembic. The Persian Ibn Sina (Avicenna) invented steam distillation about 1000 AD, a great step in chemistry. He processed petals of roses. Before that, the perfume oils originated from distillation of crushed mixtures of petals and herbs in oil as well as by mixing and squeezing.

The western culture encountered fragrances through the Crusades, and required raw materials and mixtures of the Orient. Until then, only simple lavender water from Charlemagne was known (end of the eighth century). Monks' recipes (1221) demonstrated the knowledge of perfumery in Florence, Italy. In the east, Hungarians (1370) made a perfume from scented oils, blended in alcohol, best known as *Hungary Water*. In addition, Venice became an important trading center because large amounts of new herbs, spices, and other goods came to Europe through this route. The development of skills and the technical equipment for distillation allowed the manufacturing of essential oils for the market in the fifteenth century. The emergence of perfumeries may date with the arrival of Catherine de Medici (1519–1589) to the court of Henry II. In 1580, the alchemist and apothecary Francesco Tombarelli came to Grasse (France). There, he established a laboratory for the manufacture of perfumes. Thereby, Grasse became the incubator of the European perfume industry.

In Germany, the Italian barber Giovanni Paolo Feminis created perfume water, known as *Eau de Cologne*. His nephew Johann (Giovanni) Maria Farina took over the business in 1732. By the eighteenth century, aromatic flowers and herbs were planted in the Grasse region of France, in Sicily, and in Calabria, Italy, to provide the growing perfume industry with raw materials. Even today, Italy and France are the center of European trade for scented plants.

12.3.2
Perfumes

Perfume oils provide a pleasant mixture of many fragrances, which often contain essential oils. Fragrances enhance the well-being and show a positive impact on the psyche. Depending on the composition, they exert a calming or stimulating effect on the users and the environment. Therefore, people use a perfume of their choice. Myriad scents are available. Many cosmetic products contain fragrances, or the perfume applied directly on the body delivers a sympathetic, personal fragrance. In addition, perfumes emphasize the characteristics of many consumer goods.

For the creation of a perfume for personal care, the perfumer selects about 30–200 fragrances from over 2500 natural, nature identical, and synthetic fragrances [5, 6], whereby a fragrance can consist of several components. This art is a combination of individual oils in kind and amount, creating a pleasant composition with respect to the impression, intensity, and duration of the fragrance on the skin or on a scented product ("product design"). Perfume oils are of three different fragrance types, namely, the head, heart, and base notes [7]. They differ in their volatility and aroma. The top note (head, Tête) is characterized by a light, less adherent fragrance, and only perceptible in the first 10 min. This note gives the initial impression and is thus important for selling the perfume. Thereafter, the middle note (heart, Bouquet, Coeur) appears perceptibly, which determines the character of the perfume. The base note (fixator) comprises long adherent, low volatile odorants. The scent is still noticeable after 1 h and reduces the volatility of the heart and top note. Dominated by the middle note, the following terms characterize the typical fragrance of a perfume: aldehydic, animal, aquatic, aromatic, balsamic, floral, earthy, fresh, fruity, green, woody, herbaceous, oriental, powdery, sweet, spicy, or citrus [8].

The nose perceives not only fragrances but also the flavors (taste and odor) used in food and medicines. Incorporated as liquids in food, the flavors should not affect the taste. For increasing the stability and controlling the release of flavors, the addition may occur in dry state [9], for example, with carriers or in capsules.

The world market of "Flavors & Fragrances" amounts to about €12.2 billion [10]. Major manufacturers are Givaudan (Switzerland), International Flavors & Fragrances (USA), Firmenich (Switzerland), and Symrise (Holzminden, Germany). The perfumers of these manufacturers develop in their laboratories all new feel-good fragrances for the well-known perfume brands. There are some perfume houses that produce perfumes solely for the scenting of consumer products. One of them is the Henkel Fragrance Center (Krefeld, Germany). An extensive stock of the individual fragrances with sophisticated logistics characterizes the production equipment for the perfume oils. The manufacture of each batch takes place by precisely working on the dosages [5, 6], in which the individual components are weighed and mixed in presence of an inert gas.

In consumer market, the use of fragrances splits up into four cases. Perfuming of body ensues with different concentrated, alcoholic solutions of perfume oils with a spray device or through direct application (Figure 12.6). Further possibilities exist in the use of fragrance-containing body-cleaning agents and care as well as hygiene products such as soaps, shower lotions, shampoos, bath additives, deodorants, shaving water, hair gel, skin care, and sun protection products.

Aromatherapy with essential oils can remedy sensitivity disorders (mood swings) and alleviate diseases. For this purpose, an enrichment of space/breathing air with

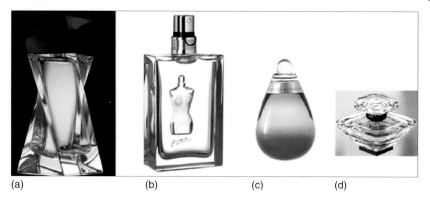

(a) (b) (c) (d)

Figure 12.6 Supply forms of perfume oils in alcoholic solutions: (a) Eau de Cologne (EdC) pump sprayer 2–5%, (b) Eau de Toilette (EdT) spray pump, 5–10%, (c) Eau de Parfum (EdP) 10–15%, and (d) Extrait de Parfum 15–30% perfume oil. (Courtesy of Douglas.)

evaporating oils or the inhalation of fragrance over a hot water bath or oral intake or rubbing/massaging into the skin is necessary [11]. Firstly, an aromatherapy massage, performed by doctors or health practitioners, achieves positive effects on the nervous system. Secondly, there exist fragrance recipes for almost all physical problems [12]. Numerous essential natural oils, alone or in admixture with other, are suitable for the alleviation of disease conditions, particularly of colds. Essential oils can help in cases of psychosomatic disorders, symptoms in the digestive tract, inflammation, pain relief in the mouth, and throat. In a much diluted form, they stimulate the function of the biliary tract successfully. Some essential oils have antiseptic and antibiotic properties, and help in skin injuries. It is also known that some oils accelerate the cure after burns.

Vaporizing perfumes, distributed with adjustable fan airflow, scent the *space* of warehouses, shops, showrooms, and sales and exhibition stands. In private rooms, controlled release allows a constant scenting. Release control is achieved by differences in the consistency of the perfume, which means the oils evaporate from the liquid, a thickened liquid, or out of solidified bars. Special spray devices enable short-term intensive air freshening. For a long-lasting slight scenting, fragrance-containing blocks in open plastic containers exist.

The perfuming of products, used at home and in trade, such as soaps, detergents, and cleaning agents, occurs predominantly with synthetic fragrance mixtures. It can be realized by incorporation of the liquids or with scented solids or by spraying. The products are used preferably in the kitchen, bathroom, and toilet areas as well as to wash and care for the laundry. Added perfumes cover up specific odors of the product and increase the attractiveness by sympathetic scenting. Fragrances improve the recognition rate. A chosen fragrance mixture takes into account the application. It should harmonize with the product (product design). Type, quantity, and form of scenting depend on the physical state and interactions with other constituents of the product. Molecular interactions [13] may affect the solubility, vapor pressure, and fragrance.

12.3.3
Extraction of Fragrances

Natural fragrances are of vegetable or animal origin (Figure 12.7). The animal scents are banned and have only historical significance. They come from glands of beaver and musk-animal or as salve-like secretions of civet. Further processing of agents usually ensues through an extraction with alcohol. The odorant Ambergris, sometimes pathologically produced in the sperm of whales, can be largely created by chemical syntheses. For imitation of this scent mixture, the chemists synthesize Ambroxan natural identical; other necessary fragrances of the mixture arise synthetically without a natural model.

Today, scenting occurs with synthetic and vegetable substances. Most natural perfumes include essential (volatile) oils, insoluble in water, and gently obtained by steam distillation (<100 °C) from the plant parts [14]. Essential oils distilled with this method are described in Table 12.1. Steam distillation generates a number of well-known fragrances from the plants: lavender from the flowering panicles with stems, clove from the leaves and buds, bay leaf and patchouli from the leaves, juniper from the berries, pine needles from the needles and twigs, stone pine from the branches and twigs, rose- and sandalwood from wood chips, tarragon and

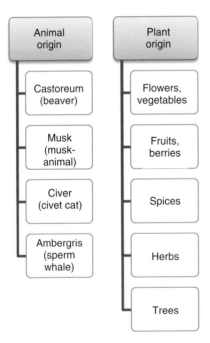

Figure 12.7 Origin of natural scents.

Table 12.1 Distillation of essential oils (fragrances) from different parts of plant.

Part of the plant	Fragrance
Blossoms	Rose, ylang ylang
Panicle with stem	Lavandin
Leaves, buds	Carnation
Leaves	Laurel, patchouli, peppermint, rhododendron
Berries	Juniper
Fruits, grains	Pepper
Needles, twigs, branches	Pine, stone pine
Branches, (pine) cones	Cypress
Wood, leaves	Camphor
Wood chips	Rose, sandalwood, cedar, amyris
Herb	Chamomile, rosemary, tarragon, basil, sage, yarrow
Grass	Lemongrass, palmarosa
Seed	Coriander, carrots, celery, anise, cardamom, cumin, fennel
Root	Ginger, iris, angelika
Resin	Myrrh, frankincense, galbanum
Bark	Cinnamon

rosemary from the herb, lemongrass, and palmarosa out of grass, coriander and carrots from the seeds, ginger and iris from the roots, and myrrh from the resin.

The technical implementation of fragrance extraction takes place in more or less simple systems (Figure 12.8). Firstly, the vegetable components are stacked dry in a container on a rack. After exposure to *water vapor*, the volatile oils evaporate out of

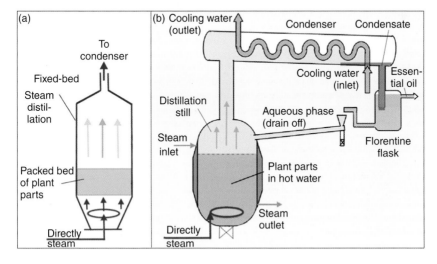

Figure 12.8 Extraction of essential oils from plant parts with steam distillation: (a) alembic with a packed bed (g/s) and (b) batch distillation with steam/evaporating water (g/l/s).

plants and condense in a heat exchanger, wherein the essential oil forms the upper phase of the condensate. An alembic (alambic or alambique) represents a small, simple distillation equipment, mostly made of copper. Large systems of steel often use a distillation still, containing suspended ground plant parts in hot water, and equipped with a recirculation system for the water phase.

By direct injection of *superheated steam* into the distillation still and/or by heating of the suspension via a heating/cooling jacket, the evaporation of the mixture takes place, consisting of water and essential oils. After condensation in a heat exchanger, the liquid forms two phases at the end of the cooler, which flow into a separation vessel. The essential oil floats on the water phase, and thence drains off. The recycled water phase contains some essential, soluble substances from the plants (hydrolate) that accumulate after repeated cycles. For instance, these agents help in aromatherapy [12].

The fragrances of some plants (stinging nettle) cannot evaporate with steam alone. In these cases, a second mixed plant supports the distillation of essential oils (*co-distillation*). The pure fragrances, gained through steam distillation, show an evaporation temperature of 130 °C (eucalyptus) to about 190 °C (sage, thyme), and in exceptional cases even more.

The hemp plant contains three different oils: first, the vegetable oils (triglycerides) from the seeds, set free by cold pressing in oil mills; secondly, the essential oils (terpenoides), isolated by steam distillation of flowers and leaves; and thirdly, the THC-containing "hash oil" (tetrahydrocannabinol) via a solvent extraction of the resin. As a fragrance for aromatherapy, the essential oil has certain significance. The seed oil is valuable, used for skin care and cure, and utilized as food oil similar to olive oil.

The rose scent arises from rose petals with steam distillation and by solvent extraction, preferably with hexane (Table 12.2), as well as from jasmine, juniper (absolute), and mimosa. After the extraction of flowers, the solvent is distilled off. A creamy mass remains, the *Concrête*. In the resolved alcohol, the waxes can be removed by reducing the solubility via low temperature treatment (−24 °C). After evaporation of alcohol, the *Essence absolue* (highest purity) emerges.

Manufacturers use hexane (boiling point, bp = 69 °C), petroleum (depending on composition: bp = 35–65 °C), or butane (bp = −0.5 °C) as water-insoluble solvents. In some cases, the extraction requires a water-soluble extraction agent. For this,

Table 12.2 Solvent extractions of plant parts.

Method	Essential oils
Extraction with ethanol (80–100%)	Vanilla (pods), honey (honeycomb), cocoa (beans), oak moss (herb), siam benzoin (resin), tolu (resin), tonka (resin)
Extraction of flowers with hexane	Champaca, Frangipani, broom, jasmine, mimosa, osmanthus, rose absolue

ethanol (bp = 78 °C) is suitable, as in the extraction of fragrances from fermented, roasted, and crushed cocoa beans or from vanilla pods. The supercritical fluid extraction of crushed flowers or plant parts with supercritical carbon dioxide for the production of essential oils represents a relatively new process used in the industry, especially for flavors and coffee.

An old extraction method, applied up to the 1980s and for extremely expensive oils (jasmine) until today, mostly in Grasse/France, works with animal fats (pork and/or beef) as extractant. The fats are spread on glass plates, and thereon the petals lay. Exchange of the placed petals happens daily. Over time, the fats solve the essential oils (*Enfleurage*). In some cases, the process accelerates by the use of molten fat at 60 °C (*Mazeration*; Enfleurage à chaud). In a "cold" extraction process, fats are saturated after 10–14 days. The release of essential oils from the pomade ensues with ethanol. The pure fragrance (*Essence absolue d'enfleurage*) remains after distillation of the alcohol [7].

Cocoa butter originates from the seeds by cold pressing. For isolating the fragrances, cold pressings of shredded fruit peels from lime, lemon, bitter orange, mandarin, grapefruit, and bergamot are necessary. After mixing with water, the squeezed liquid forms two phases. The essential oil separates in a centrifuge and flows into a tank. Galbanum is a native, fennel-related plant in Asia Minor. After cutting, a gummy liquid containing the fragrance runs from the root and the lower stem. Figure 12.9 depicts the different processes for production of fragrances from the plants.

Because the needed quantities of natural fragrances are not available in the world, the large quantities of fragrances required emerge in syntheses (natural identical or full synthetic). The raw material and synthetic routes depend on the desired scent. About 80% of the fragrances come from syntheses [5]. For example, the industry produces $12\,000\,t\,a^{-1}$ of linalool (2000) and $40\,000\,t\,a^{-1}$ of citral (BASF 2004). Citral

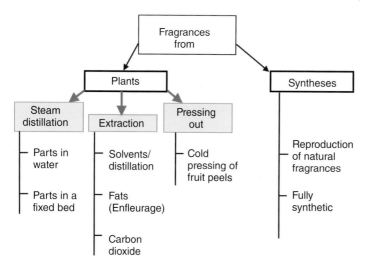

Figure 12.9 Processes for fragrance production.

is a mixture of stereoisomers geranial (citral A) and nerval (citral B) and exhibits a fresh, intense lemon fragrance. These acyclic monoterpene aldehydes provide raw materials both for synthesis of other substances and for scenting of household cleaners. Many other synthetic fragrances are not based on the terpene chemistry. These substances also serve as raw materials in a subsequent fragrance synthesis (benzaldehyde, phenylethanol, phenylacetic, and cinnamic aldehyde).

The following list gives some examples of numerous, synthetically accessible fragrances with relevance to perfumery. Aldehydes: vanillin, α-pentyl cinnamon aldehyde, cinnamon aldehyde; ketone: coumarin, ionone; alcohols: menthol, linalool. For technical or economic reasons, some scents are available only via chemical synthesis, such as green apple, peach, apple blossom and violets, lilac, lily of the valley, freesia, almond blossom, and lily [11].

12.3.4
Chemical Composition of Natural Fragrances

The chemical structure of molecules determines the scent impression. Therefore, several structure–odor relationships exists [15]. In many cases, it is possible to synthesize the targeted molecule by incorporating functional groups. In this way, the desired fragrance and required adhesion arise. The targeted modification of molecular structure is known as *molecule- or product engineering*. This modification constitutes an essential part of product design. For this reason and because odor changes may occur due to chemical reactions, the perfumer has to anticipate possible reactions in the formulation.

The results of chemical analysis depend on the plant, origin, vegetation of soil, and weather during plant growth. Natural fragrances consist mainly of a mixture of different *terpenoides*. The chemists interpret the monoterpene $C_{10}H_{16}$ as an isoprene dimer (two C_5-units), and the sesquiterpenes $C_{15}H_{24}$ as a trimeric isoprene. Terpenoides are divided into three main groups: aliphatic or acyclic (I), mono-(II), and bicyclic (III) terpenoides. Terpenes are pure hydrocarbons, or as terpenoides bear a functional group (as aldehyde, ketone, and alcohol) and may be etherified or esterified. For example, patchouli oils consist mainly of sesquiterpenols, sesqui-, and monoterpenes as well as ketones and oxides. In the following, some well-known fragrances are presented that occur in nature, generated from plants (Table 12.3).

Monoterpene hydrocarbons include myrcene (I), limonene (II), and α-pinene (III) $C_{10}H_{16}$. Limonene is the most frequently occurring monoterpene in plants. Typical terpene alcohol scents are geraniol, nerol, citronellol $C_{10}H_{18}O$ (I), menthol $C_{10}H_{20}O$ (II), and borneol (III). Frequently used terpene aldehydes and ketones are citral $C_{10}C_{16}O$ (I), menthone $C_{10}H_{18}O$ (II), and camphor $C_{10}H_{16}O$ (III). Farnesol $C_{15}H_{26}O$ (I), a sesquiterpene, is appreciated because of this muguet-like odor. Farnesol is a natural substance, found in several essential oils. On the other hand, chemists synthesize farnesol from linalool in a mixture of isomers, such as many other fragrances. The ether anethole (anise flavor) and the 1.8-cineol

Table 12.3 Main components of natural fragrances (examples).

Compound / Product	Monoterpene C_{10}	Sesquiterpene C_{15}	Terpenole (Alcohol) $C_{10,15,20}$	Terpenaldehyde/-ketone	Ester
Lavandin	+		++	+	++
Cistus	+++		+		+
Spruce needles	++		+ (Di-)		++
Sandalwood		++	+++ (Sesqui-)		
Rosemary	+++	+	+	++	+
Limette	+++		+	+	+
Patchouli	+	+	++ (Sesqui-)	+	

(II), distilled from eucalyptus, represent substances with a monoterpene structure. Other examples are geranyl acetate, linalyl acetate (I), and the phenol thymol (II).

Several fragrances that exist are only artificial. Examples are lyral, lauric ethylester, benzyl acetone, and propiophenone. In addition to natural perfumes, some odorants with volatile compounds such as silver fir, lavender, and musk are formed by copying the formula of the natural agents in chemical synthesis. Fragrances may include heterocyclic compounds (such as diallyl sulfide from garlic). Chemical structures, fragrances, and physicochemical characteristics of customary scents are described [16].

12.3.5
Possibilities in Product Design of Perfume Oils

Perfume oils represent interesting examples of product design. The variety of natural and synthetic fragrances in graduated levels of purity and from different places of origin offers the possibility of creating numerous formulations (Figure 12.10). Depending on the task, an adjustment of the scent experience takes place by the composed mixture. This is the most important element of product aesthetics for perfume oil. The composition, head, heart, and base notes, in conjunction with a fixation (such as cinnamyl acetate) control of the fragrance release times, offers the opportunity to prolong these.

For scenting of products, these properties can be important; in other cases, a uniform fragrance release over a longer period with no change in the fragrance impression stands in the focus. The development of "appropriate" product fragrances is based not only on coverage of intrinsic odors but also on delivery of interesting scents, corresponding to the product. The use of synthetic fragrances dominates the area of consumer products. Here, the scent mixtures for the human body are composed to be simpler than the natural perfume oils and consist of about 25–50 individual scents, and in some cases a few fragrances suffice.

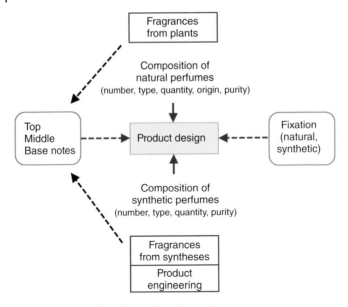

Figure 12.10 Product design of perfume oils.

For cleansers and similar products, an advertising core message guarantees hygiene and cleanness, such as "with citrus scent." Hence, in these cases the presence of one odor note suffices. Fragrances are processed predominantly as liquids. Scenting of products sometimes requires organic solubilizers (dipropylene glycol, diethylene glycol, diethyl phthalate, and isopropyl myristate). For other compositions the addition of water, mixed with emulsifiers, helps.

By chemical reactions, the fragrance impression may change over time. Thereby, perfume molecules react with other perfume or product components. Examples of reactions between fragrances include acetal formations, aldol condensations, ester cleavages, transesterifications, and (partial) oxidations of alcohols, aldehydes, and ketones to acids [12]. The reactions start at higher temperatures and in the presence of oxygen, and/or by initiation with UV radiation, as well as with moisture. Adaptation of storage temperatures, stable fragrance mixture, and inertizing with nitrogen as well as the use of UV-light impermeable packaging reduce/prevent chemical reactions. In the detergent industry, an identification of chemical changes ensues by sniffing after 3 months of storage with open and packed products at 30 °C/80% relative humidity (RH).

Product design is very different for product-dependent and application-related fragrances. Firstly, the fragrance oils can be mixed in liquids or poured into various shaped bodies. Another method is the spraying on solids. Thirdly, one or the other meltable substances allows a solidification of the oils. Figure 12.11 displays some very different solutions for the task "scenting of products." The images show the results of several laboratory experiments, which illustrate the variety. As can be seen, the processing enables usage of the perfume as pure oil, foam, or emulsion

Figure 12.11 Design of perfume oil in the laboratory: (a) oil in a water soluble ball; (b) emulsified oil in water (oil content $\sigma \sim 50\%$); (c) foamed dilute solution: $\sigma \sim 10\%$; (d) fragrance granules: $\sigma \sim 20\%$; (e) fragrance tablets compacted from granules; and (f) solidified: $\sigma \sim 50\%$.

and also as a solidified part. Furthermore, solid excipients (powder, melts) may serve as carrier for perfume oil, resulting in scented pellets, granules, or tablets. The choice of processing depends on the fragrance experience, which changes with the addition of excipients in duration and intensity.

The first step for the preparation of solid scent variants begins with a distribution of the fluid. For powders and granules with perfume oil content of up to about 20%, the spray impinges on suitable excipients in a mixer. Spraying takes place with a twin-fluid nozzle. If the customer desires a spherical product, it is possible, on the one side, to use a downstream extruder with granulator and spheronizer. Alternatively, balls with up to 70% of oil emerge by incorporation into a molten material, shaping via dropletization under cooling. The melt arises in the range of 55 °C to about 80 °C from mixtures of various fatty compounds. The perfume oil is continuously stirred into the melt in an enclosed container. The melt forms droplets in special nozzles and solidifies as spheres. The scent release from these melt granules depends on the molecular interactions between perfumes and components of the melt, and is therefore controlled by the materials as well as their proportions. An optimization of interactions can take place by measurement of vapor pressures (Section 12.3.9).

As an example of perfume oils, Figure 12.12 presents various possibilities for the *design of liquids*. First, a selection of a scent takes place for the composition. Subsequently, the discharge of the liquid from the container ensues via an opening at the bottom, or by a pressure or pumping device. The fluid escapes as spray jet or as droplets as well as foam. Secondly, a filling of oil is possible in water-soluble

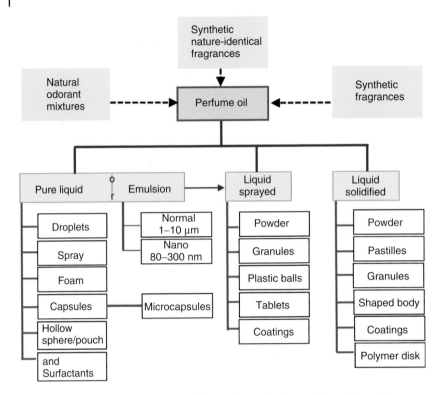

Figure 12.12 Twofold product design of the perfume oils. Top: perfume oils, made of different odorant mixtures/fragrances; lower range: liquid and solid odorants for the application.

capsules (gelatin, starch), in microcapsules [17], pouches (poly(vinyl alcohol) film), or in hollow spheres. Added lipophilic substances adjust the viscosity of the oils. Thereby, the carbon chain length and functional group, as well as the amount, allow a fine adjustment of consistency. The fragrance of the particles, core, and top layer, occurs by mixing, spraying, or coating. In some applications, water-insoluble plastic containers are in use, provided with openings for receiving the product and for fragrance release. The products are pure, gel-like and encapsulated liquids, or solidified fragrances.

12.3.6
Emulsions

For technical applications, the use of emulsions instead of lipophilic essential oil is of advantage [18]. The emulsions contain water and are not inflammable. Owing to the water content, not all shell materials are suitable for the shaped bodies. Otherwise, working with aqueous emulsions is simpler. The emulsified perfume oils need no explosion regulations regarding the storage and processing in production (zoning). After spraying the emulsion, it is not necessary to dry the product because of the small amounts of additional water (<0.6%). A positive influence on the solubility, caused by the surfactants in the emulsions, results.

Usually, the applied oil is not enough for forming a complete film on the particle surface. Therefore, negative influences on the rate of dissolution may not occur. For realization of a delay in the dissolution, a small percentage of silicone oils on the surface suffices. Further, it is possible to influence the surface properties of particles by dissolved or suspended solids, incorporated into the emulsion.

For the manufacturing of perfume emulsions with an oil content up to about 50%, two ways exist, which lead on the one hand to the normal (macro) emulsion with droplet sizes in the micron range and on the other hand to nanoemulsions with 60 to 700 nm droplets. The two types of emulsions require the use of special surfactants and technologies for the desired droplet sizes. The selection of emulsifiers for terpenoides ensues through numerous laboratory tests, using appropriate systems of several surfactants with different HLB-values. The types and quantities [19] of surfactants must be adapted to the fragrance composition.

The preparation of the two phases occurs separately, to work at different temperatures with other emulsifiers and viscosity modifiers. The oil phase, heated for a short time under effective stirring runs into the warm aqueous phase. Mixing with an Ultraturrax or with another high shear mixer generates the normal emulsion (1–10 μm). For nanoemulsions, not only more surfactants with higher contents are necessary but also an additional, energy-intensive emulsification. This happens preferably in high-pressure homogenizers (see Chapter 13) or ultrasonic reactors with multiple passes [19].

A coating with perfume oils on carriers takes place often by means of spraying. Achievable droplet sizes depend on the liquids, and in particular on the type, number, and size of the nozzles and on the operating conditions. The use of nanoemulsions allows coating the particle with "nano" oil droplets. In addition, very fine droplets arise after dissolving the fragrances in liquid carbon dioxide [20], followed by atomization at low temperatures under minimal odor formation.

12.3.7
Scented Solids

Production of fragrant solids ensues preferably via spray and solidification processes. The spray or foam method allows applying up to 25% by weight of perfume oils on fine powders. Figure 12.13 displays four possible ways for the making of perfume granules.

Solid scent particles originate usually from the following processes:

1. Fragrant powders arise by spray drying of a scented liquid or emulsion, if necessary with a coating. Spray drying leads to product losses, which may change the fragrance impression.
2. Fragrant particles arise in a mixer, whereby downstream the scented powders or agglomerates are mixed with the main stream of product. Owing to the absence of a drying stage, dry compaction processes are preferred for the

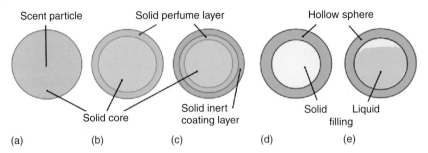

Figure 12.13 Scenting of granules: (a) fragrance evenly distributed throughout the particle; (b) fragrance in the outer coating layer; (c) fragrance in the inner coating layer; and (d,e) non-scented/fragrant hollow ball with a solid or liquid scented/not scented filling.

agglomeration. Larger (multi-layered) agglomerates emerge from powders and granules in pellet, tablet, and roller presses.

3. Coating of particles with a fragrant film, applied as spray of a solution or melt, preferably occurs in a spheronizer or mixer.
4. Coating of particles takes place with a fragrant liquid, additionally covered with an inert solid material layer for protection [21], to bind the fluid and to reduce the stickiness, and to delay the rate of evaporation.

Most of the agglomerates are fragrant by admixture of scent beads. For beads, the chemical composition and thickness of fragrant layer as well as an inert outer layer allows a controlling of fragrance release within limits. The diffusion coefficient of odorant molecules characterizes the release time. In most manufacturing processes, a partial back mixing of over- and undersized beads occurs in the agglomeration step because of the recirculation. To prevent the back mixing and limit the odor spread, an admixing of scent beads at the end of the process is recommended, just before filling. The beads may be powders, granules, or pure oil. The perfume oil may also be enclosed in water-soluble or water-insoluble shaped bodies, preferably hollow spheres or capsules.

Another possibility for the production of fragrant beads or solid layers with oil contents of 5% to more than 50% arises by a solidification of liquid perfumes [22]. Therefore, perfume oils are incorporated either in plastic or in softened and melted materials (preferably mixtures). To lower the melting point in the range of 35–120 °C and to facilitate mixing, the addition of a small percentage of solvents may be useful. Two paths are possible. The first way starts with plasticizable masses, and the second by solidification of melt (including suspensions). These methods permit the application of the extrusion and pelletizing, and for a melt the dropletization, pastillation, prilling, or casting of different shapes. Thus, the production of scented pastilles, hollow spheres, granules, gels, blocks, and candles with oil content up to 70 wt% takes place (Figure 12.14). Suitable base materials are fatty alcohols, fatty acids, surfactants, sugar, urea, paraffin, and mixed melt, investigated and optimized by experiments. In addition to the production of molded articles, a perfume-containing melt can provide excellent coating layers.

(a)   (b)   (c)

Figure 12.14 Perfume oil in a variety of forms: (a) fragrance beads, liquid in (gelatin) capsules; (b) solidified perfume oil as micropastilles; and (c) perfume oil in colored agglomerates.

The solidification of water-based emulsions, containing perfumes, occurs via the addition of thickeners (cellulose ethers and others) in combination with suitable solids. Softenable compositions, consisting of anionic surfactants in a mixture with salts, form toilet blocks by extrusion. The duration of the effect depends on the frequency of flushings in the toilet. The solubility of the stones is adjusted, so that surfactants for the cleaning and oils for the scent last for about 6 weeks.

The spatial separation of scent from reactive substances succeeds using multi-phase moldings (Figure 12.15) by installation of a film or separate tablet [23]. This procedure limits the interaction of the fragrance with the ingredients. Alternatively, separation of individual fragrance components (top, middle, and base notes) ensues by incorporation into different tablet layers. The aldehydic terpenes of the head and heart notes, as more reactive substances, come in an inert matrix. A reduction of evaporation rate for the heart and top note succeeds by incorporating the scent further inside the shaped body, separately placed. Produced by this method,

(a)   (b)

Figure 12.15 Multiphase tablets with a separation of fragrances: (a) cleaning tablet with three phases for automatic dishwashing, only one contains the odorants and (b) two-chamber shaped body with granules and a separate liquid phase; fragrance incorporated in the molding material or in the solid and/or the liquid phase.

the moldings release the characteristic fragrance for longer time. Furthermore, it prevents changing of the scent by chemical reactions.

12.3.8
Neuromarketing

Customers may recognize the products and possibly the brand manufacturers through a pleasant smell. People usually perceive a "delicate" scent unconsciously, which evokes emotions as well as creates a positive impact on the brand [24]. Measurements of brain activities in volunteers using computed tomography (CT) scanner prove that the scent acts on brain as mentioned in the statements of advertisements. In the meantime, a number of studies with some amazing results exist in this field [25]. This branch of science is called *neuromarketing*.

Space and product scenting occurs as a global feature of a distinctive brand, and to stimulate the sale [24, 26, 27]. Already, some banks, hotels, shops, and exhibition stands use the options for scenting an area. Several leading companies worldwide consider the fragrancing of technical devices for washing machines, notebooks, mobile phones, cars, and televisions. Tests for scenting run extensively on prototypes. In a few cases, fragrant articles are already in market. The design requires either conventional means such as "scent pearls" or an innovative method by the incorporation of scent into plastic parts. This will expand the product design of perfume oils considerably in the coming years.

12.3.9
Perfume Oil for Space Fragrancing

Table 12.4 shows some possibilities for scenting of spaces (halls, houses, shops, rooms, toilets, aircrafts, cars, cabinets, and vacuum cleaners). This reflects the diversity of product design already reached. In scenting of spaces and products the fragrant oils are encapsulated as a dilute solution, undiluted, sprayed, sprayed and shaped, emulsified, foamed, thickened, gel-like [28], or solidified.

For solidifying of fragrances, different plasticizable masses and melting as well as mixtures thereof are suitable. Thereby, the scent impression of a fragrance block remains the same during the application period, if the fragrances used have comparable vapor pressures. Another possibility is the application of only one artificial fragrance. After stirring in molten fatty alcohols, numerous experiments with perfume oil show the parameters for influencing the vapor pressure. The parameters examined are the ratio of liquid to solid, length of carbon chain, variation of functional groups, and use of a mixture of solids with organic/inorganic components. With knowledge of these relationships, the evaporation rates are controllable over wide ranges.

As an experiment shows, the incorporation of synthetic perfume oil in molten hexadecanol in the ratio 2 : 1 leads to a lowering of vapor pressure by a factor of 2.6 in the examined temperature range (Figure 12.16). The effect is based on molecular interactions, measurable in the solidified state as well as in the melt [29].

Table 12.4 Scenting with different products as example for the diversity of product design from liquids.

Product	Scent	Manufacture/operating principle
	Solid/liquid	
Air freshener	Liquid	Perfume oil evaporates in the air stream of a fan
Room spray	Liquid	Perfume oil sprayed into the space, with pressure or pump atomization
Fragrance pearls/granules	Solid	Perfume/fragrance on carrier material (in form of pellets, granules, extrudates, and also in textile-bags)
Fragrance beads	Solid	Fragrances on porous plastic granules (subsequently disposed)
Scent capsules	Liquid	Fragrances filled into capsules/microcapsules
Scent strips	Solid	Perfume oil on paper
Scent disks	Solid	Fragrances in plastics
Scented candles	Solid	Fragrances in paraffin melt (stirred in); pull of candles, burn down the scented candles
Fragrance lamps	Liquid	Drops of oil on a bowl of water; evaporation by heat of a burning candle beneath the bowl
Scent stones	Liquid	Odorants dripped on porous stones, the oil evaporates at room temperature due to the large surface
Scented gel/block, fragrance beads	Solid	Incorporation of fragrances in melt, gels, or plastic masses, evaporation over a long period
Scented blocks (toilet)	Solid	Incorporation of fragrances in surfactant formulations; block flushed completely in 4–8 weeks
Fragrance concentrate (toilet)	Liquid	Odorants in liquid surfactants; dosage via plastic basket
Fragrance sticks	Liquid	Rod inserted into a fragrance solution, by diffusion of perfume evaporates the oil via the rod
(Smelling salts)	Solid	Ammonium carbonate with perfume oil

During the stirring, the oil dissolves into the liquefied hexadecanol. Presumably, hydrogen bonds between the functional groups of alcohol with the various groups of fragrances are formed in solution. As shown in Figure 12.17, the fatty alcohol delays considerably the evaporation period at room temperature.

The intensity of the scent decreases by solidification. This effect is always desirable when the smell should not be intrusive, but more likely unconsciously perceived. The scent impression may change over time. The faster diffusing components evaporate rather on the surface. Therefore, the fragrance block shrinks with time, until it disappears completely.

12.3.10
Perfume Oils for Detergents

The detergent industry uses mainly alkali and *bleach stable perfume oils* for scenting the powder/granules. The corresponding fragrances originate from syntheses. The

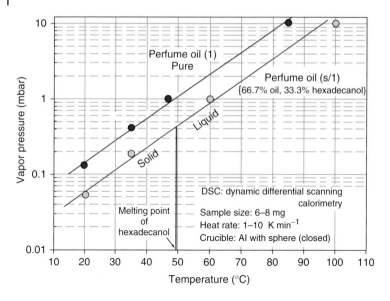

Figure 12.16 Temperature dependence of the vapor pressure of a selected perfume oil: measured both as pure liquid and dissolved in hexadecanol $\{CH_3[CH_2]_{15}OH\}$. (Method from 30.)

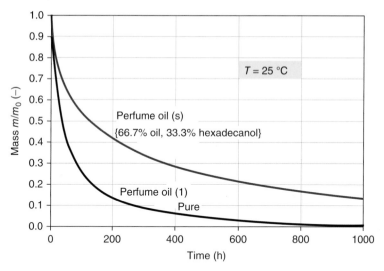

Figure 12.17 Isothermal evaporation of liquid perfume oil in comparison to oil dissolved in hexadecanol, calculated from a plurality of measured values.

detergent formulations contain only chemically inert fragrances. This is valid in the working temperature range, especially in the presence of bleaching agents. Therefore, the smell cannot change in storage and during the washing process by reactions (such as oxidation). The upper temperature limit depends on the application and lies at 40, 60, or 95 °C. Using the surfactants present, the perfume oil is emulsified in the washing process and distributed in a balance between wash water and laundry. The amount of fragrances adsorbed on the laundry increases in presence of carrier substances [31]. Therefore, the laundry smells both in the wet and dry states. When drying the laundry in fresh air, the perfume remains perceptible for some time.

Softeners, working at room temperatures, typically contain a stronger perfume. Thus, this fragrance dominates. When using a laundry dryer, the informed consumer applies the specially developed fabric softener with higher molecular weights for the odorants. These perfume oils are adapted to high temperatures and effects due to water evaporation. Therefore, a distinct fresh scent remains on the dried laundry. The fragrancing depends on both the operating conditions of the machine and the materials (cotton, wool, linen, and synthetics).

An assessment of not only the allergenic potential occurs for the individual scents (see Section 12.4), but also the skin compatibility of the entire product including the fragrance mix. This happens in particular for "sensitive" detergents, which are preferred by sensitive consumers. For some fragrances, results of *allergenic potential* are reported in the literature [32]. Formulations of harmless odorant mixtures belong to the corporate expertise. Individual scents are subject to the selection criteria, described in Figure 12.18. Some parameters are rather complex and can only be fulfilled with effort. Developed under these conditions, the fragrance mixture shows an optimized product design. Experts and consumers evaluate both this mixture and the scented products. They assess the aesthetics, especially the scent impression, as well intensity and duration of fragrance.

The perfuming of solid detergents can occur in four different ways (Figure 12.19). On the one hand, the atomization of oil on the complete product takes place and on the other, an admixing of solid "scent pearls" in form of suitable granules is possible.

The technical procedure runs as follows. With air as propellant, a two-fluid nozzle sprays the perfume oil on the powder in amounts of 0.3–1.0%. This either happens at the end of the collecting belt or in the final mixer (Figure 12.20, paths 3 and 4). Perfume oils cause a surface stickiness on the powder, which is eliminated by fine recipe ingredients or powdering [21]. Therefore, spraying and powdering run

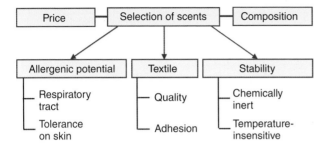

Figure 12.18 Selection of fragrances for use in detergents; some criteria for the product design.

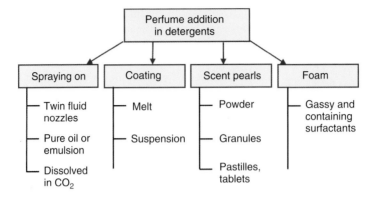

Figure 12.19 Opportunities for scenting solid detergent formulations.

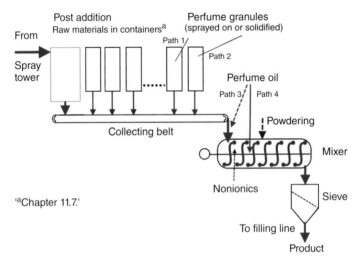

Figure 12.20 Perfuming of solid detergents in the production. Path 1: granules prepared from scenting powders; path 2: granules made of meltable materials with embedded perfume oils; path 3: perfume oil sprayed on the collecting belt dropping; and path 4: perfume oil sprayed on the detergent in final mixer.

sequentially in a continuous mixer that works directly before the filling line. For powdering, an inert powder suffices, showing particles sizes in the micron range. In the detergent sector, powdering occurs with silicas, silicates, and zeolites, but ground salts (sodium carbonate, sulfate) also reveal good results.

Another possibility represents the addition of separately produced fragrance powders, granules, or extrudates (Figure 12.20; paths 1 and 2). Added in small quantities, the "scent pearls" contain relatively high amounts of perfume. For the production, various methods exist, differing in practicability and economics.

Some interesting methods are rarely used or not at all for reasons of cost. Other processes are still going through the development phase. Instead of the pure oils, emulsions are safer and better [18]. The use of solutions in carbon dioxide [20] lies far in the distant future. Furthermore, surfactants containing oils or emulsions flow after addition of air as foam on the solid particles. With foam, a particularly uniform application and distribution is achievable [33]. Emulsions and foams will likely reach the practice only after further investigations.

12.3.11
Manufacture of Fragrance Beads

A stirrer distributes the perfume oil with the help of surfactants in an aqueous solution. The liquid, representing a solution, suspension, or emulsion, should contain dissolved and/or suspended carriers (Figure 12.21a). In closed plants, fragrant powders with uniformly distributed perfume oil (about 10 to a maximum of 25%) arise through spray drying. In some cases, pellets or granules emerge subsequently in separate facilities from the powder. Mixing of the fragrant powder with the detergent formulation occurs after dosing on the collecting belt.

The liquid with the perfume should also contain substances such as cellulose ethers, which effect an encapsulation of powder in the spray drying process. To meet the safety requirements (dust-ex., fire risk), the process runs with nitrogen in a closed cycle. The gas flows through a scrubber using the temperature reduction for dehumidifying [34]. However, a small amount of perfume oil evaporates through steam distillation. This part reaches the wash water. There, a phase separation ensues for recycling the oil. The evaporated perfumes change the odor of the scent

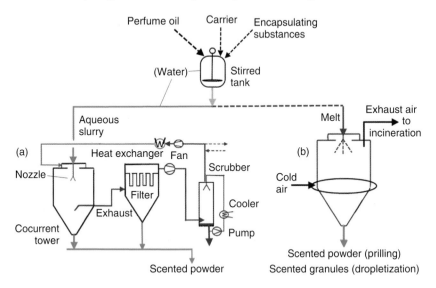

Figure 12.21 Equipment for the production of perfume powders and granules: (a) spray tower for drying of aqueous slurries with N_2 gas in circulation and (b) prilling and dropletization of melting.

particles. The manufacturer considers these changes in advance in the recipe or uses less volatile components.

Another possibility is to incorporate the fragrance oils in a mixture of liquid anhydrous excipients (~1 : 1). For this purpose, it is necessary to melt the mixture by increasing the temperature [29]. Thereafter, the melt is split up by dropletization, solidified with cold air, and converted into granules. Alternatively, in a spray tower, the powder arises by prilling the melt (Figure 12.21). The exhaust air should preferably be incinerated to avoid problems to the environment.

Figure 12.22 displays an agglomeration process, in which scent granules and extrudates emerge by using carrier substances. Because of their versatility in the incorporation of scents (to over 50%), extruders are preferred for the homogenization of solid carriers. The mass, consisting of carriers and a meltable excipient as well as the perfume oil, must not be liquefied in the extruder. Plasticization suffices. The twin-screw extruder includes several temperature zones (heatable/coolable segments) and a drilled head plate, equipped with a multiblade cutting knife

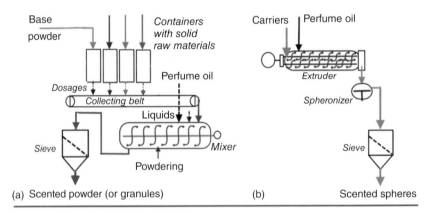

(a) Scented powder (or granules)   (b)   Scented spheres

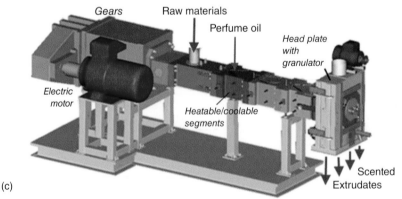

(c)

Figure 12.22 Production of fragrant agglomerates in the mixer and extruder: (a) mixing process; (b) extrusion process; and (c) a double-screw extruder with granulator [35] for the cutting of strands. (Courtesy of Lihotzky.)

(Figure 12.22c). The product passes through the holes in the head plate and leaves them in the form of spaghetti. The multiblade knife rotates directly on the head plate, and cuts the strands in the length/diameter ratio of 1 : 1. By introduction of shear forces, the product heats up and plasticizes. Therefore, the cut cylinders form a spherical shape downstream in a spheronizer by fast rolling movements. Under operating conditions, many of the softened or fusible lipophilic substances, optionally after the addition of salts, are suitable for the incorporation of perfume oils in the extruder. Subsequently, the rounded agglomerates harden in the cooling channel.

Fragrant crystals, prepared by crystallizing a melt of isomalt [31], contain fabric softener and perfume oil in solidified form. For example, bentonite, polydimethylsiloxane, or alkylamido ammonium lactate and dyes are present. The molten mixture runs into a cold mold and solidifies in the intended shape. The resulting granules are placed directly in the washing machine. They provide a pleasant scent on the laundry (Figure 12.23). In addition to the detergents, scenting of packaging exist, where perfume oils are encountered in airtight wrappings at the adhesives in the bonding area [36] and on the inside of the lid [37].

12.3.12
Personal Care and Other Products

The perfuming of skin [19] and hair care products occurs by setting of an adequate fragrance, preferred by a target group in type and strength. Creams and lotions are particularly used as skin care. To a great extent, the product influences the fragrance impression. The selected perfumes, especially the lipophilic components, can interact with some product components. In addition, other ingredients and the consistency exert a perceptible influence. In the manufacturing process, about 0.2–1% pure perfume oil flows directly into the organic phase prior to the emulsification or into the emulsion.

Figure 12.23 Scented crystals with solidified perfume oils and softening quaternary ammonium compounds.

The selection of the "best" fragrance represents a lengthy, iterative process, both in cosmetics and in the detergent sector. In the end, the determination of the best fragrance may be carried out by an odor panel (sensory assessment; OEB, Odor Evaluation Board). The decision falls usually after several tests in practice, either by the panel or in case of large products in a wider circle of 100–250 people in a household survey and test (Home Use Test, HUT [38]). The expense can be justified in large companies, because the scent, as an essential parameter of product design, affects the market success significantly.

A check for skin tolerance of fragrances, used in creams and lotions, occurs usually on 10–30 people with sensitive skin. Some products for sensitive skin, which should have a scent, contain detected skin friendly fragrances. Perfume-free products for sensitive skin are still the exception, except for eye creams and some over-the-counter (OTC) cosmetics in the pharmacy.

In addition to the personal care products, many body cleansers are perfumed. Examples are soaps, shower gels; alcoholic solutions for the face and the hand; and toothpastes. In other areas, leather, plastic, and furniture polish exude an intense fragrance. Moth repellent sachets contain drug and perfume (lavender or cedar), incorporated into granules. After spraying, the windshield washer fluid for cars unfolds a typical (alcohol) odor. Occasionally, people use fragrant stationery in different variants. Some leather products gain more attractiveness by a leather/wood fragrancing. The consumer meets other examples in daily use. Application of perfumes for increasing the sales gets more and more importance.

12.3.13
Safety

In serial examinations with patch tests, more than 15% of the German population shows sensitizations to at least one of the most important contact allergens (Information Network of Departments of Dermatology; in Germany: IVDK [39]). A major problem is that fragrance oils trigger more than 7% of personal allergies [40, 41]. Unless the substance is not identified by tests of single substances, the allergic person should avoid any skin contact with fragrances, because it is not possible to estimate the potential and the danger of allergic substances.

Fragrances are divided into four classes, from class 1 (frequently sensitizing) to class 4 (non-sensitizing). Class 4 includes, for example, almond, mandarin, and lime oil, and class 3 (rarely sensitizing) rosewood, caraway, and bergamot. The latter stands out with a unique feature: after application to the skin and simultaneous exposure to sunlight, bergamot (e.g., in eau de cologne) can trigger phototoxic reactions. The effect causes skin discoloration and occasionally bubble changes, known as *Berloque dermatitis*.

In Europe, the SCCNFP (Scientific Committee on Cosmetic Products and Non-Food Products Intended for Consumers) evaluated fragrances according to their potential for contact allergy [32]. They found 26 major allergens (Table 12.5 [42]), which are annexed to the Cosmetics Directive (Section 5a, 3 Appendix 2 A, 45, and substances from 67 to 92 [43]). Beyond a certain amount (100 ppm for rinse-off

Table 12.5 Required declaration components in perfume oils according to the cosmetic directive (strong allergens highlighted).

INCI name	Chemical description
Amyl cinnamal	2-(Phenylmethylene)heptanol
Amylcinnamyl alcohol	2-(Phenylmethylene)heptanol
Anise alcohol	4-Methoxy-benzyl alcohol
Benzyl alcohol	Benzyl alcohol
Benzyl benzoate	Benzyl benzoate
Benzyl cinnamate	Benzyl cinnamate
Benzyl salicylate	Benzyl salicylate
Cinnamal	Cinnamaldehyde, 3-phenyl-2-propenal
Cinnamyl alcohol	Cinnamyl alcohol, 3-phenyl-2-propen-1-ol
Citral	3,7-Dimethyl-2,6-octadienal
Citronellol	DL-citronellol, 3,7-dimethyl-6-octen-1-ol
Coumarin	2H-1-benzopyran-2-one
Eugenol	2-Methoxy-4-(2-propenyl)phenol
Evernia prunastri extract	Oakmoss extract
Evernia furfuracea extract	Treemoss extract
Farnesol	3,7,11-Trimethyl-2,6,10-dodecatrien-1-ol
Geraniol	2,6-Dimethyl-trans-2,6-octadien-8-ol
Hexyl cinnamal	2-(Phenylmethylene)octanal
Hydroxycitronellal	7-Hydroxycitronellal, 7-hydroxy-3,7-dimethyloctanol
Hydroxyisohexyl-3-cyclohexene-carboxaldehyde	4-(4-Hydroxy-4-methylpentyl)-3-cyclohexene-1-carboxaldehyde
Isoeugenol	2-Methoxy-4-(1-propenyl)phenol
Limonene	1-Methyl-4-(1-methylethenyl)-cyclohexene
Linalool	3,7-Dimethyl-1,6-octadien-3-ol
Methyl 2-octynoate	2-Octynoic acid, methyl ester
Alfa-isomethylionone	3-Methyl-4-(2,6,6-trimethyl-2-cyclohexene-1-yl)-3-butene-2-one
Butylphenyl-methylpropional	2-(4-*tert*-Butylbenzyl)-propionaldehyde

and 10 ppm for the leave-on fragrances), the manufacturer has to declare the use of these compounds on the product package in the INCI list (International Nomenclature of Cosmetic Ingredients) (see Section 13.2).

Applied in larger quantities on the skin, perfume oils diffuse, depending on composition, in about 15 min through the epidermis into the dermis and enter the bloodstream. Thereafter, the components are present in the whole body. Some scents can be detected later in the urine.

For respiratory intake of fragrances, however, little experimental data are available. After inhalation of perfumes and fragrances, in rare cases both irritating and allergic reactions occur, as clinical experience demonstrates. The intensity reaches from a mild mucosal irritation with cough and mucus production to severe asthmatic attacks [44].

12.4
Learnings

√ The key to product design lies in setting the desired quality and shape as well as converting into other phases and shapes.

√ Purity characterizes hydrophilic and lipophilic liquids, in addition to composition in the case of mixtures and emulsions.

√ Liquids change their properties on the introduction of surfactants and viscosity regulators as well as solids in different finenesses and quantities.

√ Liquids from nature offer a variety of product designs (example: milk).

√ Perfumes are essential oils, predominantly consisting of terpenoides. The fragrances originate from plants or from syntheses.

√ Fragrances emerge from parts of plants by steam distillation, extraction, or cold pressings.

√ Perfume oils allow a scenting of the human body, of products, and of spaces. In addition, aromatherapy utilizes fragrant oils.

√ Emulsions, foams, granules, and tablets arise from perfume oils by the addition of excipients.

√ Incorporation of perfumes ensues in different ways: admixing of scented pearls or encapsulated beads, spraying on the powders or enclosing in coating layers.

√ Perfume oil may cause allergy. Therefore, the fragrance oils divide into four classes, from class 1 (frequently sensitizing) to class 4 (non-sensitizing).

√ Annexed to the Cosmetics Directive, there are 26 fragrances with high allergenic potential, which the manufacturer of products must declare for perfumes.

References

1. Rimmel, E. (1867) *The Book of Perfumes*, 5th edn, Chapman & Hall, London, (digitized by Google).
2. Morris, E. (2002) *Fragrance: The Story of Perfume from Cleopatra to Chanel*, Dover Publications.
3. Ohloff, G. (1992) *Irdische Düfte - himmlische Lust: eine Kulturgeschichte der Duftstoffe*, Birkhäuser.
4. Ohloff, G. (2004) *Düfte- Signale der Gefühlswelt*, Helvetica Chimica Acta, Zürich.
5. Boeck, A. and Fergen, H.-U. (1994) in *Perfumes: Art, Science and Technology* (eds P.M. Müller and D. Lamparsky), Chapter 15, Chapman & Hall, London, pp. 421–440.
6. Meine, G. (2004) in *Kosmetik und Hygiene*, 3rd edn, Chapter 18 (ed. W. Umbach), Wiley-VCH Verlag GmbH, Weinheim, pp. 493–500.
7. Schwedt, G. (2008) *Betörende Düfte, sinnliche Aromen*, Chapter 4, Wiley-VCH Verlag GmbH, Weinheim, p. 178 ff.
8. Schreiber, W.L. (2013) *Kirk-Othmer Chemical Technology of Cosmetics*, Chapter 4, John Wiley & Sons, Inc., Hoboken, NJ, pp. 123–160.
9. Uhlemann, J. and Reiß, I. (2009) Produkteigenschaften und Verfahrenstechnik am Beispiel der Aromen. *Chem. Ing. Tech.*, **81** (4), 393.
10. FA Symrise (2011) Annual Report 2011, p. 47.

11. Werner, M. and von Braunschweig, R. (2006) *Praxis Aromatherapie: Grundlagen – Steckbriefe – Indikationen*, Georg Thieme Verlag, Stuttgart.

12. Wabner, D. and Beier, C. (eds) (2012) *Aromatherapie*, 2nd edn, Elsevier, Urban and Fischer Verlag, München.

13. Perring, K.D. (2006) in *The Chemistry of Fragrances* (ed. C. Sell), Chapter 11, RSC Publishing, p. 199.

14. Schwedt, G. (2008) *Betörende Düfte, sinnliche Aromen*, Chapter 3, Wiley-VCH Verlag GmbH, Weinheim, p. 99 ff.

15. Kraft, P., Baigrowicz, J.A., Denis, C., and Fráter, G. (2000) Recent developments in the chemistry of odorants note on trademarks. *Angew. Chem. Int. Ed.*, **39** (17), 2980–3010.

16. Surburg, H. and Patten, J. (2006) *Common Fragrance and Flavor Materials*, 5th edn, Wiley-VCH Verlag GmbH, Weinheim.

17. Lee, K. and Popplewell, L.-M. (2005) International Flavors & Fragrances Inc., New York, EP 1 589 092 A1, Oct. 26, 2005.

18. Rähse, W. (2005) Henkel AG & Co. KGaA, DE 103 54 564 B3, Jul. 07, 2005.

19. Rähse, W. and Dicoi, O. (2009) Produktdesign disperser Stoffe: Emulsionen für die kosmetische Industrie. *Chem. Ing. Tech.*, **81** (9), 1369–1383.

20. van Ginneken, L. and Weyten, H. (2003) in *Carbon Dioxide Recovery and Utilization* (ed. M. Aresta), Chapter 3, Kluwer Academic Publishers, Dordrecht, pp. 123–135.

21. Rähse, W. (2009) Produktdesign disperser Stoffe: Industrielles Partikelcoating. *Chem. Ing. Tech.*, **81** (3), 225–240.

22. Rähse, W. and Victor, P. (2005) Henkel AG & Co. KGaA, DE 103 57 676 A1 Jul. 21, 2005.

23. Holderbaum, T., Semrau, M., and Schaper, U.-A. (2000) Henkel AG & Co. KGaA, DE 19838127A1, Feb. 24, 2000.

24. Hehn, P. (2006) *Emotionale Markenführung mit Duft: Duftwirkung auf die Wahrnehmung und Beurteilung von Marken*, Verlag Forschungsforum.

25. Lindstrom, M. (2009) *Buyology*, Campus-Verlag, Frankfurt.

26. Knoblich, H., Schubert, B., and Scharf, A. (2003) *Marketing mit Duft*, 4th edn, Oldenbourg.

27. Rengshausen, S. (2004) *Markenschutz von Gerüchen*, V & R Unipress, Göttingen.

28. Pashkovski, E. *et al.* (2004) Colgate-Palmolive Co, New York, EP 1 379 620 A1, Feb. 19, 2004.

29. Rähse, W. and Dicoi, O. (2010) Produktdesign von Flüssigkeiten: Parfümöle in der Konsumgüterindustrie. *Chem. Ing. Tech.*, **82** (5), 583–599.

30. Tilinski, D. and Puderbach, H. (1989) Experiences with the use of DSC in the determination of vapor pressure of organic compounds. *J. Therm. Anal.*, **35**, 503–513.

31. Rähse, W. *et al.* (2007) Henkel AG & Co. KGaA, DE 10 2006 016 579 A1, Nov. 10, 2007.

32. Herman, S. (2002) *Fragrance Applications: A Survival Guide*, Allured Publishing Corporation, Carol Stream, IL.

33. Rähse, W., Larson, B., and Semrau, M. (2002) Henkel AG & Co. KGaA, DE 101 24 430 A1, Nov. 28, 2002.

34. Rähse, W. and Dicoi, O. (2009) Produktdesign disperser Stoffe: Industrielle Sprühtrocknung. *Chem. Ing. Tech.*, **81** (6), 699–716.

35. Rähse, W. (2007) *Produktdesign in der chemischen Industrie*, Springer-Verlag, Berlin.

36. Weber, R., Reins, J., Boeck, A., Schaper, U.-A., and Berg, M. (1986) Henkel AG & Co. KGaA, EP 0114301, Dez. 4, 1986.

37. Barthel, W. and Lahn, W. (2000) Henkel AG & Co. KGaA, DE 198 58 858 A1, June 21, 2000.

38. Lawless, H.T. and Heymann, H. (2010) *Sensory Evaluation of Food: Principles and Practices*, Springer, New York.

39. Schnuch, A., Uter, W., Lessmann, H., Arnold, R., and Geier, J. (2008) Klinische Epidemiologie der Kontaktallergien. Das Register und das Überwachungssystem des Informationsverbundes Dermatologischer Kliniken (IVDK). *Allergo J.*, **17**, 611–624.

40. Schnuch, A., Uter, W., Geier, J., Brasch, J., and Frosch, P.J. (2005) Überwachung der Kontaktallergie: zur

"Wächterfunktion" des IVDK. *Allergo J.*, **14**, 618–629.

41. Uter, W., Balzer, C., Geier, J., Schnuch, A., and Frosch, P.J. (2005) Ergebnisse der Epikutantestung mit patienteneigenen Parfüms, Deos und Rasierwässern. Ergebnisse des IVDK 1998–2002. *Dermatol. Beruf Umwelt*, **53**, 25–36.

42. Schnuch, A., Uter, W., Geier, J., Lessmann, H., and Frosch, P.J. (2007) Sensitization to 26 fragrances to be labelled according to current European regulation. Results of the IVDK and review of the literature. *Contact Dermat.*, **57**, 1–10.

43. Bundesministerium der Justiz (1977) Verordnung über kosmetische Mittel (Kosmetik- Verordnung), Bundesministerium der Justiz, 16.12.1977; letzte Änderung 11.6.2009, *//bundesrecht.juris.de* (accessed 27 November 2013).

44. Schnuch, A., Oppel, E., Oppel, T., Römmelt, H., Kramer, M., Riu, E., Darsow, U., Przybilla, B., Nowak, D., and Jörres, R.A. (2010) Experimental inhalation of fragrance allergens in predisposed subjects: effects on skin and airways. *Br. J. Dermat.*, **162**, 598–606.

13
Design of Skin Care Products

Summary

Chemists or pharmacists formulate tailor-made products for the skin with caring or healing effects. The products differ in number, concentrations, and types of active ingredients, according to the task and application. They are classified as cosmetics, cosmeceuticals (active cosmetics), and pharmaceuticals. The usual scented cosmetics serve for beautification. Cosmeceuticals are dermatological products for sustaining and restoring a healthy skin. Many pharmaceuticals, available on prescription, heal diseased skin. A skin cream consists of several lipophilic and hydrophilic substances, forming an emulsion. The ingredients are divided into excipients, additives, and active materials. Active cosmetics contain various ingredients in effective concentrations, ensuring the stressed skin with an optimal supply of moisture, lipids, and vitamins. Special substances, for example, for the reduction of skin aging or for the stimulation of skin renewal, are important components of the formulation. The emulsions, produced discontinuously in stirred vessels with rotor–stator emulsifiers, flow metered via a filling line, preferably in hygienic dispensers, which allow taking out the cream without air access.

The discussion of product design of two immiscible liquids that form an emulsion with the aid of surfactants takes place using the example of skin care products. This involves creams and lotions for the care of the face, body, as well as the hands and feet. Care creams are a part of cosmetics.

13.1
History of Cosmetics

The term *cosmetic*, from the ancient Greek word meaning "order or decorate," refers to the body and beauty care. This includes the maintenance, restoration, and enhancement of the beauty of the human body. The first sign of cosmetics dated back about 10 000 BC. The Mesolithic people applied grease and castor oil to soften the skin. With plant dyes, they painted tattoos. About 7000 years later (3000 BC), Egyptian parchment described the use of creams to sooth the skin and to reduce

Industrial Product Design of Solids and Liquids: A Practical Guide, First Edition. Wilfried Rähse.
© 2014 Wiley-VCH Verlag GmbH & Co. KGaA. Published 2014 by Wiley-VCH Verlag GmbH & Co. KGaA.

wrinkles [1]. In the ancient Near East, men applied oils to their hair and beard. The women used eye paints, rouge, powders, and ointments for the body.

About 50 BC, Cleopatra was known as a *beauty*, also because of her intensive use of cosmetics. She possessed many products from nature. These were beeswax, honey, and natural oils, as well as products made from fruits, vegetables, herbs, and seeds, besides eggs and milk. She bathed in goat milk for skin regeneration. At that time, there were mirrors, makeup, makeup containers, combs, wash dishes, wigs, as well tweezers and blades for removal of annoying hair. Vermilion and red ocher were used for coloring the lips and cheeks, henna for the hair, skin, toenails and fingernails, and the malachite green, gray galena, and finely ground antimony for eyes and eyeliner.

The Greek physician Galen (about 200 AD) developed the first cold cream from beeswax, olive oil, and water [1]. The Romans introduced communal baths for noble persons. In the Middle Ages, they used hair dye and makeup, in addition to natural skin care and herbal remedies. During those times, only a pale complexion was considered beautiful. With lead white, they achieved a flawless pallor. This substance and others are highly toxic and often caused abscesses that did not heal.

During the Renaissance, the Venetians dyed their hair using plant colors, fixed with clay and baked in the sun. When it came to Elizabeth I of England (about 1580) and Catherine de Medici in France, dyeing of the cheeks and lips became popular again. The red lip color came from cochineal, a red dye from the cochineal scale insect. In the eighteenth century, bismuth oxide, mercury oxide, tin oxide, and talc were used to whiten the skin. Red makeup for the lips and cheeks emerged from safflower, cochineal, redwood, sandalwood, and vermilion. In addition, the hair was treated with greasy pomades. Hair powder consisted mostly of wheat or rice starch, partly colored.

At present, there are ranges of cosmetic products that have been tested for their safe use. The aim has not changed in thousands of years. Cosmetics mean increasing attractiveness by beautifying the body and face.

13.2
Regulations of Cosmetic Products

The cosmetic products cover the areas of skin and hair, mouth and teeth; there are smell and color cosmetics as well (Table 13.1). In Europe, the permitted substances and their amounts as well as the use of these products are defined by the Cosmetics Directive; in the United States, this comes under the Food and Drug Administration. The Directive guarantees free movement of cosmetic products within the European market if these products are not dangerous to human health under normal or foreseeable conditions of use. The Directive includes prohibited substances, and substances with restrictions in the amount or specific conditions of use, apart from authorized colorings, preservatives, and UV filters. A search for substances is possible on the Internet via the cosmetic ingredients database "Cosing." It is mandatory for the labeling on the packaging of cosmetic products

Table 13.1   Typical cosmetic products.

Definition in the Cosmetics Directive, first part of Article 1 (quote)

A "cosmetic product" shall mean any substance or mixture intended to be placed in contact with the various external parts of the human body (epidermis, hair system, nails, lips, and external genital organs) or with the teeth and the mucous membranes of the oral cavity with a view exclusively or mainly to cleaning them, perfuming them, changing their appearance and/or correcting body odors and/or protecting them or keeping them in good condition.

Category	Products
Skin cleansing and care	Soap, cleansing milk, cleaning water, perfume, and bubble baths
	Eye and face cream, body lotion and creams, hand and foot cream, and gels
	Shaving cream, shaving soap, and aftershave
	Depilatories
	Sunscreen milk, sunscreen lotion, and water-repellent lotion
Dental and oral care	Toothpaste, dental floss, toothbrush, and mouthwash
	Dentures: cleaning and adhesion
Hair treatment	Shampoo, styling, conditioner, permanent wave, hair gel, spray, coloring, powder, hair care, and pomade
Decorative cosmetics	Makeup, rouge, mascara, eye shadow, eyeliner, lipstick, nail polish, artificial nails, and foundations
	Self-tanners
Scent and smell	Perfume, eau de toilette, eau de cologne, deodorant, and antitranspirant

Table 13.2   Labeling of cosmetic products according to the Council Directive 76/768/EEC (upper part: Quotes from the Regulation).

Name and address of the manufacturer or of person responsible for marketing the product
The nominal contents at the time of packaging, by weight or by volume
Date of minimum durability indicated for products with a minimum durability of <30 months
Period of time after opening for which the product can be used for products with a minimum durability of more than 30 months (indicated with the symbol representing an open pot of cream)
Function of the product and particular precautions for use
Batch number

List of ingredients: INCI, International Nomenclature of Cosmetic Ingredients

Order of the ingredients according to their mass proportions (highest percentage first)
Ingredients <1% in any order
Perfume – 26 fragrance allergens must be declared from a certain amount (see Section 12.3.13)
The CI (color index) number specifies the dyes

to have certain information, which has to be understandably and legibly written on the packaging (Table 13.2).

Furthermore, the Cosmetics Directive requires the elaboration of a qualified security assessment of the formulation from an expert and demands a registration of the product with the relevant authorities. Currently, the production is carried out, of course, according to the good manufacturing practice (GMP) standard.

13.3
Product Design

The manufacture of designed products, according to the customer's wishes, requires trained teams of chemists, chemical engineers, and marketing people. The product includes the packaging. It is possible to change product properties within wide limits or to set superior features. Novel forms, applicators, and increased performances, developed in cooperation with the customer, characterize product design (Figure 13.1). This new way of thinking starts from a desired product, but not from an existing process. In addition, consumer awareness of the brand (not discussed here) especially provides a key for sales [2].

As a rule, the known or assumed performance of the product decides the purchase by the consumer. Convenience and aesthetics are of secondary importance. There are a few exceptions. This includes cosmetic products for which the customer shops mostly according to the aesthetics and awareness of the brand (Table 13.3).

Many measurable parameters characterize fine solid particles, such as particle sizes and distributions, form factors, colors, surfaces and release properties, fragrances, and bulk densities. Suited, coordinated, and directed processes may define their product design [3]. In addition, liquid products, in particular liquid

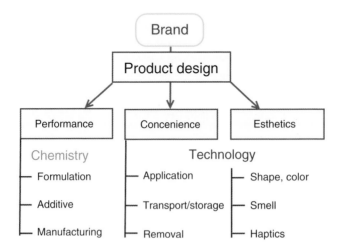

Figure 13.1 Characteristics and composition of the product design for cosmetic products (without packaging).

Table 13.3 Purchasing behavior of consumers.

Ranking	Household products	Cosmetic products
1	Performance	Esthetics (fragrance, packaging, and design)
2	Brand (awareness and image)	Brand (awareness and image)
3	Convenience (handling, storage, and durability)	Convenience (jar and packaging)
4	Esthetics	Performance (beautification)

mixtures, are subject to the laws of product design. To illustrate the diversity of product design, liquid skin care products are particularly suitable, because there are numerous ways to fix the product performance, handling convenience, and aesthetics. The following is a detailed discussion of skin care to prove the statements made, on the basis of two papers in German [4, 5].

For the design settings, the recipe is responsible, which is composed of auxiliaries, additives, and active ingredients. Figure 13.2 lists the medical and physical–chemical parameters that play a role in product development, and are adjustable by appropriate substances [6].

Further degrees of freedom exist by producing different physical–chemical emulsions. The possible emulsions consist of oil or water droplets, in the form of a (macro-)emulsion with droplet sizes in the micrometer range or as a mini- (or nano-)emulsion with droplet sizes of around 100–600 nm [7]. In addition, the selection of an appropriate packaging optimizes, in another way, the product design.

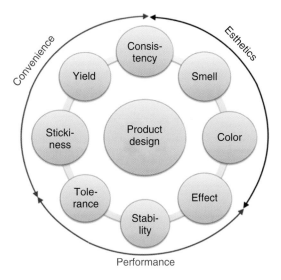

Figure 13.2 Elements of product design for cosmetic products.

13.4
Skin Care

Dermatological products are developed for the care and health and healing of the skin, hair, and nails, and are classified, according to the task, as cosmetics, cosmeceuticals, and pharmaceuticals. Cosmeceuticals play a major role in medical skin care. The skin represents an area of about $2\,m^2$, weighs 3.5–10 kg, and is 0.3–5 mm thick. The largest organ in the human body, it characterizes appearance and fulfills numerous functions such as protection, regulating the heat balance, and the perception of contact stimuli.

The drying out and roughness of the skin prevents the hydro-lipid coat, which is also called the *acid mantle*, that must be available on the skin in the right composition and thickness. The skin is daily exposed to many stress factors (too much sun, cleaned too intense, too dry ambient air, and environmental influences). The strain under prolonged exposure damages the film on the skin. Therefore, the skin care is of great importance for the well-being of people, at present and in the future.

Table 13.4 Features of dermatological creams.

Orientation	Cosmetics (skin care)	Cosmeceuticals (medical skin care)	Dermatics (medicament)
Manufacturer	Cosmetic industry	Cosmetic industry	Pharmaceutical industry
Tasks	Beauty	Health	Healing of skin diseases
	Care	Care/prevention/concomitant therapy	Eczema
	Moisture	Skin rejuvenation/moisture	Dermatitis/psoriasis
	Scent	Itching	Seborrheic keratosis
		Acne	Microbial infestation and inflammation
		Sun damage	Rashes
		Medical treatment of diseased skin	Wounds
Active ingredients			
Number	1 (advertised)	>2	Mostly 1
Concentrations	Low	Medium to high	Effective/age dependent (children and adults)
Additives			
Perfume	Yes	No	No
Dye	In some cases, yes	No	No
Emulsion type (preferred)	o/w	o/w	w/o
Water (%)	70 to >80	50–70	60–80?
Container	(Pump) bottles crucibles	Dispensers	Tubes

The cosmetic industry provides suitable products for all aspects of skin care, as shown in Table 13.4. After application, the cream remains for the most part on the surface, smoothing the skin. Two cosmetic product lines are available: perfumed creams for beauty (cosmetics in the strict sense) differ from those for maintaining health (cosmeceuticals). The classification in cosmetics, cosmeceuticals, and pharmaceuticals is done according to the task and need. Furthermore, there are products to relieve symptoms (cosmeceuticals and dermatics) as well as for healing damaged skin (dermatics and pharmaceuticals). A precise definition is not possible as there are overlapping areas.

Pharmaceutical products are used to treat and heal sick skin. The healing ingredients remain not only on the skin surface but also penetrate into the epidermis. Dermatics treat primarily pathological disorders of the skin such as psoriasis, eczema, inflammation, cancer, rashes, deficiencies, and many others [8]. They are dermatological therapeutics. The galenic composition supports the active material, transport into the skin, and release (duration) of the therapeutic substance.

13.4.1
Cosmetic Products for Beautification

The care of healthy skin around the eyes and lips, on the face, hands, and the total body is possible with commercially available creams and lotions. The corresponding products represent the great diversity found in the cosmetic industry. The nature of the skin depends on, among other things, genetic factors, gender, type, age, and weight. The formulations used also differ, depending on the thickness of the skin (feet, hands, and face) and the skin condition (dry, normal, and bold) and, also on the intended use (skin texture, sun protection, and anti-aging). Cleaning, revitalizing, perfuming, and deodorizing are the main functions of cosmetics. In addition, there are color cosmetics, which underline the beauty with rouge, eye shadow, mascara, lipstick, and nail polish, and further with hair styling products.

One performance parameter stands out in the development of usual cosmetics. Mostly, people wish to improve skin moisture or the surface structure and barrier function; older people want effective ingredients against skin aging (Figure 13.3). Improvements in structure play a role in the case of dry skin. A cleaning agent reduces problems associated with oily skin. A weak level of sun protection could be beneficial in products used daily, because UV light causes increased skin aging. Bright sunlight requires special products with high sun protection factors (SPFs). The SPF of the lotion depends on skin type and the intensity of solar radiation.

Typical commercial products for the skin have a pleasant consistency and provide short-term care. From a physical point of view, the creams preferably consist of oil in water (o/w) emulsions. Normal (usual) cosmetics contain only a few active ingredients, formulated in low concentrations, but in relation with many auxiliary materials. Traditional products contain about 25–40 ingredients, including many substances for preserving, emulsifying, and thickening, apart from the active and base materials.

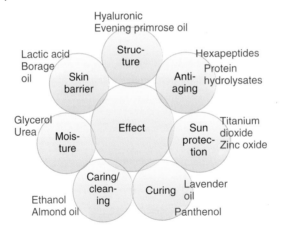

Figure 13.3 Benefits of skin care products (cosmetics and cosmeceuticals) with examples of suitable ingredients.

Almost all cosmetic care products are scented, and spread a pleasant fragrance. The effect lasts only for a short time. A product sale takes place mostly by the pleasant scent and, less frequently, an interesting substance (silk proteins and urea), or because of a particular function (smoothing and moisturizing). According to the EU Cosmetics Directive [9], first the products must do what is written on the packaging (moisturizer, reduces wrinkles, etc.) and second, the products require a declaration of nanoparticles (solids), even if they are safe, on the packaging.

13.4.2
Active Cosmetics for Healthy Skin

Depending on the recipe, cosmetics cure slight dermatological diseases [10]. For such products, between cosmetic care and healing with medicines, dermatologists in the United States created the term *cosmeceuticals*. The word, formed from cosmetics and pharmaceuticals, means "active cosmetics." These designed skin care products ensure not only beauty but also a healthy condition of the skin with essential ingredients [11].

Cosmeceutical creams or lotions are, preferably, o/w emulsions. They combine effective cosmetic ingredients with "healing" natural oils and vitamins/provitamins in amounts that provide a rich supply to the skin and show visible success. By using selected ingredients, physiological processes occur more than usual and support the skin's own regeneration. To avoid undesired effects, it is necessary to reduce the auxiliaries to a minimum. This affects not only the amounts of emulsifiers and preservatives but also waxes, synthetic oils, and other synthetic adjuvants.

To minimize the risk of allergic reactions, cosmeceuticals should not contain perfumes or dyes. To increase acceptance, some creams contain intensively tested fragrances that do not cause allergic reactions. The possible addition of (auxiliary) materials and their maximum levels, as well as the identification of substances on the packaging (INCI, International Nomenclature of Cosmetic Ingredients), are subject to the EU cosmetics regulation [9].

Cosmeceuticals contain biologically active substances in effective concentrations. The first group of these substances comes from plants/plant parts, obtained through different processes such as grinding/pressing and extracting with solvents or with the aid of superheated steam. Depending on the origin of applied substances, the mode of action on the skin surface can be described as healing, caring, moisturizing, lipid-supplying, smoothing, refreshing, rejuvenating, revitalizing, promoting circulation, antiseptic, anti-inflammatory, antioxidant, astringent, and UV-protective. The second group includes the vitamins and pro-vitamins. These substances act on and in the skin in many ways. Some can work with synergistic effects, especially the antioxidants (vitamins C and E), the stimulators of circulation (vitamin B_3), regulators of sebum production (vitamin B_6), anti-aging (vitamin A), and healing ingredients (vitamin B_5). The third group of substances, proteins and peptides, derived from microorganisms or synthesized in analogy with nature, provides improvements in skin hydration, firmness, elasticity, and skin protection. Examples represent products that are able to reduce small facial wrinkles [12], improve peripheral circulation, reduce hair growth in unwanted places, or fight a fungal infection [13].

Cosmeceuticals [14] not only serve for the maintenance of health but also often for the support and supplement of a therapy. These problems require compositions other than those normally used in the cosmetic industry. The active substances work simultaneously in effective concentrations, partially with specific synergistic actions. Among the most important materials are some unsaturated natural oils that boost, depending on the amount, the sheen on the skin. This is undesirable for usual, daily cosmetic products, but is acceptable for active cosmetics. The oils, in combination with moisturizers and vitamins, enable medical actions besides the usual care. The primary target is to maintain or restore a healthy skin. The inclusion of novel peptides [12, 15] provides firming effects. A medical application of cosmetics has demonstrated the calming impact on a dry, itchy skin, bringing the skin back to a healthy state. Studies of volunteers and/or the descriptions of many applications document the improvements achieved by cosmeceuticals.

For cosmeceuticals, there are no legally binding rules. They are subject to the same laws that apply to the usual cosmetics. Table 13.5 gives some aspects of the delimitation between normal and active cosmetics used for therapeutic agents.

13.4.3
Differences between Cosmeceuticals and Drugs

Cosmeceuticals are cosmetic products and not medicines. The GMP cosmetic guidelines represent the basis of manufacturing. Marketing requires a qualified security assessment of the recipe, executed in accordance with the EU cosmetic regulations. Offices of the federal states publish the formal procedures for marketing. For pharmaceuticals, however, there are legal provisions. The Medicines Act [16] contains the definition, requirements, production, licensing, registration,

Table 13.5 Legal basis of dermatological products in Europe [[6]].

	Cosmetics	**Cosmeceuticals**	**Dermatics (pharmaceuticals)**
Legal basis	LMBG	No legal basis; functions proposed by the U.S. dermatologists	AMG
	Food and Commodities Act		Medicines Act
Tasks	Externally applied Care and cleaning	Externally applied Improve and prevent and eliminate forms of minimal dermatological diseases	Inside and outside Heal, alleviate symptoms of diseases and pathological symptoms
	Appearance	Physiological effects are limited to the skin	Render harmless pathogens, parasites, and foreign substances
	Odor	No side effects No systemic effect	Systemic[a] effects possible

[a] Possible effects throughout the body via the blood circulation.

clinical testing, delivery assurance, and quality control aspects. For marketing, authorization (Section 21 AMG) is needed. The permit is granted by the competent federal authority or the Commission of the European Communities or the Council of the European Union.

After AMG Section 1 (abridged) pharmaceuticals are defined as "substances or preparations of substances that are intended for use in the human body and show properties for treating or alleviating or preventing diseases or pathological symptoms...." The results of clinical studies represent the basis for the approval process of drugs, especially the proof of claimed effects (indications proof). Furthermore, the product must be safe in application. The emulsions or ointments have a healing effect in the case of skin lichen, eczema, inflammations, skin rashes, and nutritional deficiencies. Pharmaceuticals contain not only the actual active ingredients but also different consistency regulators, emulsifiers, and preservatives. They usually represent "water in oil (w/o) emulsion." A fatty layer on the skin and a very slow penetration of water-soluble active ingredients characterizes this type of emulsion after application. Single-phase systems are predominantly known as an *ointment*, which consists, for example, of a base substance (Vaseline®) and an active ingredient. Dermatologists prescribe recipes to treat and heal diseased skin with pharmaceuticals, while cosmeceuticals are on sale without prescription.

To check the effectiveness of therapeutic agents, they usually only contain one medicinal substance, at optimized concentration. Normally, a special galenic with suitable excipients guarantees the pharmaceutical availability of the active ingredient [17]. The pharmacological potency and duration of action depend on the concentration, additives, and the formulation (pharmaceutical chemistry and technology). The active substance exists as a pure or chemically modified natural material, or is synthesized completely, or produced biotechnologically. Except for the α-hydroxy and salicylic acid and urea, the drugs contain completely different agents as formulated in the cosmeceuticals. Examples may be potent cortisone, antibiotics, antifungals, and retinoids.

13.4.4
Natural Cosmetics Label

Normal cosmetics use more than a 1000 different active ingredients and excipients. These agents come from plants, animals, or out of syntheses. The formulation of animal products, except milk and honey, are renounced by responsible manufacturers. The cosmetic industry needs no animal testing; they make no sense and cause additional costs. Also, the Cosmetic Directive bans animal experiments. Powdered raw materials (drugs) contain many germs. Therefore, a treatment is necessary, for example, a radiation, such as used for spices. Alternatively, an enhanced preservation of the cream is possible, but not desirable. The addition of these powdered plant parts should be avoided completely, also because they show no real contribution to skin care. It is better to use extracted active ingredients.

The raw materials, especially the preservatives, differ in efficacy, biodegradability, and their side effects, such as their allergic potential. Therefore, it is important to give the right recommendations. Serious manufacturers of cosmetics use well-known effective substances and extracts from nature, preferably from plant parts, but also salts, for example, from the Dead Sea. In some cases, a chemical modification occurs. Glycerol originates from natural oils by ester cleavage. Starting from wood, the cellulose ethers arise by a sequence of different reactions.

With the variety of raw materials, it is certainly useful to provide guidance to the formulators. These tasks are assumed by the certification companies, which assign an organic, bio or eco or natural cosmetics label. A realistic classification, based on the effects on the skin, might be a reasonable measure. However, some go far beyond this target. They split up the chemistry into good (permitted) and bad (not allowed), and allow methylations but ban sulfonations. As bioethanol, ethanol is good, bad from chemical synthesis, although in high purity it is the same.

It is misleading that the certifier does not discuss either the purity of substances or the interactions with other ingredients. Propylene glycol is available in four grades, from the technical to pharmaceutical quality. Sodium lauryl sulfate (SLS) may cause sensitization. The presence of betaines suppresses this property of SLS, depending on the mass ratio. Many manufacturers of hair shampoos therefore use this (prohibited) combination.

Whether the biodegradation of ingredients represents a useful criterion for the exclusion of cosmetic ingredients should be discussed intensively. Ingredients that have not been biologically degraded often adsorb on biosludge and reach the environment even if only in very small amounts. An improvement of some certification rules might go a long way in increasing the quality of the products and the credibility of the recommendations.

A final comment: for the skin, the known therapeutics cannot be certified, owing to the base components and the active ingredients. Certifiers ban lanolin and Vaseline, rightly in the case of ordinary cosmetics, but most dermatica contain one of these substances.

13.5
Emulsions

The optimal supply with lipophilic and hydrophilic active substances in larger quantities requires the formation of an emulsion. An emulsion offers the best physical attributes for skin care products to act on. The preparations are, preferably, o/w emulsions because in this way the 1–10 μm large oil drops can quickly penetrate into the skin, without leaving an oily film. In contrast, on applying w/o emulsions, the oil remains for several minutes, often as a sticky film, which is undesirable in cosmetics. Polymers can control the release of substances in both phases. The water as the continuous phase partially evaporates on the skin, leaving a thin film with the nondiffusible substances in this environment and reinforcing the hydro-lipid film (acid mantle). With a buffer in the water phase, the pH on the skin is set to 5. The substances smooth the skin and provide protection against pathogens with a suitable, partially water-/oil-soluble preservative.

13.5.1
Basics (Definition, Structure, and Classification)

Emulsions are thermodynamically unstable systems of two immiscible liquids. A surfactant distributes the first one evenly in the form of droplets in the other, continuous phase. The liquid is the external or continuous phase, while the internal or disperse phases are the droplets. In the case of oil, droplets arise as an o/w emulsion; otherwise, a w/o emulsion with water droplets is formed [18]. According to the Bancroft rule [19], the HLB (hydrophilic–lipophilic balance) number of the emulsifier (Figure 13.4) decides the type of emulsion. The HLB value indicates the hydrophilic–lipophilic balance of the system and can be determined, analogously according to Griffin [20], from the molecular weight (MW) of the hydrophilic portion (MG_{hydro}) in relation to the total MW.

Emulsifiers with HLB values between 3 and 6 (to 8), which are dissoluble in oils, form (w/o) emulsions. Moreover, surfactants with HLB values of 11–18 produce the (o/w) emulsions. HLB values of 8–11 characterize the transition states between the lipophilic and hydrophilic behavior [21]. In this region arises an o/w or w/o emulsion, depending on the type of production and temperature, but preferably the o/w-type. The percentage distribution of the phase components (water and oil) does not affect the type. W/o droplets are often smaller than o/w droplets (Figure 13.5).

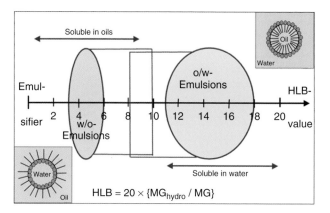

Figure 13.4 Control of the emulsion type by the HLB-value of the emulsifier (insets: lipophilic parts are shown by lines and the hydrophilic groups by circles).

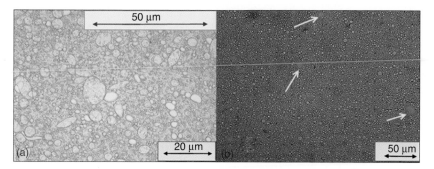

Figure 13.5 Microscopic images of commercial skin care products (a) o/w emulsion with small and large droplets and (b) w/o emulsion type with some large drops.

The stability of the system depends crucially on the emulsifier. Some nonionic surfactants stabilize emulsions very effectively and are therefore preferred. The choice of emulsifier depends on the chemistry of the oil, because oil and emulsifier should be compatible. High-quality natural oils such as borage, hemp, and evening primrose consist mainly of C18 triglycerides. Clearly, therefore, it is best to use a nature-based C18 emulsifier for the manufacture of an o/w emulsion.

Ethoxylated fatty alcohols are one of the most effective nonionic surfactants, creating optimal results with about 30 molecules of ethylene oxide (EO) (MW about 1600, HLB = 16.6). Other emulsifiers are in most cases less effective (higher quantity is required, larger droplets). A macro-emulsion, containing high-quality natural oils, is stable for several years with 0.7% of a fatty alcohol with 30-EO units. Such a stabilized cream, about 4 weeks old at the time of measurement, is shown in Figure 13.6. With this small amount of surfactant in combination with natural oils, the argument is unfounded that skin drying is caused by EO-containing nonionic surfactants. A real assessment is possible by testing the complete recipe in comparison with other less effective surfactants, which require an addition of 2–5%.

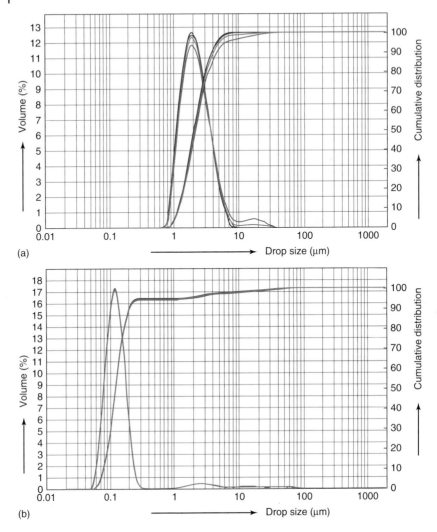

(a)

(b)

Figure 13.6 Differences between macro- and mini-emulsions (a) droplet size distribution of a nonionic surfactant-stabilized o/w skin cream with a mean diameter d_{50} of 1.8 μm and (b) stable cream with nano-droplets ($d_{50} = 120$ nm; from 50 to 300 nm) after three runs through a high-pressure homogenizer; measured three times after dilution by laser diffraction/equipment: Malvern.

While the drop size distributions of commercial products are usually in the micron range, nanosized droplets of about 80–600 nm characterize a mini-emulsion. To generate such a nanoemulsion, two surfactants are required, namely, a hydrophilic and a lipophilic emulsifier. The preferred surfactant combination with HLB values of 16.6 and 5 results in small, stable droplets. The temperature stability of the emulsion depends on the MW of the emulsifiers, the higher the better. The best results are found with a polymeric silicone-based emulsifier, which is even suitable for multiple emulsions, namely, cetyl PEG/PPG-10/1-dimethicone (PEG, poly(ethylene glycol); PPG, poly(propylene glycol)) (Abil$^{®}$ EM 90 from Evonik) in quantities of 0.5% – especially for a stable o/w mini-emulsion applied in combination with about 2% Eumulgin$^{®}$ B 3 (COGNIS/BASF). The droplet size distribution also depends on the homogenizer and the operating conditions. A dual-piston high-pressure homogenizer (Niro Soavi) gives, after three runs, a close and reproducible drop distribution with $d_{10} = 80$ nm, $d_{50} = 122$ nm, and $d_{90} = 200$ nm (Figure 13.6b).

Micro emulsions are clear and transparent. Because of the high amounts of emulsifiers (about 15–35%) needed for the production, use for skin care seems less appropriate; similar restrictions apply to multiple emulsions. A multiple emulsion can be produced by the emulsification of an o/w emulsion in oil, or a w/o emulsion in water. The additional costs for multiple emulsions (higher amounts of surfactant, manufacturing, and fewer degrees of freedom in the formulation) result in no benefit for the skin care at present.

13.5.2
Stability of Emulsions

An essential quality feature of skin care products is the time to use after manufacture and after opening the package. The stability of the emulsion [18], the constancy of color, flavor, and consistency (such as immediately after preparation) are the guidelines for product design. Several steps, regarding the recipe, manufacturing and filling conditions, and the packaging material (Figure 13.7), can optimize the lifetime of the product.

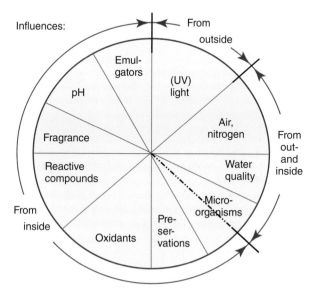

Figure 13.7 Influences on the stability of skin care products.

For oxidation-sensitive ingredients such as vitamins and natural oils, protection is possible by introducing chemical groups and by addition of antioxidants. Vitamin A palmitate and vitamin C phosphate are examples of the chemical modification of sensitive components. Tocopherols (especially in the presence of citric acid) protect the precious natural oils, with a high content of γ-linolenic acid (18-3), against rancidity. Both smell and color are inconspicuous. The use of substances with oxidizing power leads to significant changes in color, smell, and effect.

The quality of water is of crucial importance because almost all infections in the product come from the water. Ion exchange columns remove heavy metals and lime-forming ions before use. Membrane plants, supported by UV lamps, sterilize the water. Preservatives in the product not only inhibit the growth of microorganisms but also destroy them. An agent used in low concentration is sufficient, when working in the optimum pH range. Sensitive ingredients also need a stable, constant pH. Various reactions such as hydrolysis, oxidation, esterification, and ester cleavage can alter the pH, thus reducing the effectiveness of preservation and creating a more pleasant environment for microorganisms. Therefore, the use of a buffer system (pH 5.0–5.5) has certain advantages.

During manufacture, deviations in the temperature profiles, in the order of recipe ingredients, or in the dissolution of substances as well as an insufficient energy input lead to a destabilization of the emulsion. Nonoptimal type and amount of emulsifiers, reactions supported by heavy metals, UV radiation and or oxygen, temperature fluctuations, changes in pH, and microbial contaminations over time contribute to the disintegration. On the one hand, the stability of an emulsion depends on creaming/sedimentation as well of the Ostwald ripening [22]. This describes the disappearance of small droplets, because the bigger ones are more stable. In addition, after some time further aggregations and coalescences stimulate the destabilization.

Careful selection of formulation components, emulsifiers, and preservatives, as well as setting and maintaining a constant pH-value by a buffer system, guarantees the stability of the cream/lotion. Production and filling run under optimized conditions in clean, disinfected equipment, taking into account the recipe-dependent temperature limits. The flushing of all containers with nitrogen is recommended during production and bottling. Furthermore, a packaging (bottle) that is impermeable to UV light, water vapor, and oxygen, and equipped with a removal system without contact to the air and to the product, is needed.

Depending on the chemistry and technology, the distribution expresses the quality of the emulsion. The lower the d_{50} and the narrower the monomodal distribution the better should be the stability. The stability is checked with a centrifuge test (40 °C, 30 000 g, and 5 min). All narrow distributed emulsions that pass the test remain stable over a long period (some years).

In a few cases, the formulation contains fine solids (powder). The particles settle mainly on the surface of the droplets and assist in the stabilization (Pickering emulsions). These and other types of emulsions, similar to the PIT (phase inversion temperature), the micro and multiple emulsions, are fully described in the literature [23, 24] and are not discussed here. They do not play a major role in skin care.

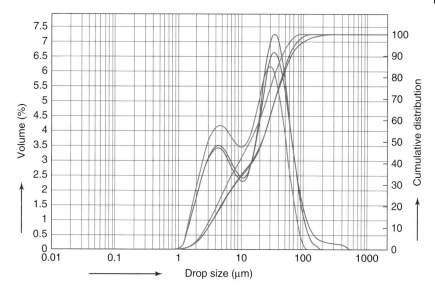

Figure 13.8 Droplet size distributions of a commercially available o/w skin cream, few weeks after production, triple measure after dilution by laser diffraction/equipment: Malvern.

Well-known competing products from the market (creams) have been measured with laser diffraction, to gain an impression of the emulsion stability of a few weeks/months after manufacture. Remarkably, the four investigated samples of different producers show bimodal distributions. The corresponding o/w droplet size distribution is shown in Figure 13.8, and the associated micrograph (Figure 13.4a) shows the visible drops. In addition, a measured w/o market product also shows a bimodal distribution (Figure 13.4b). The larger droplets in the emulsions are in the range of 20–60 µm (maximum 100 µm); they point to an already incipient destabilization. In other images, some big bubbles were considerably more visible. Amazingly, several market products begin to disintegrate after only a few months. Probably, the emulsifiers are inadequately stabilized. The consumer notices the segregation only after complete disintegration, and thereafter pumps only water out of the bottle.

13.5.3
Preparation of Emulsions in the Laboratory

A manufacturing process is developed in a laboratory/pilot plant before transfer to production in accordance with the scale-up conditions. The following laboratory instructions are used to manufacture the emulsion, often described in this or a similar form, but with higher temperatures. First, in a stirred vessel successively dissolve the hydrophilic components at about 45–50 °C in sterile water. Depending on the ingredients and the consistency of the added viscosity regulators, the dissolution process may vary and definitely can take an hour or more to complete.

Next, acids or bases are added to adjust the pH value to 5.0–5.5. AHAs (α-hydroxy acids) such as lactic or citric acid are suitable for this purpose.

To limit the required temperature, it is recommended to formulate the recipes without high-melting organic substances, and also without the use of hard paraffin and waxes. In a separate vessel under nitrogen atmosphere, the lipophilic components dissolve at 55–60 °C up to give a clear phase. The perfume oil runs in at the end or during the emulsification. Immediately after dissolving all components, the oil phase flows into the nitrogen-layered aqueous solution at 45–50 °C. Here, the homogenizer runs for several minutes with varying speeds. A temperature control system ensures that the temperature in the liquid remains constantly at 45 °C. After homogenization and cooling down, the addition of shear and temperature-sensitive materials as aqueous solutions (e.g., peptides) under slow stirring completes the procedure. Finally, the product is filled, at about 25 °C, under nitrogen into the dispensers.

The manufacture of mini-emulsions from macro-emulsions requires, after addition of another surfactant, more than one run through a high-pressure homogenizer at 45–50 °C.

13.6
Structure of Skin Care Creams

Creams/lotions are thickened emulsions; they consist of excipients, additives, and active ingredients. The nature, number, concentrations, and interactions of these substances determine the product design (performance, handling, and aesthetics), and can be influenced greatly.

13.6.1
Excipients

Excipients are substances needed for the manufacturing, application to the skin, and stabilization. The number of excipients listed in Table 13.6 shows that a stable cream needs at least 9, and in some cases up to 20, agents. These include the thickeners for the two phases, emulsifiers, solvents, and spreading agents. In addition, in a cream, pH regulators are required to set the pH of the skin to 5. A buffer system (acid/base) preferably executes the pH constancy. Furthermore, the buffer, the absence of oxygen, moderate temperatures, and UV-light-impermeable dispensers prevent undesirable chemical reactions. The strength of the preservatives depends on the concentration and on the pH value.

The emulsifier system is crucial for the stability of an emulsion, and for the effect on the skin. This persists in o/w emulsions of one or two emulsifiers (surfactants) with HLB values of about 12–17 as well as viscosity regulators for both phases. Nonionic surfactants such as ethoxylated fatty alcohols are effective already at low concentrations (0.5–1.5%). Other emulsifiers need 2.0–5.0% for stability and a low limit of allowable amounts of salt.

Table 13.6 Excipients of a cream according to their functions.

Excipients	Number of common substances	Number of required/recommended substances	Used concentration ranges[a] (%)	Typical substances
Consistency (viscosity control)				
Organic phase	>60	1	2–5	FAs
Aqueous phase	>40	1	0.2–3	Acrylates, cellulose ethers, and xanthan
Emulsifiers				
o/w macro-emulsions	>80	1	0.7–2	FA ethoxylates
w/o macro-emulsions	>50	1	1–3	Sorbitan oleate
Mini(nano)-emulsions	? (<20[b])	2	0.5–3	
Preservatives	56	1	0.1–0.4	Sorbic, salicylic acid, and parabens
Complexing	12	0 or 1	0.2–0.6	Sodium citrate
Antioxidants	15	1	0.2–2.5	Vitamin E
pH control	31	2	0.1–1	Citric acid and sodium citrate
Solvents and solubilizer	>14	1	0.5–4	Ethanol and propylene glycol
Spreading	>17	1	0.5–3	Silicones
Liposomes	>5	0 or 1	<2	Lecithin

FAs, fatty alcohols.
[a] Material and application dependent.
[b] Estimated.

The stabilization of the emulsion happens not only with emulsifiers but also with the thickeners. It is recommended that there are at least two substances as viscosity regulators, a lipophilic and a hydrophilic one. The viscosity of the water phase can be adjusted with synthetic polymers, like modified polyacrylates, as well as numerous natural or semi-synthetic polymers such as starch, xanthan, and cellulose ethers. The hyaluronic acid thickens the water phase extremely, so even tiny amounts are enough to produce, with this well-known skin ingredient, an increase in viscosity. Lipophilic compounds of high MW, as long-chain fatty alcohols, set the viscosity of the oil phase. The final viscosity reached in the emulsification process depends on the optimum amount of thickening agents as a function of type and amount of the ingredients, which is estimated in a series of experiments. The amount and nature of excipients determine the structure of the cream. Furthermore, extensive experience is of help in choosing the most suitable active substances, of the best combinations for the viscosity and for the spreadability and haptic feeling.

The stabilization of polyunsaturated natural lipids requires an addition of antioxidants. Even a slight oxidation of these natural oils results in a clearly perceptible, rancid odor. This undesired reaction, catalyzed by heavy metal ions, should be additionally suppressed by the exclusion of oxygen and the use of complexing agents. After applying the cream, the larger water-soluble molecules remain on the skin. For their transport into the epidermis, proteins encapsulated in liposomes are suitable. Dissolved lecithin in water forms hollow spheres (vesicles), in which the proteins can be stored. The loaded vesicles are able to diffuse along the hair shaft and through the pores into the skin and transport active ingredients to the epidermis.

It is assumed that the classes of nonionic fatty alcohol ethoxylates make the skin more permeable and support the introduction of (harmful) substances, further mobilizing the skin fats. A quote from Paracelsus (1493–1541) may be the right comment to these findings: "All things are poison and nothing is without poison, only the dose makes that a thing is no poison." In particular, the low concentrations, and also interactions and influences of the environment (pH and salts), thickening agent, and lipids, impinge on the effect of surfactants on the skin. Alcohol ethoxylates have the advantage of emulsifying very well at low concentrations, even in the presence of high salt loads. During the application of such cosmeceuticals, no adverse effects on the skin are observed or measured.

13.6.2
Preservations

A cream is manufactured at a low level of bacterial counts. It is not sterile, and so germs are present in minimal amounts. These few germs are usually not a problem. (Caution: with organic solids, microorganisms can be introduced into the water phase.) After opening the package until the complete consumption some months later, bacteria contaminate the content, which renders it useless. Therefore, the utilization of a preservative is required. First, the preservative prevents further multiplication of the introduced microorganisms. Formulated in higher concentrations, it is able to reduce the number of germs. The required concentration of preservative lies mainly in the range of 0.1–0.35%, and depends particularly on the efficacy of the recipe and the packaging. The permissible limits are set from the European cosmetics regulation [9].

In cosmeceuticals, some preservatives are preferred such as the parahydroxybenzoic esters (PHB-ester, parabens), sorbic acid, and phenoxyethanol. Phenoxyethanol represents an ether of phenol with ethylene glycol, which has only a moderate activity spectrum, especially against gram-negative bacteria, and is therefore either formulated in relatively high concentrations (about 0.8%; allowed: <1%) or in combination with other agents. The esterified hydroxybenzoic acid with a methyl, ethyl, propyl, butyl, or benzyl group, works effectively in mixtures (0.1–0.8%), preferably in a combination of methyl- with propyl-4-hydroxybenzoate. The optimal effect lies in the pH range 6–8. The propyl ester is more effective against molds, and is also hydrolytically stable. From the perspective of chemistry, on the one hand, hydrolysis occurs under basic conditions with a pH shift and, on the other hand, the strong oil solubility (=inactivation) reduces possible applications.

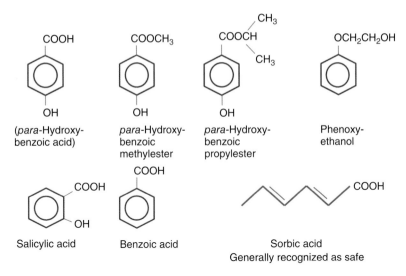

Figure 13.9 Customary preservatives for cosmetic creams and lotions.

Parabens rarely cause allergies but show, compared to sorbic acid, a higher allergy potential. Analyses of breast tumor tissues of women revealed traces of benzyl esters. They are suspected to cause cancer and may no longer be used [25]. Furthermore, butyl ester possibly triggers hormonal changes in fetuses. Methylparaben may promote skin aging in sunlight (UV rays). According to present knowledge, documented in numerous studies, the parabens preserve safely. Nevertheless, they provide a degree of risk under intense usage. Several customary preservatives are shown in Figure 13.9, including the precursor for the PHB-ester.

In the pH range of skin from <5.0 to 5.5, the physiologically acceptable sorbic acid (hexadienoic) applies as the best preservative. This acid disassembles in the body similar to the fatty acids via β-oxidation [26]. For food, sorbic has been given the GRAS (generally recognized as safe) status in the United States. Above pH 5.5, equilibrium exists between the undissociated acid and the salt, with too little free acid, so that the effectiveness decreases. The hexadienoic acid, which is poorly soluble in oil and in water, dissolves satisfyingly in organic solvents. These are present in cosmetic formulations as solubilizers and, therefore, sorbic acid is particularly suitable for cosmetic creams. Sufficient sorbic acid-preserved cosmeceuticals pass the microbiological stress test described in the European Pharmacopoeia [27]. The bacterial count in the inoculated samples is reduced by the preservative within 7 days by at least three orders of magnitude.

Sorbic acid is sensitive to oxidation and builds up gradually in the presence of air over time. Three measures will help suppress the reactions: the exclusion of oxygen, the addition of antioxidants, and stabilizers. Therefore, the cream runs in a container under nitrogen. By using "airless" dispensers, there is no gas atmosphere above the cream. No growth of microorganisms happens in buffered products, even in a long-time application.

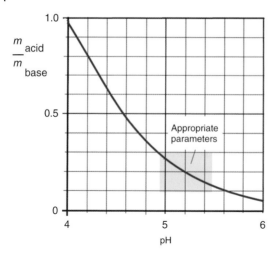

Figure 13.10 Mass ratio of citric acid monohydrate to sodium citrate dihydrate, the best suitable buffer system for skin creams.

The water-soluble potassium sorbate in combination with the buffer system citric acid/citrate (Figure 13.10) preserves well at a pH of about 5. The salt releases sorbic acid, which is stabilized by the buffer system. The buffer provides not only for a constant pH of the cream but also for the correct pH on the skin for a long time, and is important for the effectiveness and durability of the ingredients.

13.6.3
Additives

Chapter 12, especially Sections 12.3.2 and 12.3.3, deals with the perfume oil as an important additive in skin care products. Many customers tolerate fragrance oils, and choose cosmetic products for an individual scent impression. Sensitive people search for well-tolerated and fragrance-free products. Additives (Table 13.7) increase the risk of an allergic reaction on sensitive skin. Therefore, substances for scent and coloring are typical ingredients in "normal" cosmetic products, but are usually undesirable in cosmeceuticals.

Table 13.7 Additives for fragrance and color.

Additives	Number of common substances	Number of required/ recommended substances
Fragrances including essential oils	>2500	0
Perfumes	>1000	0
Dyes	145	0

A number of essential fragrant oils show positive effects on the skin. Examples are chamomile, carrot, rose, rosemary, sage, and cedar. Only special products contain these oils. Their typical, intense smell is often not perceived as pleasant, despite good efficacy. In contrast to cosmetics, the cosmeceuticals normally contain no additives. After several positive experiences, all creams should be free of substances in solid or semisolid consistency, such as natural and synthetic waxes, tallow, solid vegetable fats, paraffin, or Vaseline.

13.6.4
Cosmetic Active Ingredients

The active ingredients are divided into groups (Table 13.8). In groups 1 and 2 are the natural oils and vitamins/provitamins, in 3 and 4 the moisturizing agents and the AHAs. Other added substances (group 5) enhance or generate desired effects. Examples include itch-relieving, anti-inflammatory, keratolytic or especially rejuvenating substances.

Depending on the concentrations, these active materials determine the effect on the skin together with the remaining components. Examples are the AHAs, urea, vitamin A and natural product extracts, and the peptides. In relatively low concentrations, the substances are very useful; in high concentrations, they may lead to undesirable effects. The patch test [28] checked the recipe for compatibility (contact allergy) in susceptible individuals. Low but effective concentrations and/or avoidance of critical ingredients lead to a positive safety assessment [29].

13.6.5
Typical Effects of Cosmetics

Products for the skin protect and care for the body and face, eyes and lips, and hands and feet. The cosmetic industry offers most of these specialties for children, teenagers, women, and men. Some examples of advertised benefits of cosmetic skin care products, especially facial care, are listed in Table 13.9.

Cosmeceuticals have higher concentrations of active ingredients, and they operate fruitfully in combination with drug therapy. In general, they are suitable for all people. Table 13.10 shows some examples of product applications that clarify the differences from cosmetic care.

13.7
Essential Active Substances from a Medical Point of View

Each skin cream should contain at least three essential active ingredients in effective concentrations to maintain a healthy skin: each one selected out of the group of natural oils, moisturizers, and vitamins. The following subsections describe the effect of selected substances on the skin.

Table 13.8 Substances with specific effects on and in the skin.

Group	Active ingredients	Number of required/ common substances	Number of recommended substances	Used concentration ranges (%)	Typical substances
1	Lipids				
	Natural oils	>40	1–3	7–20	Hemp, evening primrose, borage, argan, sunflower, and almond
	Synthetic oils	12	1	<3	Silicone (paraffine)
	Natural or hardened waxes	>20	0	<2	Beeswax and lanolin
	Synthetic waxes	42	0	<2	Artificial jojoba
2	Vitamins				
	Provitamin	13	2 or 3	0.2–2.5	Vitamin A, (C), E, B_3, and B_5
	Dexpanthenol	—	—	1.5–5	D-Panthenol
3a	NMF	16	2–5	0.2–6	Urea, lactate, amino acids, and glycerol
3b	Further NMF, salts, and proteins	>14	2 or 3	0.2–4	Dead sea salts, sorbitol, protein hydrolysates, hyaluronic, and glycose
4	AHA (α-hydroxy acids, fruit acids, and salts)	9	2–4	<1	Lactic, malic, glycolic, and citric acid
5a	Plant extracts	>120	0–2	<2	Allantoin, aloe vera, flavonoide, ginseng, linden blossom, hamamelis, marine algae, marigold, and pineapple
5b	Caffeine substances	8	0	<0.3	Green tea and cocoa
5c	Essential oils	>125	0 or 1	0.3–0.8	Chamomile, grape-fruit, jasmine, lavender, squalene, nettles, geranium, sage, and sandalwood
5d	Biotechnological synthesis	>3	0–2	<0.1[a]	Antarcticine, ubiquinone (Q10), and hyaluronic
5e	Synthetic peptides	>20	0–2	<0.1[a]	Hexa- and tetra-peptides

NMF, natural moisturizing factors.
[a] Active matter.

Table 13.9   Cosmetic skin care, o/w and w/o creams and lotions ("milk").

Skin type	Body part	Advertised benefits
All	Body Face Hands Feet	General skin care and sun protection
All	Eyes	Care in the areas around the eyes
Normal skin (also sensitive and mixed skin)	Face (body)	Moisture, vitalization; cleaning with body milk; (matting) care with UV protection; and care with self-tanners
Dry skin	Face	Against rough, dry skin feeling, and moisture supply
	Body Hands	
Older skin	Face (body)	For firmer body and/or facial skin; against wrinkles, and reduces wrinkle depth
Impure skin	Face	For clear and beautiful skin

Table 13.10   Effects of cosmeceuticals [30].

Skin type	Body part	Offered effect
All (also for sensitive and dry skin)	Body	Care by simultaneous supply of moisture, lipids, and (pro)vitamins
	Face Hands Feet	
Acne	Face	Reduces the impact
All	Nose	Cares and reduces snoring
Neurodermitic areas	Body	Care by simultaneous supply of moisture, lipids, and specific (pro)vitamins; contains substances to destroy microorganisms
	Face Hands Feet	
All	Feet	Destroys fungi
All	Face	Reduces lady beard
Older skin	Face	Care and firming by the simultaneous supply of moisture, lipids, (pro)vitamins, and several anti-aging ingredients
Older skin	Body	For the prevention of pressure ulcers

13.7.1
Linoleic and Linolenic

First, among the essential ingredients are natural oils that contain bound linoleic and γ-linolenic fatty acids. The human body cannot produce either of these fatty acids. Therefore, they must be supplied from outside. A deficiency leads to disruption of the barrier function of the skin with a significant increase in transepidermal water loss. Linoleic is a di-unsaturated (C18-2, ω-6) fatty acid that is an essential part of the epidermis. The top layer of the skin, the stratum corneum, contains ceramides, free fatty acids, and phospholipids. Ceramides exist as lipid bilayers and regulate the water balance in the skin. Linoleic represents the largest share of the essential ceramide 1. Second, linoleic acid supports the elimination of skin irritation after topical application (contact dermatitis) and reduces light damage to the skin and age spots.

As important is the triple-unsaturated (C18-3, ω-6) γ-linolenic acid, which is missing in neurodermitic skin, and represents a raw material for the group of tissue hormones (prostaglandins [6]). A lack of these hormones, which are involved in cell metabolism, leads to rough, dry, cracked, and itchy skin. Natural oils such as borage, evening primrose, hemp, and black currant, and the oil from the seeds of pomegranate, supply the skin with linoleic and γ-linolenic. One of these natural oils (triglycerides) in amounts of 5–20% should be included in good skin cosmetics. The composition of fatty acids bound to hemp oil corresponds most closely to the skin's fatty acids.

13.7.2
Urea

In dermatology, one of the most important natural moisturizing factors is urea. It binds water in the upper layers of the skin, reduces the transepidermal water loss, and contributes to the elasticity of the horny layer. Urea acts as an anti-inflammatory, an antibacterial, and in higher concentrations also as an antipruritic and keratolytic. On application to open wounds, this substance – as acids – initiates a short burn. Urea is nontoxic and well tolerated.

Essentially, compared to healthy skin, urea is lacking in dry skin by up to 50%, in psoriatic skin up to 60%, and in neurodermitic up to 70%. These missing amounts must be supplied from outside to the skin. Not only in supporting a therapy but also as a precaution, active creams contain urea with 2–4%, and thus are also suitable for children.

13.7.3
Panthenol

The water-soluble provitamin D-panthenol [31] penetrates into the skin and reacts in the epidermis to give pantothenic (vitamin B_5). Pantothenic is the main component of coenzyme A, which controls in the skin some metabolic reactions of fats, carbohydrates, and proteins. Therefore, topically applied panthenol enhances the formation of new skin cells and thus measurably promotes regeneration of the skin. Furthermore, this "panacea" improves moisture retention and elasticity of the skin, soothes itchiness, and shows anti-inflammatory and wound healing properties. For use in cosmetics, apply the recommended amount of 1.5–3%, and in specialty products for wound healing apply up to 5%. Panthenol is stable at pH 5 up to 45 °C.

13.8
Penetration into the Skin

To optimize skin care products, it is essential to study the penetration of lipophilic and hydrophilic substances, to examine the evidence, and to derive rules for the formulation. In this way, the product design is controlled by measurements.

13.8.1
Skin Structure

An explanatory picture of the structure of the skin is given in the cross sections shown in Figure 13.11. The thickness of each layer depends on location, age, weight, and gender. Values for skin thickness, such as those found on the forearm, characterize typical skin areas. Depending on the activity of the fat and sweat glands, and on the current cell condition, the skin surface is covered by a film made of water, electrolytes, polar substances, lipids, urea, and amino acids. This film shows acidic properties and, therefore, is called the *acid mantle*, or because of its moisture-preserving function it is also known as the *natural moisturizing factor*. It provides an additional protective barrier against microorganisms.

The cornea consists of a network of lipid double membranes and dead cells (keratinized corneocytes) and renews permanently from the basal cells of the epidermis. The basal membrane delimits to the directly underlying dermis. Figure 13.12

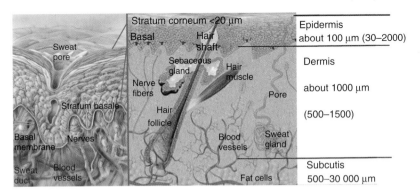

Figure 13.11 Structure of skin. (Unlabeled photos courtesy of COGNIS (Skin Care Forum 29 and 34).)

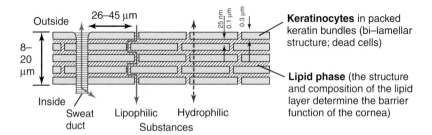

Figure 13.12 Proposed models for the structure of the stratum corneum [32].

shows a conceptual model of the construction of the stratum corneum. This concept shows that lipophilic substances easily penetrate through the lipid phase ("mortar") into the lower layers of the epidermis. The water-soluble substances must diffuse through the corneocytes ("stones") and meet, thereby, again on lipid phases, making it difficult to penetrate deeper. A simpler option is the diffusion through sweat-filled channels and pores.

13.8.2
Applying the Emulsion

The emulsion is applied to the cornea. The surfactants of the cream emulsify the acid mantle, so that the agents are directly in contact with the skin and diffuse more easily. The processes on the skin that expire after application of the emulsion, especially three actions, are shown schematically in Figure 13.13.

Process 1 stands for the stationary state of the skin, showing the acid mantle and a slow drying. In the second process, the applied cream forms a layer on the skin, heated to body temperature. Notably, the surface area is increased greatly by applying the cream. Parts of the film dry by circulated air, whereby the emulsion breaks down. Dehydration (process 3), negatively affects diffusion into the skin. Without changes on the skin, the new steady state of the acid mantle is reached in about 12–24 h, depending on the recipe. Movement, sweating, rubbing of laundry on the skin, and cleaning the body all disturb or remove the acid mantle temporarily.

13.8.3
Proof of Performance

Comparison of different skin parameters before and after application and over a long period proves the performance (power) of the skin cream/lotion. Any distortion of results is avoided through a statistical design with about 15–25 (in special cases, 50) persons. First, a performed test for skin tolerance excludes primary irritant properties. This happens either in patch tests or in the open application tests. The test aims to exclude the possibility of allergic reactions over a period of 1 to about 5 weeks, under realistic conditions or intensified by a provocation test. This test is particularly sensitive and is important for allergic skin types [28].

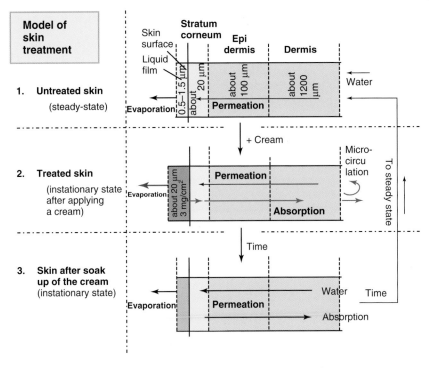

Figure 13.13 Processes that occur on the skin after applying an emulsion.

Various standardized measurement methods are available (Table 13.11) to assess the barrier function, moisture, surface structure, and lipid content in the skin. On one hand, the values obtained allow a precise statement about the skin condition. On the other hand, the results disclose improvements after use of the care product with statistical certainty.

Figure 13.14 shows an example of the determination of skin moisture after a single application by a capacitive measurement over a period of 48 h with over 20 volunteers. Suitable water-soluble substances improve the skin moisture significantly by binding/adsorption of water in the epidermis. Charge-free, low MW substances can penetrate into the skin, especially in the case of a hydrated cornea, although the cornea constitutes a barrier to water. The diffusion paths run transepidermal, transfollicular along the hair follicles, or transglandulär through pores.

The fine chemicals manufacturer and commodity traders usually give some evidence of the effects of formulated substances, for concentrations and incompatibilities with other substances, pH values, and temperatures. The desired effect is targeted and the recipe is optimized toward the measured skin parameters ("product design"), including the exclusive use of natural products. For each desired effect, there are several more or less effective alternative substances from nature.

Table 13.11 Determination of skin characterizing parameters with different apparatuses [33].

Measuring device	Measurement
Corneometer	Skin moisture
Sebumeter	Skin surface lipids
Cutometer	Skin elasticity
Foits (fast optical *in-vivo* topometry of human skin)	Skin surface, structure, and wrinkles
TEWL	Barrier function
pH	pH on the skin surface
Ultrasonic	Skin thickness
3D image analysis of silicone replicas	Microrelief (fold width and depth)
CLSM and OCT [34]	Structural changes with age (fibrous structure)

TEWL, transepidermal water loss; CLSM, confocal laser scanning microscopy; OCT, optical coherence tomography.

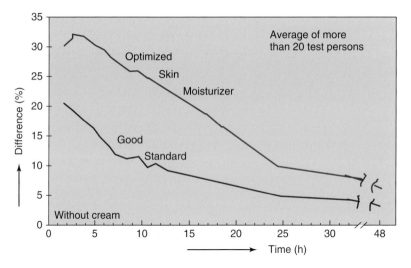

Figure 13.14 Measurement of skin moisture as differences (percentage) with a corneometer: application of the cream at time $t = 0$; capacitive measurements of the treated skin minus measured values of the untreated skin.

13.8.4
Penetration of Lipophilic Substances

A specialized NMR (nuclear magnetic resonance) spectroscopy [35]), conceived and performed by Dicoi *in vivo* on volunteers, tracks the diffusion of precious natural oils into the skin. Particularly suited is the mobile NMR-Mouse® [36] in conjunction with a newly developed and assembled device to hold up the arm.

The equipment changes the position of the measuring arrangement in small steps according to the desired measuring depth. The measuring areas used are on the left and right forearm. Each measurement field produces slightly different values for the untreated skin. The measuring system produces data approximately every 15–30 μm over the skin depth. Each measurement takes about 1 min (65 s) to complete. More recently, both the step size and the measurement times have been reduced.

Evaluations of the NMR measurements succeed by calculating the proton densities, which are based on the maximum value of the untreated skin in the epidermis. Furthermore, the response time t_2 of the signal identifies, after calibration, the substance (water: 12.2 ms and natural oil: 88.1 ms). The relevant test results come from curves of the treated minus untreated skin. One example of supplying the skin with pure natural oils (mixture of almond oil with borage) is displayed in Figure 13.15. After 10 min, the supernatant oil on the skin was removed. The measured values [4] indicate that these oils diffuse very quickly in considerable amounts into the epidermis. About 40 min after application, there is still a significant effect in the epidermis. In the meantime, most parts of the oil have penetrated into the deeper layers of the dermis.

Figure 13.16 shows, graphically, typical responses, first for the untreated skin and second for the skin after applying a miniemulsion. After a 60 min move in of the oil, the curve indicates that the emulsion is significantly more effective than the pure oil. By adjusting the viscosity of the oil droplets, they diffuse significantly slower through the epidermis. In this way, time remains for action and for filling of the fat cells.

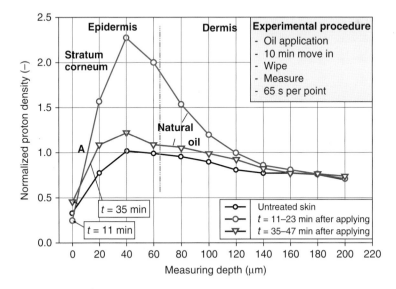

Figure 13.15 Normalized proton densities as a function of the measured depth in the skin after treatment with a mixture of natural oils (triglycerides).

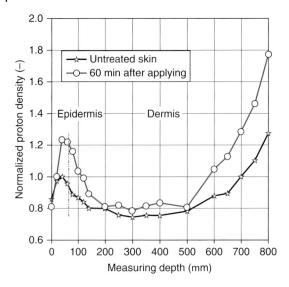

Figure 13.16 Treatment of the skin with an o/w mini-emulsion; differences between the two curves caused by diffused natural oils on the left forearm are clearly visible in the stratum basal of the epidermis and in the deeper layers of the dermis. (Measuring depth zero for the first value can be ~25 µm inside the skin.)

In sections, calculated according to Fick's law, the curves emerged under the assumption of a thermodynamic equilibrium at the basal membrane. Already after a few minutes, high values of the oil appear in the epidermis near the basal layer and in the bottom layers of the dermis (reticular dermis and elastic fiber network layer). In the lower layers of the dermis, the values increase for some hours before the oil phase values slowly drop to normal levels. The measurements show the same reproducible patterns, in the case of mini-emulsions ($d_{50} = 120$ nm) with open, as in practical application, as well as with covered application without water evaporation for theoretical considerations. Comparable values for the untreated skin originated from magnetic resonance imaging (MRI) measurements [37].

The examined inter- and intracellular water contents in the epidermis depend on the measuring depth in the skin [38]. The maximum values are in the stratum spinosum and basal, with inter- and intracellular water contents of comparable sizes.

On covering the skin area after applying the cream, for several formulations both macro-emulsions (d_{50} about 2 µm) and mini-emulsions (d_{50} about 120 nm) show comparable values of the oil content in the skin. In practical applications, both emulsions disintegrate quickly, but at different speeds, owing to water evaporation. (The velocity of the water evaporation, also measurable by using the NMR-Mouse [39], was not determined.) After applying a macro-emulsion, the rapid evaporation leads to a reduced diffusion and a normal oil content in the skin (only one formulation tested and, therefore, the result is not meaningful). The macro-emulsion seems to produce significantly increased oil values only in the

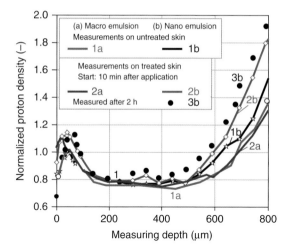

Figure 13.17 Treatment of the skin with a macro-emulsion (curves 1a, 2a) and a mini-emulsion (1b, 2b, and 3b): the measurements were made 10 min after applying the cream on the left and right forearm; measuring time per curve about 25 min.

stratum corneum, but not in the dermis (Figure 13.17), as observed after applying mini-emulsions. Applying larger quantities causes probably a stronger increase of the oil concentration in the skin. The effects are mass dependent. Measurements on w/o emulsions are pending; owing to the thickening of the oil phase, slow mass transfer is expected.

13.9
Targeted Product Design in the Course of Development

The three cases examined differ dramatically in practical application: pure oil is unpleasant on the skin, and penetration through the skin too fast for healing actions. Better effects are produced with creams. According to the measurements, macro-emulsions preferably reside in the upper layers of the skin, while natural oils of a mini-emulsion penetrate into deeper areas (dermis). The application of creams is easy; both water- and fat-soluble substances diffuse in a controlled manner into the skin without a greasy feeling. The lipophilic phase should contain key ingredients for the skin, such as di- and triglycerides with high levels of double or triple unsaturated acids (ω-6 fatty acids and C18), possibly also the acids themselves, and squalene (triterpenes). Added oil-soluble vitamins, particularly radical suppressors, support the reduction of the visible effects of skin aging. (The penetration behavior of major water-soluble substances cannot be assessed because of lack of appropriate *in vivo* investigation methods.)

The task of developing specific effects for skin care products can be solved chemically because there are unlimited possibilities to mix components both in the type and quantity. According to Figure 13.18, preoptimization of the technology starts with a basic formula in combination with proven emulsifiers and viscosity regulators in the first cycle. During further optimization, it is advisable to examine from time to time the emulsification by measuring the droplet size distribution, by testing the stability with a centrifuge, and checking the preservation by microbiological tests.

Physical measurements on the skin and the simultaneous application tests allow feedback concerning the use of active ingredients in the type and quantity (cycle 2,

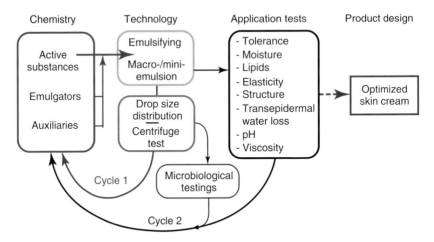

Figure 13.18 Cycles for optimizing product designs.

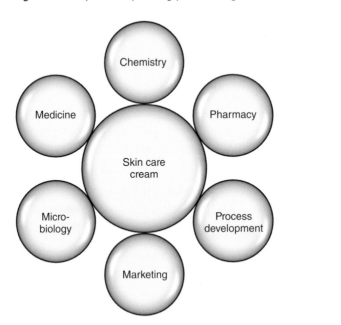

Figure 13.19 Cooperation with various departments in developing new formulations for skin care products.

Figure 13.18). Furthermore, there is the possibility of setting the properties of the developed skin care product. Using this optimization method in combination with known active ingredients, several skin properties are ultimately better than the market standard in terms of, for example, skin moisture, elasticity, barrier function, and surface structure. Targeted product designs presuppose laboratory equipment, measuring apparatus, and internal and external resources to test the skin properties of probands. In addition, such development is not possible without the knowledge and skills of different specialists (Figure 13.19) working synergistically together in this process. In the end, the product goes to market, and for the next cream the same optimization process is started.

13.10
Production of Skin Care Products

For the safe production of cosmetic products, especially emulsions for skin care, four points must be followed precisely, concerning the plant design, materials, and the process procedure:

1) Note the GMP regulations.
2) Ensure hygiene.
3) Select suitable materials with treated surfaces.
4) Optimize the plant design.

Modern production facilities are arranged according to the rules of GMP. The GMP guidelines [40] ensure quality and hygiene in pharmaceutical, food, and cosmetic industries. They refer in particular to raw materials, plant equipment, rooms, air/water, packaging, and finished products as well as to the staff. Laboratory controls and documentations secure the traceability of each batch.

In the cosmetics industry, where the cosmetic-GMP is valid, production takes place in largely sanitized facilities (pharmaceutical and biotechnological industries: sterile). Personnel entry into the production area is only possible after passing locks, in compliance with the GMP requirements of cleanliness and protective clothing. Owing to structural measures for the premises and for the production plant, cleaning and disinfecting inside and outside are generally easy. For disinfection of the inside surfaces against germs, bacteria, fungi, spores, and viruses, several liquids that contain chemicals such as peracetic acid, hydrogen peroxide, sodium hypochlorite and alcohols, quaternary ammonium compounds, and glyoxal, in acidic or alkaline solution, or mixtures thereof are suitable.

Guidelines for hygienic processing in the food industry are published by the "European Hygienic Engineering and Design Group (EHEDG)," founded 1989. The American 3-A Sanitary Standards organization has a similar mission and common interests, as well as the NSF International. EHEDG actively supports European legislation, which demands hygienic processing and packaging of food in hygienic machinery and in hygienic premises (EC Directive 2006/42/EC for Machinery, EN 1672-2, and EN ISO 14159 Hygiene requirement).

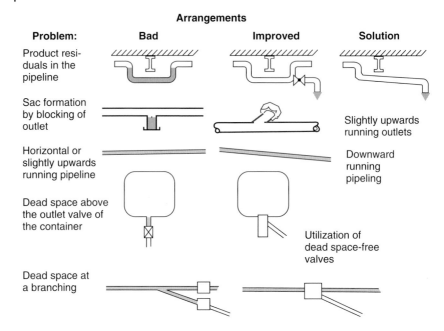

Figure 13.20 Hygienic requirements for the installations.

In the food, pharmaceutical, and in the biotechnological (see Chapter 10.5) as well as in the cosmetic industries, these instructions help find and install suitable equipment and facilities. The most essential point is the prevention of microbiological infections. This target requires a number of costly measures concerning the premises, machinery, materials, and design:

- All installations must be free of dead spaces (Figure 13.20), and show a product flow from the top downwards. Preferably, they are completely drainable at the deepest point.
- Only stainless steel with smoothed surfaces in all parts should be used (Figure 13.21).
- There should be precise (automatic) welding of stainless steel to meet the hygienic requirements.
- Simple cleaning and disinfection of the plant in all parts is a must, preferably with an automatized facility (Figure 13.22).
- The premises should be tiled or sealed with epoxy resin.
- The design of sensors should be hygienic.
- Hygienic water should be provided by membrane filtration and UV-irradiation.
- The air should be cleaned by sterile filter (recommendation).

Owing to the high hygiene standards, the plant must consist of stainless steel with highly polished walls inside, dead space-free valves, and be equipped with spray nozzles for disinfecting and cleaning (CIP, cleaning in place) [41]. The system should be emptied easily and completely. Directly before each batch, a disinfection

Rotating spray devices

Thyphoon   Twister   Tempest

Double seat valve   Single seat valve   Mixproof butterfly valve

Centrifugal pumps (self-priming)

Tank bottom valve   Mixproof shuttle valve   Modulating control valve   Spray balls

Figure 13.21 Dead space-free valves and pumps for hygienic production, and equipment for automatic cleaning of tanks. (Courtesy of GEA/Tuchenhagen.)

cycle with 2-propanol is run through the entire production equipment. In modern plants and filling lines, filters remove germs from the air, at the intake of outside air and/or in the room by air circulation. Sensitive recipes require a low excess pressure of nitrogen during manufacturing and filling. The filling takes place preferably in "closed" systems, under purging with nitrogen, or in special cases with germ-free air. A thorough cleaning and disinfection of the entire system including the filling line follows after completion of each batch. The batch-wise manufacturing of creams and lotions is standard in the cosmetic industry.

The whole chain of production, from raw materials to the finished cream in a dispenser, expires in compliance with the cosmetic-GMP guidelines [40, 42] under careful documentation of all substeps. The heart of the production plant is a temperature-controlled vessel for emulsifying, which is equipped with agitators and with a rotor–stator machine. One stirrer works close to the wall with a scraper. Most often used is the anchor and helical ribbon stirrer or special structures, partially heated/cooled. If the vessel contains two stirrers, the second one runs counterclockwise, is smaller, and located centrally or in the field of half the radius. Common vessel sizes are 50–4000 l, and in exceptional cases up to $10 \, m^3$. All wetted parts consist of polished stainless steel (material numbers AISI 304 and 316, see Chapter 16, and Section 10.5.1), to reduce surface roughness.

In the manufacturing, the water phase with all water-soluble substances, and the oil phase with the lipophilic components (free of waxes) are heated separately under

Figure 13.22 Cleaning in place: (a) preparation of cleaning solutions in a small CIP-plant; (b) principle of the cleaning during production; and (c) a large-scale production plant showing the numerous dead space-free valves; (1) concentrated solutions; (2) solutions in the concentration of application; (3) heat exchanger; (4) valves; and (5) heated solutions to the emulsification plant. (Courtesy of GEA/Tuchenhagen.)

stirring in vacuum at temperatures of about 45–60 °C. The addition of temperature-sensitive substances occurs after cooling of the emulsion. O/w emulsions arise by an energy-intensive division of the oil phase in an aqueous phase, in the simplest case by direct energy dispersion in a stirred tank. Here, driven by a metering pump, the lipid phase flows directly into the rotor–stator homogenizer, which works deep inside the vessel within the two-phase mixture. This arrangement,

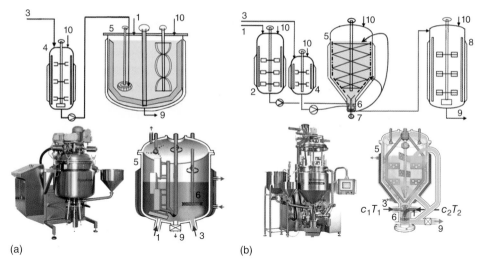

Figure 13.23 Discontinuous production of cosmetic emulsions: (a) "all in one" emulsifying vessel and (b) flanged emulsifier with recirculation of the product; (1) raw material supply of water and hydrophilic substances, (2) stirred tank for dissolving the hydrophilic substances and controlling the temperature, (3) raw material supply of lipophilic substances, (4) stirred tank for dissolving lipophilic substances and controlling the temperature, (5) stirred emulsifying vessel, (6) rotor–stator machine, (7) three-way valve, (8) temperature-controlled stirred tank as a reservoir for filling, (9) to the filling line, and (10) nitrogen supply to remove air and/or vacuum connection. (Images Courtesy of FrymaKoruma.)

shown in Figure 13.23a, is the easiest and, except for an o/w-cream or lotion, sufficient execution of production. The vessel employed for emulsifying also serves to regulate the product temperature and in some cases also as a reservoir for the filling plant. For rapid emulsification, addition of the oil to the water phase directly within a separate multistage rotor–stator machine is more effective, because this action reduces the residence (manufacturing) time by increasing the local power density. In these plants, the different designed emulsifiers [43] work either under the vessel in the outlet flange area (Figure 13.23b), or are externally mounted in a cycle. According to the level in the vessel, the emulsion flows, selectable, through the small and/or large circle. To decouple the energy input from the product flow, a metering pump in front of the external mounted rotor–stator machine is advisable.

The lower the throughput and the higher the viscosity of the continuous phase, the more energy dissipates by the gear rims of the homogenizer, lowering the danger of coalescence. The stirring arrangements for high viscosities in conjunction with commercially available emulsifiers normally meet the requirements for the manufacture of an o/w-skin cream or lotion, with viscosities of about 5 ± 3 Pa s, measured at room temperature. In practice (Figure 13.24), most creams are produced in facilities that are analogous to Figure 13.23b. To clean the apparatus, the lid opens hydraulically and lifts the stirrer (Figure 13.24a).

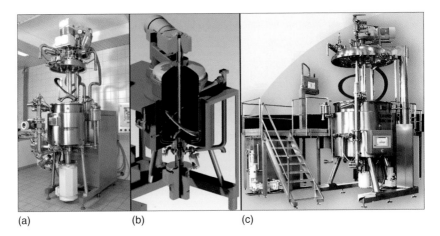

Figure 13.24 Plants for the manufacture of creams/lotions: (a) the top-driven agitator; (b) the bottom-driven rotor–stator dispergator and the cycle; and (c) 2000 l production plant. (Courtesy of Ekato Systems GmbH.)

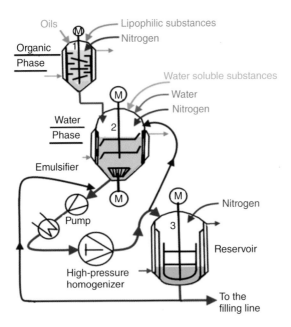

Figure 13.25 Batch plant with a high-pressure homogenizer to produce o/w mini-emulsions; (1) stirred vessel for the preparation of the organic phase, (2) aqueous dissolving and emulsifying tank, which is also the reservoir for the first and third run through the homogenizer, and (3) reservoir for the second run and for the filling line.

The production time depends on the batch size, recipe, and technology. A batch, starting with weighing and dissolving the substances up to discharging into the stirred reservoir of the filling line, lasts about 5–10 h, including the heating and cooling. The actual homogenization in the stirred tank takes about 5–25% of the total completion time. To realize a mini-(nano)emulsion [44], the macro-emulsion should preferably run three times (in some cases, five) through a high-pressure homogenizer [45]. A product cooler in front of the homogenizer controls the

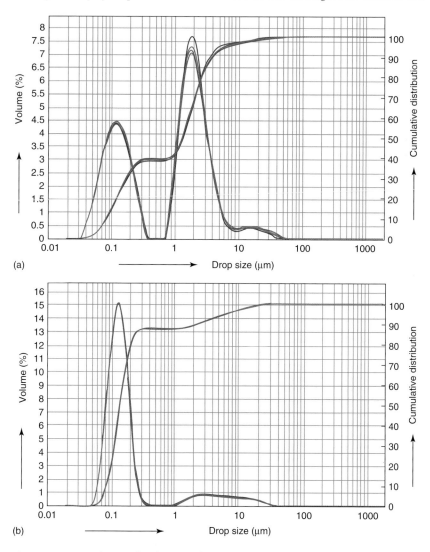

Figure 13.26 Droplet size distributions of a skin care product after (a) one and (b) two runs through a high-pressure homogenizer (the same medium with triple pass; see Figure 13.6); three times measured after dilution by laser diffraction/equipment: Malvern.

temperature to about 45 °C (Figure 13.25). To produce a nano-emulsion, the three runs through the homogenizer require, depending on plant design, an additional 2–4 h, and a total schedule of about 6–14 h.

For the nano-emulsion, two alternately used stirred tanks ensure the processing of the batch. The first and the third, which run through the high-pressure homogenizer, start from the emulsifying vessel, while in the second pass through the homogenizer the emulsion flows back from the reservoir of the filling line to the emulsifying tank. After the third stage, the cream runs from the stirred reservoir to the filling line, where the bottling happens.

Figure 13.26 displays the drop size distributions after the first and second pass through a 600 bar homogenizer. Owing to the high pressure differences, the microorganisms burst [46]; the product is completely sterile in every case. After the first run, numerous nano-droplets arise, and the drops remain in the size of a macro-emulsion. Already in the second run, both the extremely fine droplets below 50 nm and most of the droplets outside the micro-range disappear. The third passage leads to a close, stable distribution with very few or no micrometer drops. On measuring the volume of the drops, a few big drops show a large impact on the distribution. No significant progress (Figure 13.27) arises if, first, the pressure increases from 600 to 800 bar. Second, the measured values for three passes are within the error limits.

Figure 13.28 shows the working principle in a valve of a high-pressure homogenizer. The pressurized (macro-)emulsion is pressed through a narrow gap. Here, the pressure drops off, whereby the emulsified drops break up into very small (nano)droplets. The gap (pressure) is adjustable and regulates the throughput. An industrial high-pressure homogenizer with five pistons, used up to 1500 bar, is shown in Figure 13.29. For the considered working pressure of 600 bar, a throughput of $9500 \, l\,h^{-1}$ is possible for one run, which means about $3000 \, l\,h^{-1}$ for the process under consideration of three passes.

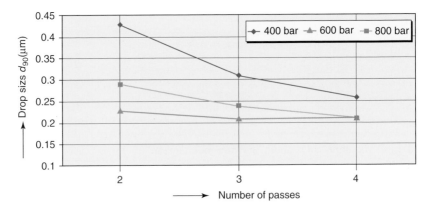

Figure 13.27 Diameter d_{90}-drop size as a measure of the width of the droplet size distributions after several runs through a high-pressure homogenizer (series of measurements on different days with freshly prepared macro-emulsion).

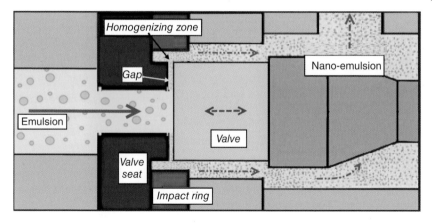

Figure 13.28 Principle of the homogenizing at a valve with adjustable gap in a high-pressure homogenizer.

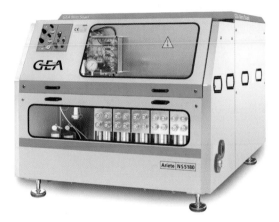

Figure 13.29 Industrial high-pressure homogenizator for the production of nano-emulsions. (Courtesy of GEA Niro Soavi.)

There are several technical possibilities for producing an emulsion [18]. In the cosmetics industry, emulsification takes place preferably in rotor–stator machines. The alternative emulsification by ultrasound with the same recipe leads to droplet size distributions that lie somewhere between the two possibilities, "Ultra-Turrax and high pressure" (d_{50} about 350 nm, but depend on the surfactants). The emulsification techniques, many more or less only applied in pilot plants, can be summarized by the terms *rotor–stator machines, colloid mills, pressure valves, jet disperser, membranes,* and *ultrasound and high-pressure machines/equipment*. Owing to the rotor–stator principle, different mills emulsify well. The specific energy input is $\sim 10^4 - 10^{10}$ J m^{-3}. Lower values correspond to macro-emulsions, and higher values to mini-emulsions.

13.11
Bottles and Prices of Cosmetic Creams

In the skin care market, there are many suppliers with well-known brands from perfumery and fashion. Their products are centered primarily on odor. Even very expensive creams use formulations with known components, as all manufacturers have access to the same commodity market and literature. In general, the products consist of a base material, according to the intended use. In good creams, a very expensive material causes special effects, and supports the advertising.

The following calculation is based on 50 ml of a facial cream. Depending on the formulation and purchased quantities, the cost of raw materials for the base material including the manufacturing lies approximately at €3–8 kg^{-1}. High-grade synthesized hexapeptides with skin-firming effect cost in solution about €2000 kg^{-1}. However, the peptide generates an additional cost of about €70 kg^{-1} creme, using the average recommended amount of 3.5%. High-quality glass crucibles cost €1.00–3.00 (€2.00) including coloring and labeling. Thus, the calculation of a 50 ml jar with high-quality content achieves a value of €5.80 {0.3 + 3.50 + 2.00 (3/2013)}, when ordering many thousand pieces, and about €6.80 for smaller quantities. For simple recipes, the raw material costs amount to €0.15 for 50 ml. Plastic crucibles or bottles (from China) are available for €0.05–0.1. The production costs, including the filling, determine only a small portion of the total costs. They depend strongly on the quantity and lie at €0.05–0.3 per bottle. An estimate for a simple care cream by ordering more than ten thousand 50 ml bottles shows €0.35 (0.15 + 0.07 + 0.13). That means, there is a broad price range from <0.35 to about €10 for the production of 50 ml units. The sale prices in shops lie between <€2 and €300. In addition, there are the allocated development costs. Marketing costs represent the largest item, and often exceed the production costs by far.

Figure 13.30 shows well-known brands. High-price creams are filled preferably in glass jars, although these receptacles have hygiene issues.

In the cosmetic creams market, the following container types (Figure 13.31) are common: for small quantities (10–50 ml) the tube, little bottle, or crucible; for medium size (25–200 ml) the pump bottle (pump dispenser); and for large size (100–500 ml) bottles with a round opening, also available as an upside standing bottle. The large bottles are sealed with a flip cap or a screw cap. A dosage is possible by shaking or squeezing the bottle. In the airless dispensers, a new generation of hygienic bottles, a piston, or a bag delivers the cream to the outlet. A vacuum forces the movement of the product.

Precious formulations contain high levels of polyunsaturated oils. These oils need protection against chemical reactions. UV radiation initiate not only oxidations in the cream, which take place even in the presence of only small amounts of oxygen, but other chemical reactions (ester cleavage and condensation) also occur. Therefore, the use of UV-light-impermeable bottles is necessary. Depending on the type of container and an overpack, for cosmeceuticals transparent or translucent materials such as glass, polyethylene terephthalate (PET) or polyethylene (PE)/polypropylene (PP) are suitable only in exceptional cases, because the

Figure 13.30 Creams and one oil with famous brands: price guide: 30 and 50 ml jars for €50–300 unit^{-1}. (Courtesy of Douglas.)

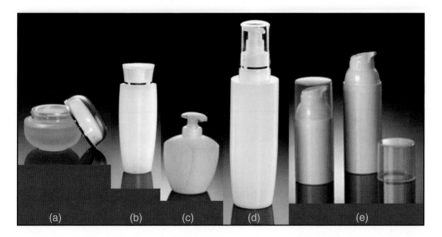

Figure 13.31 Bottles for cosmetic creams: (a) crucible made of glass with a plastic lid, (b) bottle with a screw cap, (c) bulbous pump bottle, (d) slim pump bottle, and (e) airless dispensers. (Courtesy of Pohli.)

sensitive ingredients must survive a storage and consumption time of 2 years and more.

Against UV rays, coloration with plastic dissolvable paints or with incorporated pigments protects the containers. Pigments are suitable fine solids such as titanium dioxide. One of the best materials for the bottles is the white, TiO_2-containing, opaque PP, because it is chemically neutral, physiologically harmless, and shows a smooth, haptically and visually attractive surface. For the properties of the plastics used, gas permeability plays an important role. Depending on the temperature, through high-density polyethylene (HD-PE) only 20–33% of the oxygen is diffused, when compared with an equally strong wall of low-density polyethylene (LD-PE) [47]. PP exhibits comparable values to HD-PE. In both plastics, HD-PE and PP, the water vapor permeability is low. Owing to the exact dimensions of the piston, in the PP airless dispensers thicker walls are necessary, so the gas permeability drops accordingly.

After a careful filling of the almost germ-free cream under nitrogen into the containers, the microorganisms do not multiply until use, even in weakly preserved products. After opening the containers, there are different hazards that infect the cream with germs. High-priced creams, filled in attractively shaped crucibles, meet the ambient air on a large surface. In addition, a major disadvantage is the removal of the cream with the finger. Microorganisms not only reach the cream over the air, but especially through a contact with the fingertips. Therefore, creams in jars must contain higher contents of preservatives. Similar conditions, but with reduced contaminations, occur in the big bottles. In the smaller pump bottles, only the ambient air contacts the cream during ordinary use. Therefore, from a hygienic point of view, pump dispensers are better, compared to crucibles and to big bottles.

The best choice for hygiene and convenience provides the airless dispensers, shown in Figure 13.32. Available in sizes from 15 to 150 in round or oval shape, they offer the greatest protection against contamination during both storage and application. The filling of airless dispensers takes place free of air at room temperature, in most cases from above. Thereafter, the head with the valves is

(a) (b) (c)

Figure 13.32 Airless dispensers: (a) explanation of the function via the cross section: (1) cap (PP), (2,3) head (PP), (4) upper valve (EVA, ethylene vinyl acetate), (5) bellows (LD-PE), (6) lower valve (POM, polyoxymethylene), (7) container for the cream (PP), (8) piston (HDPE), (9) bottom (PP); (b,c) 100 and 50 ml market product after filling from the top: cosmeceutical skin care creams in airless dispensers. (Courtesy of Pohli (a) and ATS License GmbH [30] (b,c).)

pressed on the filled container. The alternative bottom filling requires a separate sealing station. During the cream removal, the piston moves slowly upward. To stop the removal, not only the valves but, in the case of some types, the lips at the outlet also close. A contact of the cream inside the dispenser with the air is virtually impossible. Such dispensers are optimally suited for preservative-free formulations.

13.12
Design of all Elements

For cosmetics, the product design includes the cream formulation and the dispenser. The essential elements are product performance, convenient handling, and aesthetics. The optimal supply of appropriate active substances through the skin is given the highest priority in development. In addition, the brand plays a major role (Table 13.12). The other listed points of an optimal design, including the need for essential marketing elements such as brand name and logo, colors, and lettering, are carefully considered and executed. They contribute to a consistent product appearance, with importance not only for simple cosmetic products but also for cosmeceuticals.

An unpleasant odor, a disagreeable color, or a greasy consistency may be tolerated in therapeutic agents, but are surely not accepted by the users of cosmeceuticals. However, the use of alternative materials or changed processes overcomes such product failings. Optimal aesthetics and dosage with an airless dispenser increase the frequency of application and differentiate products in the market.

Table 13.12   Design elements of active cosmetics.

Elements of design	Cream	Dispenser
Performance (responsible: the composition)	Effect Tolerance Stability	Hygiene
Convenience (responsible: the auxiliaries and the container)	Consistency Productivity (range)	Removal Dosage Form
Esthetics (responsible: the composition and the container)	Color Gloss Odor Stickiness	Color Form Surface Easy pumping
Brand/brand family (responsible: the composition, the container and the label)	Typical ingredients Exclusion of substances (without perfume, dyes, etc.)	Name and lettering Logo
		Label, text, and layout Colors

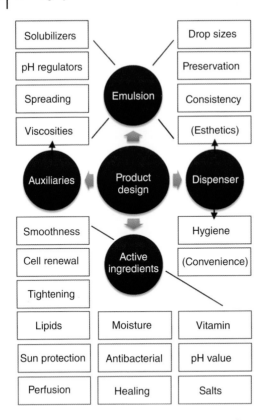

Figure 13.33 Product performance parameters of active cosmetics for the skin.

After applying an active skin cream (Figure 13.33) twice a day for some days, most people show a visible improvement in the appearance of their skin, and also of diseased skin. The results of multiple measurements (before application, after an hour, hours, days, weeks, and months) prove the effectiveness of a skin cream. Typical measurement parameters are the moisture and lipid content, skin texture, thickness, elasticity, and barrier function of the skin surface [48]. The chemist is usually responsible for the development. He or she tailors the recipes with suitable ingredients to solve a specific skin problem, taking into account the medical effects of the active substances. Thus, the gap between cosmetic cream and ointments for the therapeutic benefit is closing.

13.13
Learnings

√ In Europe, the Cosmetics Directive (in the United States, the FDA) regulates the ingredients, manufacturing, labeling, and marketing.

√ Skin care creams and lotions are emulsions.

√ The emulsions consist of several excipients (emulsifier, preservation, and thickener) and active ingredients (lipids, vitamins, and moisturizers), possibly additional substances (perfume).

√ The cosmetic industry produces oil-in-water emulsions for common products, and for specialties the water-in-oil emulsions.

√ The droplets of skin care products lie in the range of one to some microns, in the case of the rarely encountered nanoproducts about 100–250 nm.

√ The area between ordinary cosmetics and dermatics covers the cosmeceuticals. These effective cosmetic products supply the skin with essential lipids, moisturizers, vitamins, and salts.

√ Cosmeceuticals contain more than one active ingredient in higher concentrations. They serve the well-being and maintain the health of skin.

√ Several measurements provide information about the condition of the skin and the positive effect of skin care creams.

√ Lipophilic substances, such as lipids and essential oils, penetrate immediately into the skin.

√ The Good Manufacturing Practice for Cosmetics (GMPC) dictates the premises and facilities, the processing and cleaning/disinfection.

√ The manufacturing ensues batch-wise in polished stainless steel facilities, containing a stirred vessel with a rotor–stator emulsifier.

√ For practical and hygienic reasons, the filling of creams occurs into airless dispensers that exclude contact of the product with air. These bottles are suitable for preservation-free products.

References

1. Stepanovs, J. (1999) in *Skin Saver Remedies* (ed. W. Harald), Tietze Publishing, Australia.
2. Rähse, W. and Hoffmann, S. (2003) Product design – interaction between chemistry, technology and marketing to meet customer needs. *Chem. Eng. Technol.*, **26** (9), 1–10.
3. Rähse, W. (2007) *Produktdesign in der Chemischen Industrie*, Springer, Berlin.
4. Rähse, W. and Dicoi, O. (2009) Produktdesign disperser Stoffe: Emulsionen für die kosmetische Industrie. *Chem. Ing. Tech.*, **81** (9), 1369–1383.
5. Rähse, W. (2011) Produktdesign von Cosmeceuticals am Beispiel der Hautcreme. *Chem. Ing. Tech.*, **83** (10), 1651–1662.
6. Ellsässer, S. (2008) *Körperpflegekunde und Kosmetik*, 2nd edn, Springer-Verlag, Berlin.
7. Bouchemal, K., Briancon, S., Perrier, E., and Fessi, H. (2004) Nano-emulsion formulation using spontaneous emulsification: solvent, oil and surfactant optimization. *Int. J. Pharm.*, **280**, 241–251.
8. Gabard, B. (2000) *Dermatopharmacology of Topical Preparations: A Product Development Oriented Approach*, Springer, Berlin.
9. European Commission (2009) EU Cosmetic Regulation – Consumer Affairs– Text of the Cosmetics Directive, *http://ec.europa.eu/consumers/sectors/cosmetics/documents/directive/* (accessed 7 March 2013).

10. Draelos, Z.D. (2005) *Cosmeceuticals: Procedures in Cosmetic Dermatology*, Elsevier Saunders, Philadelphia, PA.

11. Walters, K.A. and Roberts, M.S. (eds) (2007) *Dermatologic, Cosmeceutic, and Cosmetic Development: Therapeutic and Novel Approaches*, Taylor & Francis, New York.

12. Draelos, Z.D. (2007) The latest cosmeceutical approaches for anti-aging. *J. Cosmet. Determatol.*, **6** (s1), 2–6.

13. Zuzarte, M. *et al.* (2011) Chemical composition and antifungal activity of the essential oils of Lavandula viridis L'Hér. *J. Med. Microbiol.*, **60**, 612–618.

14. Talakoub, L., Neuhaus, I.M., and Yu, S.S. (2009) in *Cosmetic Dermatology* (eds M. Alam, H.B. Gladstone, and R.C. Tung), Saunders, Philadelphia, PA, p. 7.

15. Lupo, M.P. and Cole, A.L. (2007) Cosmeceutical peptides. *Dermatol. Ther.*, **20** (5), 343–349.

16. AMG (1976/2005) Gesetz über den Verkehr mit Arzneimitteln, *http://www.gesetze-iminternet.de/amg 1976/index.html* (accessed 7 March 2013).

17. Lüllmann, H., Mohr, K., and Hein, L. (2006) *Pharmakologie und Toxikologie*, 16th edn, Chapter 2.6, p. 39, Chapter 22, Georg Thieme Verlag, Stuttgart, pp. 258–262.

18. Schuchmann, H.P. (2007) in *Product-Design and Engineering*, vol. 1 (eds U. Bröckel, W. Meier, and G. Wagner), Wiley-VCH Verlag GmbH, Weinheim, pp. 63–93.

19. Mollet, H. and Grubenmann, A. (1999) *Formulierungstechnik*, Wiley-VCH Verlag GmbH, Weinheim.

20. Griffin, W.C. (1949) Classification of surface active agents by HLB. *J. Soc. Cosmet. Chem.*, **1**, 311–326.

21. Mollet, H. and Grubenmann, A. (2008) *Formulation Technology: Emulsions, Suspensions, Solid Forms*, Wiley-VCH Verlag GmbH, Weinheim.

22. Ostwald, W. (1900) *Über die vermeintliche Isomerie des roten und gelben* Quecksilberoxyds und die Oberflächenspannung fester Korper. *Z. Phys. Chem. Stoechiom.*, **34**, 495.

23. von Rybinski, W. (2005) in *Emulgiertechnik* (ed. H. Schubert), Behrs Verlag, Hamburg, pp. 469–485.

24. Umbach, W. (ed) (2004) *Kosmetik und Hygiene*, 3rd edn, Wiley-VCH Verlag GmbH, Weinheim.

25. Andersen, K.E., White, I.R., and Goossens, A. (2011) in *Contact Dermatitis* (eds J.D. Johansen, P.J. Frosch, and J.-P. Lepoittevin), Chapter 31, Springer-Verlag, Berlin, p. 560.

26. Lück, E. and Jager, M. (1995) *Chemische Konservierungsstoffe*, 3rd, Chapter 19 edn, Springer-Verlag, Berlin, p. 158 ff.

27. Sigg, J. (2005) in *Emulgiertechnik* (ed. H. Schubert), Chapter 21, Behrs Verlag, Hamburg, p. 595 ff.

28. Oakley, A. (2008) DermaNet NZ: Standard Series of Patch Test Allergens, *http://dermnetnz.org/dermatitis/standardpatch.html* (accessed 7 March 2013).

29. Rogiers, V. and Pauwels, M. (2008) Safety assessment of cosmetics in Europe, in *Current Problems in Dermatology*, vol. 36 (ed. P. Itin), Karger, Basel.

30. ATS License GmbH (2012) Cosmeceuticals: Product Portfolio, *http://www.ats-license.com/* (accessed 27 November 2013).

31. Ebner, F., Heller, A., Rippke, F., and Tausch, I. (2002) Topical use of dexpanthenol in skin disorders. *Am. J. Clin. Dermatol.*, **3** (6), 427–433.

32. Heymann, E. (2003) *Haut, Haar und Kosmetik*, 2nd edn, Hans Huber Verlag, Bern.

33. Fluhr, J.W. (ed) (2011) *Practical Aspects of Cosmetic Testing*, 1st edn, Springer, Berlin.

34. Neerken, S., Lucassen, G.W., Bisschop, M.A., Lenderink, E., and Nuijs, T. (2004) Characterization of age-related effects in human skin: a comparative study that applies confocal laser scanning microscopy and optical coherence tomography. *J. Biomed. Opt.*, **9**, 274–281.

35. Blümich, B. (2005) *Essential NMR for Scientists and Engineers*, Springer, Berlin.

36. Blümich, B. and Blümler, P. (1999) Verfahren zur Erzeugung von Messsignalen in Magnetfeldern mit einem

NMRmouse-Gerät. DE 199 39 626.4, Aug. 20, 1999.

37. Agache, P. and Humbert, P. (eds) (2004) *Measuring the Skin*, Springer, Berlin.

38. Kemenade, P.M. (1998) Water and ion transport through intact and damaged skin. PhD thesis. CIP-Data Library, Technische Universiteit Eindhoven.

39. Dicoi, O., Walzel, P., Blümich, B., and Rähse, W. (2004) Untersuchung des Trocknungsverhaltens von Feststoffen mit kernmagnetischer Resonanz. *Chem. Ing. Tech.*, **76** (1–2), 94–99.

40. HAS (Health Sciences Authority) (2008) GMP Guidelines for the Manufacturers of Cosmetic Products, December 2008.

41. Gea Process Engineering Inc. (2011) CIP Cleaning-In-Place/SIP Sterilization-In-Place, Internet 2011, *http://www.niroinc.com/gea liquid processing/cleaning in place sip.asp* (accessed 7 March 2013).

42. EFfCI (The European Federation for Cosmetic Ingredients) (2005) GMP Guide for Cosmetic Ingredients, 2005 Revision 2010, *http://www.effci.org/assets/snippets/ file-download/GMP/Archive%20%20GMP% 20Guidelines/EFfCI GMP 2010.pdf* (accessed 7 March 2013).

43. Jafari, S.M., Assadpoor, E., He, Y., and Bhandari, B. (2008) Re-coalescence of emulsion droplets during high-energy emulsification. *Food Hydrocolloids*, **22** (7), 1191–1202.

44. Jafari, S.M., He, Y., and Bhandari, B. (2007) Optimization of nano-emulsions production by microfluidization. *Eur. Food Res. Technol.*, **225** (5–6), 733–741.

45. Gea Niro Soavi North America (2012) High-Pressure Homogenization Technology, *http://www.nirosoavi.com/highpressure-homogenization-technology.asp* (accessed 7 March 2013).

46. Christi, Y. and Moo-Young, M. (1986) Disruption of microbial cells for intracellular products. *Enzyme Microb. Technol.*, **8**, 194–204.

47. Nentwig, J. (2006) *Kunststoff-Folien: Herstellung- Eigenschaften- Anwendung*, Carl Hanser, Munich, p. 100.

48. Förster, T. (2004) in *Kosmetik und Hygiene*, 3rd edn (ed W. Umbach), Wiley-VCH Verlag GmbH, Weinheim, pp. 85–92.

14
Influencing the Product Design by Chemical Reactions and the Manufacturing Process

Summary

In chemical conversions, the raw materials and the synthesis route as well as the chemical reactor(s) with the appropriate operating conditions influence the product design. For the raw material, several origins often exist, for example, via petroleum or other natural resources. The product quality depends in particular on the removal of reaction heat and the residence time distribution. For highly exothermic reactions, the falling-film or loop reactor and the narrow fixed bed (catalytic) reactor are suitable. Furthermore, the fluidized bed with an internal heat exchanger allows such reactions. The types are explained with examples. The final product design, characterized by the particle size, shape, haptic, and smell, arises after isolation and purification in adjusted follow-up treatments (rounding, coating, coloring, and perfuming).

14.1
General Remarks

As shown for polymers, the selective control of the chemical reaction and the design of the reactor allow affecting the properties of the products as well as the formation of byproducts. An example represents polyethylene (PE) (see Section 9.4). The targeted formation of polymers with different densities and degrees of branching ensures chemical reaction engineering. This also applies to the usual chemical reactions, although the possibilities there are considerably limited. Furthermore, the downstream steps are crucial for the quality. The final product results from shaping procedures. Altogether, many parameters constitute the design of chemical products (Figure 14.1). A well-directed influence of the three parameters – performance, convenience, and aesthetics – is possible in limits, especially for solid and multiphase products. The appearance belongs to product design, in addition to the particular formulation or chemical modification. The shape, size, consistency, dissolution, bulk density, color, odor, stability, feel, and finish characterize the aesthetics and influence partly the convenience.

The physics and chemistry of the raw materials determine the product design (Chapter 15), besides the manufacturing process as well as the operation conditions

Industrial Product Design of Solids and Liquids: A Practical Guide, First Edition. Wilfried Rähse.
© 2014 Wiley-VCH Verlag GmbH & Co. KGaA. Published 2014 by Wiley-VCH Verlag GmbH & Co. KGaA.

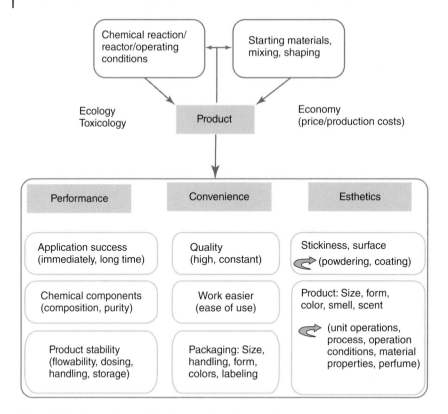

Figure 14.1 Composition of the design of chemical products.

[1–3]. Fine adjustment occurs by the machinery/equipment (machines and apparatus, M&A) and their sequence in the process. In some cases, the optimization requires special treatments. Further discussion takes place with a focus on the chemistry influencing the final product:

- Raw material basis, pretreatments;
- Synthesis route;
- Chemical reactor(s), reaction procedure, and reaction category;
- Isolating, concentrating, and purifications;
- Unit operations and alternatives;
- Posttreatments and their operating conditions;
- Arrangement and design of machinery/equipment;
- Recycling of undersize and oversize, filter dusts, and possibly products out of spec.

14.2
Elements of the Manufacturing Process

Each individual element of the manufacturing process influences the product design. Figure 14.2 displays schematically the emergence of chemical products. The

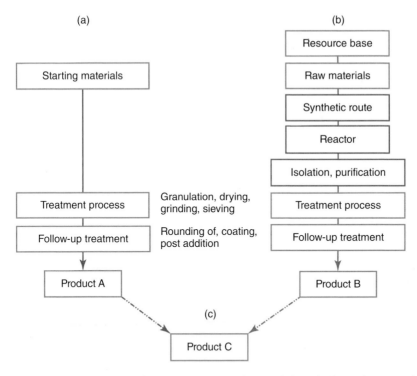

Figure 14.2 Different manufacturing routes (a) without and (b) with chemical reaction (b); the final product may arise (c) by mixing the two.

products originate by mixing and shaping of starting materials, and additionally via chemical reactions. From suitable raw materials, tailored to their use, the desired product arises. The chemistry allows several ways of synthesis in different reactors. Normally in downstream process steps, isolation and concentration as well purification [4] takes place, sometimes also drying, shaping, and additional treatments such as scenting, coloring, or surface coating. Or the chemical substance, mixed with other ingredients of the formulation, runs into an extruder, pelletizer, or granulator. In product design, solvents and by-products only play a role if they remain in the product. In contrast, the recycling streams are of great influence with respect to quality.

14.3
Raw Materials and Synthesis Routes

The product design of a chemical product depends on the chosen synthesis route, and also crucially on the resource base. Listed in Table 14.1 are eight examples that illustrate the statement.

Firstly, the production of vitamin E occurs via synthesis and secondly from natural oils. Only the natural-based manufacturing process leads to the complete right (R+)

Table 14.1 Influences of the raw material and the synthesis route (without polymers)

Product	Raw material base/synthesis route	Characterizing feature
Vitamin E (α-tocopherol)	Petroleum/synthesis	Racemate
	Plant oils	$+(R)$ enantiomer
Surfactants	Petroleum/ethoxylation	C_{13-15}
	Plant oils/ethoxylation	$C_{12,14}$ and $C_{16,18}$
Fatty alcohol	Soya oil/transesterification and hydrogenation	About 85% C_{18}
	Coconut oil/transesterification and hydrogenation	About 85% C_{8-16}
(Human)Insulin	Porcine pancreas	Molecules almost identical
	Biotechnological synthesis	Molecules identical
Fuel	Petroleum/distillation	Gasoline, diesel
	Rape oil/methylation	(Bio)Diesel
Cellulose ether	Cotton, linters	High degrees of polymerization
	Wood/spruce and beech	Average degrees of polymerization
Sodium carbonate ("soda")	Soda lakes	Differences in purity, particle sizes, and bulk densities
	Solvay process	
Titanium dioxide ("white pigment")	Sulfate process	By-products: SO_2 and waste acid
	Chloride process	High purity and short life of the reactor

rotating molecules, which are able to reduce the free radicals in the human body. In contrast, the synthesis leads to racemates (left and right rotating molecules), so only 50% of the product can be effective. In addition, the racemate needs stabilization as acetate (DL-α-tocopheryl acetate). A similar example represents the manufacture of insulin. Currently, the biotechnological products replace the previously used isolates from porcine pancreas. Only this way allows the preparation of insulin that is completely identical to human insulin.

Surfactants, such as the nonionic surfactants, are accessible by synthesis from petroleum components. Petroleum contains all C numbers; therefore, the hydrocarbons comprise even- and odd-numbered carbon chains. For washing, the used synthetic nonionic surfactants have carbon chain distribution between C_{13} and C_{15}. The plant-based products (coconut, palm and palm kernel oil, soybean, and sunflower oil) contain only even-numbered carbon chains. Thus, the natural-based surfactants have $C_{12,14,16,18}$ units. Because the plants provide different raw materials, the surfactants differ in performance among each other and to synthetic nonionic surfactants. In addition, the natural oils show different carbon chains, depending on the plant. While soybean oil has largely bound C_{18} fatty acids at the glycerol, there is a broad distribution in the coconut oil with a maximum at C_{12}.

Rapeseed methyl ester (RME) is a substitute for diesel. The crops, grown on the rape fields, are pressed after harvest. In a transesterification with methanol, the refined oil reacts to glycerol and RME [5]. In Europe, the diesel contains small amounts of RME. The fuel works fine without any change in the vehicle for many years.

For cellulose ethers, having an average degree of polymerization (DP) to 1750, mainly ground cellulose ex spruce wood is in use as raw material. High-viscosity types need cotton linters, because this raw material shows average degrees of polymerization well over 2000, depending on the region. The quality of the etherification (selectivity) depends not only on the raw material but also on the type of mill, the alkalization, and the reactor. The fiber structure of the products differs with the raw material (linters, spruce, and beech wood).

In nature, soda is mined at the edge of soda lakes. In addition, sodium carbonate is synthesized in different variants of the Solvay process. In all cases, the customer gets soda, but the quality differs from one to the other. A distinguishing feature represents the heavy metal content, the other particle size distribution and bulk density. Even in this simple case of an inorganic salt, the products are not identical. For example, the soda for the manufacturing of sodium percarbonate [6] must be almost free of heavy metals (<5 ppm), because otherwise the reactant hydrogen peroxide decomposes.

The final example concerns the manufacturing of a product, starting with the same material, but using different synthesis routes. Titanium dioxide from ilmenite or rutile is an important pigment worldwide, with about 4–5 million $t\,a^{-1}$. The two methods in parallel are the sulfate- (via $Ti(SO_4)_2$) and the chloride (via $TiCl_4$) processes, which should be comparable in the economy [7]. By the ban on dumping of waste acid in the North Sea, the sulfate process was converted. Since 1990, the reprocessing of the waste acid occurs completely.

These examples demonstrate that the chemistry of the raw materials and the method of synthesis as well as the reactor and the operation conditions determine the product properties. The same applies to the various stages of the preparation process and the finishing stages. In some cases, posttreatment may have a decisive influence on the design. It consists of an adaptation of the particle sizes and shapes as well as the surface finish (Chapters 5 and 6). Measurable physical, chemical, and disperse properties characterize the product design exactly.

14.4
Chemical Reactor and Reaction Sequence

In a number of cases, the chemical reactor affects the product quality [8]. Table 14.2 shows some examples. A perfectly chosen and operated reactor enables fully isothermal reactions, avoids back mixing, and allows a reactor dependent macro-/micro-mixing. As known, the optimization of the heat and mass transfer

Table 14.2 Product design as a function of the chemical reactor

Production	Main problem	Reactors	Product characteristics
FAS: From fatty alcohol and sulfur trioxide (gas)	Heat removal, formation of by-products	Falling-film reactor (falling film on the inner surface of the apparatus)	Color (light yellow)
		Cascade of stirred vessels	Color (dark brown)
Fatty alcohol sulfate salts (FAS-Na): From fatty alcohol sulfates and sodium hydroxide	Heat removal	Loop reactor with tubular heat exchanger	Color and odor; active matter in the solution
FA: From the methyl ester by high-pressure hydrogenation	Heat removal	Multitube reactor with catalyst	Color
PA: From o-xylene and oxygen (air)	Heat removal, selectivity	Multitube reactor with catalyst	Color
Zeolite 4A: From sodium aluminate and sodium silicate solution	Back mixing: by-product formation, particle size distribution	Tubular reactor	Little by-product, optimal particle size range
		Multistage stirred column	
		Batch-stirred tank reactor	

FAS, fatty alcohol sulfate; FA, fatty alcohol; PA, phthalic anhydride.

guarantees good results, taking into account the maximum temperature in combination with the residence time (distribution). Inadequate heat removal leads to product changes, which manifests themselves in darker color and amplified odor. This is often observed in the presence of oxygen.

In exothermic reactions, the local heat dissipation represents the most difficult task. Discolorations may occur by too high temperatures ("hot spots"). These changes are not acceptable. Usually, temperature peaks cause yield losses, because the selectivity is temperature dependent. Numerous methods exist for the control of heat: high exchange area per unit volume, internal and external heat exchangers, intercooling, low cooling water temperature, dilution of the reactants, dilution of the catalyst, intensifying the flow, evaporative cooling, changing the type of reaction from masses to suspension, and so on.

The sulfonation of fatty alcohols [9] with an air/sulfur trioxide mixture (liquid/gas reaction) required a long-term optimization to the present-day falling-film process. About 30 years ago, many manufacturers used four-stage cascades with internal and external water cooling. In the first three reactors, the reaction started and proceeded up to 50–70% conversion. In the last stirred vessel, the diluted components reacted slower. Despite strong water cooling, it was not possible to keep the temperature at a level of about 40 °C. The resulting fatty alcohol sulfate got a dark brown color because of local hot spots. Only modern reactors allow nearly isothermal operations. In a plurality of vertically arranged tubes, the liquid flows as a falling film along the inner surface of the pipes. The outer tube surface is intensively cooled. The gas reacts spontaneously at the large surface of the liquid. It results in a yellow/brownish product (fatty alcohol sulfate).

Some neutralization reactions give rise to a great deal of heat. An example is the continuous reaction of fatty alcohol sulfates (sulfuric esters) with concentrated sodium hydroxide solution, a liquid/liquid conversion. For thermal management, the reactants are diluted in a loop reactor with the product and pumped with high speed through a tubular heat exchanger. The speedy mixing happens in a high-speed homogenizer. The machine works according to the rotor–stator principle and pumps the liquid through the loop. Only a relatively small proportion of the product (~63% active matter) flows continuously out of the circuit (Figure 14.3), while the large volume runs through the loop.

In catalyzed hydrogenation and oxidation reactions, gases or a gas with a liquid react in a catalyst bed, fixed or fluidized. In fixed beds, it is possible to reduce the excess temperatures, but not to eliminate them. For this purpose, the diameter of the tubes often is limited to a maximum of 1 in. at a length of 3–6 m. The

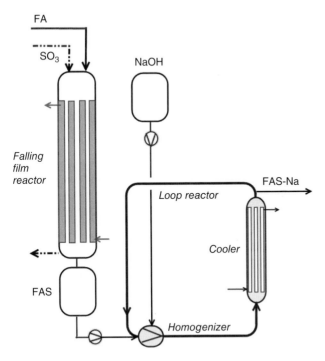

Figure 14.3 Continuous sulfation of fatty alcohols with sulfur trioxide in a falling-film reactor and the neutralization of the originating fatty alcohol sulfate in a loop reactor.

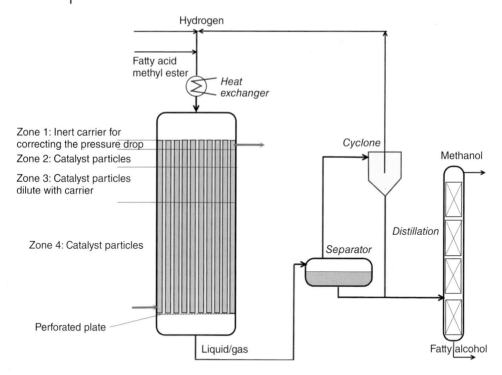

Figure 14.4 Multitubular trickle bed reactor with 6 m tubes, filled with catalyst particles (copper chromite), for the hydrogenation of fatty acid methyl esters to fatty alcohols [10].

tubes are filled with catalyst and inert material (carrier material in the region of the excess temperatures) for dilution at all critical places. Figure 14.4 displays as example of the hydrogenation of fatty acid methyl esters, running at 200–250 °C and 200–300 bar, with excess hydrogen. This catalytic liquid/gas reaction represents a conversion in multitubular trickle beds, which suppresses hot spots. Excess temperatures, arising in shaft reactors, result in undesired by-products. The liquid reaction products, methanol and fatty alcohol, split up in a distillation column before further purification occurs.

More heat leads to the oxidation of *o*-xylene to phthalic anhydride. After evaporation, the gaseous hydrocarbon reacts with oxygen of the air at vanadium pentoxide (V_2O_5) carrier catalysts, modified with titanium dioxide. The multitubular reactor contains many tubes, 3 m long and 1 in. in diameter. Especially in oxidations, the local dilution of the catalyst allows minimization of side reactions.

The manufacturing of zeolite occurs under plug flow conditions. Experiments in the laboratory showed that back mixing promotes the side reactions, and shifts the grain size in the unwanted range of larger particles. Therefore, the selected reactor and the operation conditions must prevent back mixing. For detergents, the production of zeolites (type 4A) happens batch-wise in a stirred tank reactor or continuously in very long plug flow reactors as well as in an arrangement of two stirred columns, divided in compartments. Trials demonstrate that in these

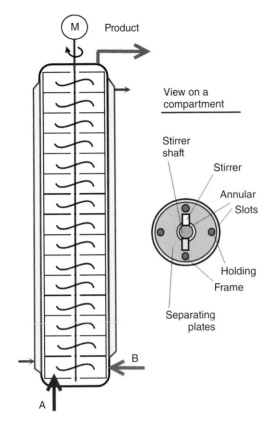

Figure 14.5 Stirred column, divided in 15 compartments, with plug flow behavior for precipitations/crystallizations; flow over at the agitator shaft and on the edge of the compartments.

compartments stirred columns under certain flow conditions allow residence time distributions similar to plug flow reactors. They offer a space-saving alternative to tubular reactors in continuous processing.

In the first column, the precipitation of the zeolite occurs by mixing of sodium aluminate with silicate solution. The plug flow, realized by high axial velocities through 15 compartments (Figure 14.5), suppresses the side reaction to sodalite, and allows an adjustment of the crystallite size in the range of 3–5 μm without large portions of oversize. There are three advantages of the stirred column: compact design, easy setting of the residence time distribution, and forced tangential flow. The efficiency requires short residence times and small areas for the overflow. The Reynold numbers for the axial and radial flow (agitation) describe the flow behavior. The optimal conditions follow from measurements of the residence time distribution and should amount to $Re_a = 1500$–2000 (d, diameter of the column) and $Re_r = 50\,000$–$70\,000$, but depend on the stirrer and the viscosity.

Polymerizations, in particular, illustrate the influence of the reactor (Section 9.4), because the quality depends on the residence time distribution under the operating

Table 14.3 Variants of the product properties by different reaction types (polymerizations)

Polymerization process in	Advantages/ problems	Reactor type	Product
Solution	High dilution, good control of the heat	All types	"Crystals"
Suspension	Catalyst; high dilution, good control of the heat	Stirred tank reactor and cascade	Powder
Emulsion	Droplets as micro reactors	Stirred tank reactor and cascade	Direct production of fine powders
Mass	No work-up process, difficult heat management	Extruder	Granules
		Thin layers on a belt	Powders from grinding

conditions (pressure, temperature, and catalyst) and the possibility of adding some raw material later (copolymerization) [3]. Furthermore, the amounts of the required product decide the choice of a batch or continuous stirred tank reactor, cascade, or plug flow reactor or loop. An example of the influence of the different types of reaction represents the polymerization and copolymerization of styrene. Styrene is one of the most important raw materials for the production of plastics, such as for the polystyrene insulation materials (Styrofoam), acrylonitrile–butadiene–styrene (ABS), styrene–acrylonitrile, and styrene–butadiene rubber (SBR).

Polymerizations run in solution, suspension, emulsion, or in mass (Table 14.3), batch-wise or continuously. Tower method with pre-reactor, tubular reactors, stirred tank, and cascades, alone or in combination, and special designs are in use [11]. Although formally an identical product (polystyrene) always results, the characteristic product properties differ significantly (DP, transparency, impact strength, and density). Furthermore, the particles are not identical in shape and size distribution. Therefore, the manufacturer must find the "right" facilities such as the operation and reactor assembly for each application of the customers.

14.5
Entire Procedure

The chosen manufacturing process and the involved M&A, further the sequence of operations and the recycled quantities, just as the operating conditions determine the product design. In processing of solids, the pretreatment of raw materials (grinding or agglomeration and sieving) influence the quality of the final products (e.g., tablets). Suitable process control allows an optimizing of the quality by

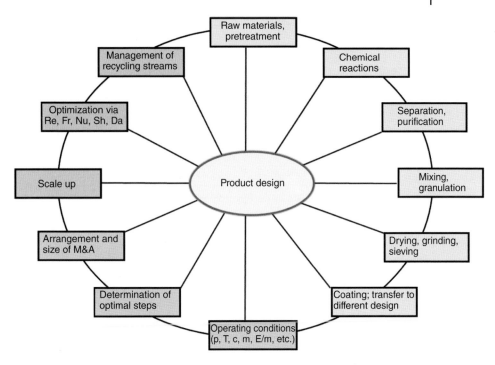

Figure 14.6 Opportunities to influence the product design by process parameters.

adapting the recycle streams in type and amount. If necessary, an additional process step must be inserted. Especially with solid-phase and multiphase procedures, many different operations, and designs are also possible. An optional posttreatment such as a coating, its implementation in detail, the choice of the optimal unit operations plays an important role in product design. In the field of consumer goods, some selectable design parameters exist that in the rule define marketing, such as color or scenting.

Figure 14.6 gives an overview of the dependencies. The manufacturer has the choice: many chemical products or chemical-based consumer products are producible by at least two different processes and many variants of the unit operations. Therefore, it is necessary to determine in advance exactly how the product should look (specification list). The selection of suitable M&As, besides their design and sequence and the operating conditions, show a major influence on the design of products.

The example of sodium percarbonate illustrates the influence of the manufacturing process on the product design. Two methods are in industrial use, namely, first, the crystallization process (dissolved sodium carbonate and hydrogen peroxide) and, second, the fluidized bed spray granulation, where hydrogen peroxide is sprayed on soda under drying conditions. As already shown in Figure 1.4, the design differs significantly. The flowability of the crystallized product is similar to that of the sugar crystals. Changes in the size distribution are limited. However, spray granulation allows the production of very round and dense granules with selectable sizes in a broad range. The products from the two procedures are chemically identical but show differences in the bulk densities, shape factors, particle size distributions, and solubilities.

Another example is the grinding of cellulose ethers. From the small fibers of the cellulose ethers arises in impact mills a product, which is milled only to a small extent, and has a low bulk density. This "woolly" material is not acceptable. The process development showed that compaction before grinding improved the process significantly. Particularly suitable mills compact the fibers and cut them at the same time.

For basic chemicals, process alternatives are common. The production of caustic soda from sodium chloride in the electrolytic chloralkali process occurs after three different procedures: amalgam, diaphragm, and membrane process. The amalgam

Table 14.4 Unit operations and advantageous alternatives.

Unit operation	Operation equipment	Alternative	Advantage of the alternative
Thermal concentration in liquid phase	Normal pressure and vacuum Distillation Evaporator (circulation, vertical tubes, thin film, and falling film)	Membrane procedures Ultrafiltration Nanofiltration Reverse osmosis	Gentle to the product, no thermal stress
Emulsifying	Rotor/stator machine Ultrasound	High-pressure homogenizator	Smaller droplets
Grinding of solids	In gaseous phase Impact mills Ball Roller Vibratory	In liquid phase Bead mills	No thermal stress and smaller particle sizes
Deodorizing	Packed column	Spray in an empty vessel (Section 10.5.3)	Easy cleaning and disinfection
Drying of substances from solutions	Spray tower Fluidized bed dryer Drum dryer	Spray agglomeration in fluidized beds Thin film drying	Suitable for difficult material to be dried
Sieving, screening (<1 mm)	Round or rectangular planar sieves (also drums) Tumbling Vibrating	Classifier in different designs, with four wheels (Section 7.8.1)	Sharper separation
Mixing continuously (liquids)	Inline homogenizer (rotor/stator principle)	Static mixer	No moving parts, for liquids and gases
Heating/cooling of a stirred tank reactor	Jacket Internal coil	External heat exchanger in a loop	Larger exchange surfaces possible
Solids out of a silo	Rotary feeder Vibrating bottom	Rotating square rods	Less blockages
Coating	Solution	Melt	No drying necessary

process, owing to the use of mercury, is no longer used or installed. In all cases, the chemically identical product originates. In the sulfation reaction, a hydroxyl compound (as fatty alcohol) reacts with a sulfonating agent (−C−O−S− formation). Apart from the aforementioned sulfur dioxide, sulfuric acid and fuming sulfuric acid are suitable, and in special cases chlorosulfonic and amido sulfonic. Similar is the case of sulfonation (−C−S− formation). Here, too, sulfur trioxide in a gas/liquid reaction and fuming sulfuric acid in the liquid phase leads to the sulfonic acid group (−SO_3H) of hydrocarbons. The reaction products of the sulfation and sulfonation serve as surfactants in detergents and cleaning agents.

For each task, there are quality-critical alternatives: instead of evaporation, perhaps a chromatography or deodorizing or a suitable membrane process alone or in combination are better solutions. The choice of the best procedure depends on the simplest way to produce the desired shape. Table 14.4 illustrates some unit operations and possible alternatives, which offer benefits.

14.6
Choice of Machines and Apparatuses

The individual steps of the process consists M&A connected in series. According to the brochures, many selected alternatives deal with the same unit operation. Practice from long years of experiences shows that there are great differences in performance and economy. It is crucial to choose the best solution and a competent machinery manufacturer. In process development, the selection of the process and suitable M&A manufacturers is based on the analysis of the qualities, benefits, and costs. Table 14.5 describes some parameters for the selection of an extruder. Many possibilities of the design impede the overview. Therefore, the customer should use a specification list for the purchase. Imposition of such a list requires long-standing expertise and meaningful pilot tests.

The variety of screw geometries in total or the different screw elements plugged together allows for an optimal adaptation to the problem. Finding the right arrangement and process control require extensive knowledge of materials and equipment, because the multiplicity of parameters leads to an almost incalculable number of variants. In addition, a closer study during the selection of the optimal machine shows that some manufacturing-specific features must be considered. These include appropriate granulating and plate-changing devices. For each task, there are several alternatives, but usually only one or two variants, which lead to superior product qualities and trouble-free operation.

Extrusion is one of the design technologies in use, apart from other agglomeration processes. Figure 14.7 shows the most essential. Several are particularly suitable for a special product design. Agglomerates arise from powder by granulating and from solutions by drying. But the moldings from melt are interesting because the tools allow almost any shape. Many aesthetic shapes originate from the extrusion [13] and granulation [15] as well as from spray agglomeration [16] and tableting [17].

Table 14.5   Extruder as an example of different embodiments of a machine [12–14]

Extruder types
  SSE
  TSE, corotating
    high speed
    low speed
  TSE, counter-rotating
    high speed
    low speed
  Closely intermeshing extruders, co- and counter-rotating
  Ko-Kneaders
  Planetary extruder
  Series connections (twin screw or planetary in combination with an SSE)
The co-rotating extruder operates with screw elements, which differ from those of
  counter-rotating extruders
High-speed extruders use other screw elements compared to low-speed machines
Design of the extruder/operating parameters
  Design of the screws (straight, tapered, gradient, distances, angles, gap widths and
  depths, etc.) and the jacks, depending on the zones
  Different screw elements plugged on a shaft (elements for shearing, kneading, mixing,
  conveying, and pressurizing)
        Heatable/coolable hollow screws or full material screws
        By electric bands different heatable sections
        Geometrical dimensions
        Energy input
        Mass flow rate
        Side feeder (additional supply of raw material)
        Automation
        Temperature profile
        Pressure profile
        Design of the extruder head plate
        Design of the holes
        Granulation by cutting

SSE, single-screw extruder; TSE, twin-screw extruder.

The choice of process often depends on the raw materials and the final product as
well on economy.

In particular, the drying process determines the product design. Each type of
dryer provides a particular application [18], such as a gentle drying in vacuum
or under inert conditions. Therefore, the drying takes place in nitrogen or in
superheated steam with energy recovery. Table 14.6 shows some examples of
suitable dryers. For dryers such as "fluidized beds," a variety of types (feed,
bottom plate, geometries, heat exchangers, and flow) and operations (bed height
without/with vibration, air temperatures, and quantities) exist.

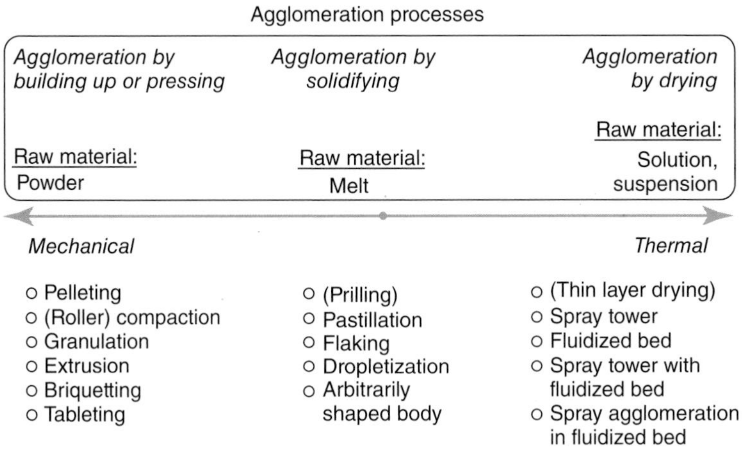

Agglomeration processes

Agglomeration by building up or pressing	Agglomeration by solidifying	Agglomeration by drying
		Raw material:
Raw material:	Raw material:	Solution,
Powder	Melt	suspension

Mechanical ←——————————————————————————→ Thermal

Mechanical		Thermal
○ Pelleting	○ (Prilling)	○ (Thin layer drying)
○ (Roller) compaction	○ Pastillation	○ Spray tower
○ Granulation	○ Flaking	○ Fluidized bed
○ Extrusion	○ Dropletization	○ Spray tower with
○ Briquetting	○ Arbitrarily	fluidized bed
○ Tableting	shaped body	○ Spray agglomeration in fluidized bed

Figure 14.7 Shaping processes by agglomeration (the thermal methods are agglomerations, assessed from the final product).

On the behavior of the solid material in the dryer is based the selection. Plants are preferred, which not only dry but also shape the products or preserve the dispersed condition (fluidized bed). The design and scale-up usually occur by the characteristic numbers. Specific features that are in the know-how of the manufacturer facilitate trouble-free operation. For the order, in addition to the technical solutions, price and service are the key criteria.

For the hot gas, heating via contact surfaces or radiation supplies the evaporation heat. The heating can take place in two ways at the same time, for example, in a fluidized bed dryer directly over the gas and indirectly through internal heat exchangers. For drying, air or air mixed with flue gases is common. In some cases, nitrogen or superheated steam represents an inert drying gas. The product leaves the dryer in the form of powder or granules and flows directly into the packaging. In the case of flaky or lumpy shapes, a compaction with sieving or a grinding/sieving process is necessary before the sale.

Advantageously, two basic operations may occur in a dryer. Examples illustrate some complex processes, such as drying/shaping in spray granulation [19], or simultaneously drying/coating or spray-drying in combination with an agglomeration in the integrated fluidized bed [16]. Furthermore, mill drying influences the choice of the operating parameters, and the recycle streams, possibly coming from the integrated sifter, limit the product design. Thus, it may be advisable to distribute individual basic operations on two or more machines/equipment. For solids, examples may be granulation and drying, extrusion and cutting, and milling with an external sifter. Granulations run often in several stages with multiple recycling streams.

The development process occurs in pilot plants, which must contain the important steps, simulating the later production. For the scale-up of solid processes, a minimum size of the pilot plant is necessary. The particles should be present

Table 14.6   Unit operation "drying" with alternatives for continuous dryers

---

(1) *Drying of moist solids*

Drying of powders and granules
    Fluidized bed
    Paddle dryer
    Mixer with heating/vacuum
    Airlift
    Flow dryer
    Pneumatic dryer
    Tray dryer
    Belt dryer
    Drum, tumble dryer
    Screw dryer
    Plate, disk dryer
    Roller dryer
    Microwave dryer

(2) *Drying and grinding of moist solids*

Drying (e.g., filter cakes) in different mills
    Impact mill
    Ball mill
    Table mill
    Air turbulence mill

(3) *Water evaporation and drying of solutions/suspensions*

Drying after/without preconcentration in membrane plants or with thermal methods
    Spray tower
    Flash dryer
    Mixer (spraying/granulating/drying)
    Fluidized bed (spraying/granulating/drying)
    Spray agglomeration in fluidized bed
    Different mills (see 2)
    Roller dryer
    Paste dryer
    Thin film dryer
    Freeze dryer

---

in their final size and in a representative amount and quality. Ensuring quality requires the inclusion of recycle streams in the pilot scale. Only then is it possible to produce comparable to the production facility.

14.7
Operating Conditions

The influence of important physical parameters (pressure, temperature, time, and energy) on product design ensues by way of some examples. The production of

PE succeeded, as discussed in Section 9.4, either at low (4–40 bar) or at high (500–1500 bar) pressures. The HDPE (high-density polyethylene) with improved elasticity and flexural modulus originates at rather low pressures. The products are not only different in their density but also in other characteristics as thermal expansion or tear strength. The procedure, characterized by synthesis pressure, influences decisively the quality. It runs in solution, suspension, or gas phase [11].

In rectifications, temperature-sensitive liquid mixtures also split up without quality losses. The adjustment of maximum temperature takes place by applying a vacuum/high vacuum. Further, a short residence time ensures stability. In the case of drugs, low temperatures in freeze drying exclude damage by thermal effects. Temperature and time are the main parameters for sterilization of products and equipment or in the pasteurization of milk. Enzymatic processes run almost at room temperature, so that fat splitting performs even at low temperatures.

In many biotechnological processes (see Section 10.5) with aerated, stirred tanks (fermenters), the yield depends on the oxygen supply of the microorganisms and thus on the input energy (2 to $>10\,\text{kW m}^{-3}$). The result of grinding (particle size distribution) shows a great energy dependency. In particular, the production of nanoparticles (down to about 10 nm [20]) needs high-energy inputs in bead mills (Section 7.6). The permeate flux in membrane plants depends on the superficial velocity and therefore on the supply of pumping energy. The residence time as well as the introduced specific energy affect the result of homogenization in extruders, mixers, and dispersers, and reflect in the product design.

Among other factors, the importance of operating conditions can be explained by the examples of the spray drying and tableting. Spray drying with single- or twin-fluid nozzles decides not only the pressure and the particle size distribution but also the operating temperature and the design of the nozzle. For reproducible production of tablets, including the tableting aid, grain size distribution and the moisture of the raw material are of great importance. In addition, the pressure, filling, tableting speed, and temperature must be observed carefully (Section 5.1). Especially for tablets, recycling provides a problem, because the ground material must match the bulk density, particle size, and shape in order to avoid segregation.

14.8
Drying Gas

The choice of process, the operating conditions, and equipment depend on the task. From the wastewater of a biotechnological plant (enzyme production), an organic slow-release fertilizer is producible. After decanting the sludge, the removal of residual water occurs in a disk dryer to a powdery substance, which could be used as fertilizer. However, the strong smell of the product precludes marketing directly to the consumer. Therefore, the developer tried to remove the odorous substances

Figure 14.8 Fertilizer granules from the wastewater of an enzyme factory, made via spray granulation in a fluidized bed (Figure 14.9).

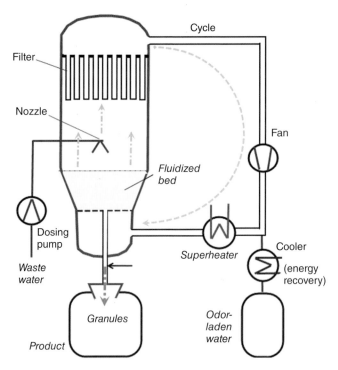

Figure 14.9 Production of fertilizer granules from the wastewater of a biotechnological production in a fluidized bed, using superheated steam in a closed system (principle of the spray agglomeration is shown in Figures 8.23 and 8.24).

with steam, using the effect of water distillation. In fact, the drying performs well with superheated steam at an inlet temperature of about 200–350 °C in a spray granulation [21]. The resulting fertilizer is granulated spherically and generates hardly any odor, that is, the granules exhibit significantly improved product design, as shown in Figure 14.8.

To satisfy the mass balance, the water evaporated out of the spray must be separated. This excess steam with a temperature of about 120–130 °C condensed in an energy recovery system (see Figure 8.6). Because of the complete condensation, there is no exhaust gas, that is, no water vapor and no smells in the environment. The noncondensable fraction, containing the residual odorous air, runs through the burner of the superheater. In this way, the organic constituents decompose into carbon dioxide and water. Whenever products withstand a short-term thermal stress of about 115 °C without problems, the manufacturer should use steam drying, especially for ex-proof production. Table 8.2 demonstrates the manifold advantages of drying with superheated steam instead of hot air.

Drying with superheated steam operates on the one hand at normal pressure. Examples are the spray drying of detergents, the fluidized bed granulation of fertilizers, or the tray drying of wood. Second, numerous processes work under pressure, such as fluidized beds for drying lignite or of beet chips. Other applications are the dewatering of biomass or waste wood (Figure 14.9).

14.9
Learnings

√ Raw material, the synthesis route, and the chemical reactor(s), as well as the operation conditions influence the product design.

√ The design of the chemical reactor must meet all requirements, especially due to the heat removal/supply and residence time distribution.

√ For the lifetime of the reactors, the selection of the material is essential.

√ Falling-film, loop reactor, and narrow fixed bed reactor, further the fluidized bed with an internal heat exchanger allow the execution of highly exothermic reactions.

√ Stirred columns, divided in many compartments, may be a space-saving substitute to plug flow reactors.

√ For most of the unit operations, some alternatives exist. Such procedures are preferred, as they bring benefits to the design and/or to the process.

√ Agglomerations occur during building up and pressing of solids, solidifying of melts, and drying of solutions.

√ Needed for unit operations and reactions, extremely complex individual machines are constructed (e.g., extruders). Pilot trials and expertise help install the "best" solution.

√ Operation conditions may influence the product design, especially in the case of polymerizations.

References

1. Borho, K., Polke, R., Wintermantel, K., Schubert, H., and Sommer, K. (1991) Produkteigenschaften und Verfahrenstechnik. *Chem. Ing. Tech.*, **63** (8), 792–808.

2. Thoenes, D. (1994) *Chemical Reactor Development: From Laboratory Synthesis to Industrial Production*, Springer, Berlin.

3. Winnacker, K. and Küchler, L. (1981) *Chemische Technologie*, 4 Aufl., Bd. 5, Carl Hanser Verlag, München, S. 353 ff.

4. Cussler, E.L. and Moggridge, G.D. (2011) *Chemical Product Design*, 2nd edn, Cambridge University Press, Cambridge.

5. Kaltschmitt, M., Hartmann, H., and Hofbauer, H. (2009) *Energie aus Biomasse: Grundlagen, Techniken und Verfahren*, 2nd edn, Springer, Berlin.

6. Acton, Q.A. (ed) (2013) *Boron Compounds–Advances in Research and Application*, Scholarly Editions, Atlanta, pp. 127–130.

7. Winkler, J. (2003) *Titandioxid*, Vincentz Verlag, Hannover.

8. Nauman, E.B. (2008) *Chemical Reactor Design, Optimization, and Scaleup*, John Wiley & Sons, Inc., Hoboken, NJ.

9. Behler, A. *et al.* (2001) in *Reactions and Synthesis in Surfactant Systems* (ed. J. Texter), Chapter 1, Marcel Dekker, New York, pp. 1–44.

10. Behr, A., Agar, D.W., and Jörissen, J. (2010) *Einführung in die Technische Chemie*, Spektrum Akademischer Verlag, Heidelberg.

11. Keim, W. (ed) (2006, 2012) *Kunststoffe: Synthese, Herstellungsverfahren, Apparaturen*, Wiley-VCH Verlag GmbH, Weinheim.

12. Rauwendaal, C. (2001) *Polymer Extrusion*, 4th edn, Carl Hanser Verlag, München.

13. Hensen, F. (ed) (1997) *Plastics Extrusion Technology*, 2nd edn, Carl Hanser Verlag, München.

14. Kohlgrüber, K. (ed) (2008) *Co-Rotating Twin-screw Extruders: Fundamentals, Technology, and Applications*, Carl Hanser Verlag, München.

15. Salman, A.D., Hounslow, M.J., and Seville, J.P.K. (eds) (2008) *Granulation*, Elsevier, Amsterdam.

16. Pietsch, W. (2002) *Agglomeration Processes: Phenomena, Technologies, Equipment*, Wiley-VCH Verlag GmbH, Weinheim.

17. Mohos, F. (2010) *Confectionery and Chocolate Engineering: Principles and Applications*, John Wiley & Sons, Ltd, Chichester.

18. Mujumdar, A.S. (ed) (2007) *Handbook of Industrial Drying*, 3rd edn, CRC Press.

19. Uhlemann, H. and Mörl, L. (2000) *Wirbelschicht-Sprühgranulation*, Springer-Verlag, Berlin.

20. Breitung-Faes, S. and Kwade, A. (2007) *Einsatz unterschiedlicher Rührwerkskugelmühlen für die Erzeugung von Nanopartikeln. Chem. Ing. Tech.*, **79** (3), 241–248.

21. van Deventer, H.C. (2004) Industrial Superheated Steam Drying. TNO-report R 2004/239 (via Internet), Forschungsvereinigung für Luft- und Trocknungstechnik, Frankfurt/Main, Heft L 202.

15
Design of Disperse Solids by Chemical Reactions

Summary

In extraction and processing of materials from natural resources, reactions between ground raw materials and fluid media (gases, liquids) occur. Starting materials are mineral and fossil materials such as ores and coal, and renewable resources. In model ideas ("shrinking core"), the starting particles shrink during reaction. In one case, the solid phase forms gases and disappears completely. In the other case, the solid retains its size during the conversion, forming a product shell around the core. The starting particle inside becomes smaller, until it is completely consumed by the end of the reaction. The design of shaped particles depends on the source, type, amount of impurities, pretreatments, and the disperse properties of the raw materials. However, raw materials such as powders, granules, or fibers, show a completely different composition after reaction, but largely maintain the size and shape of the starting material. Other disperse properties such as bulk density, density, and flowability change with the reaction. Minor components, contaminants, and residues of the raw material remain in the product. A further influence on the product design (cleaning, shaping) requires downstream processes. If, however, the product is a melt, product design becomes relatively simple, involving cleaning, additional materials, or forming by tools and in air.

15.1
Importance of Solid-State Reactions

Industrially relevant reactions of liquids and gases with solids occur in metallurgy and chemistry. These uncatalyzed reactions take place during mining of natural resources as well as during their concentration, purification, and finishing. This preferentially affects mineral and fossil materials (ores and coal) as well as renewable resources (wood and agricultural products). The wide application of solid-state reactions in the exploration of natural resources explains the overriding importance of these reaction types. The capacity amounts to several hundred million tons per year.

Industrial Product Design of Solids and Liquids: A Practical Guide, First Edition. Wilfried Rähse.
© 2014 Wiley-VCH Verlag GmbH & Co. KGaA. Published 2014 by Wiley-VCH Verlag GmbH & Co. KGaA.

This chapter discusses the technically relevant fluid/solid (g/s, l/s) reactions. The product design [1], that is, the necessary treatment of the raw materials regarding breaking down, purification, and enrichment, is of further interest. In addition, possibilities of influence [2] of the reactions on the purity, size, shape, and structure of formed particles are of significance. Various examples demonstrate whether and how chemical reactions affect product design.

15.2
Theoretical Bases

On grinding minerals as well as most other solid raw materials, finely divided particles obtained prior to any reaction. This improves the accessibility for fluid reactants and increases reaction rate (greater surface area). Usually, an inorganic solid reacts with fluids (liquid and/or gas). Reaction products may be solid, fluid, or solid and fluid. The conversion of inserted particles changes the chemical structure, often also the crystal structure, and other characteristic properties. In other cases, the solids disappear completely because the products formed are gaseous or liquid. The starting materials originate from natural resources. Therefore, after the reaction, the remains left are gait, impurities, by-products, ash, or slag.

Figure 15.1 displays the formal equations of some reactions from fluids with solids. According to conventional ideas, the particles shrink during the progress

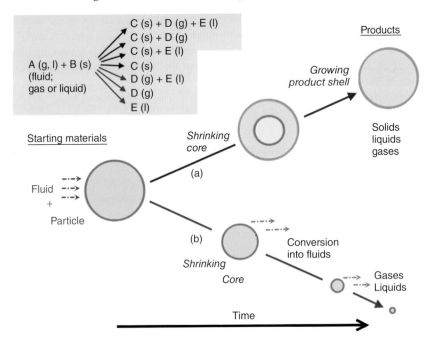

Figure 15.1   Shrinking-core model for spherical particles: shrinking solid cores as starting material for formation of products in gas/solid or in liquid/solid reactions.

of a reaction. Starting from the particle surface, in the first case (the upper route a) a solid product forms inside a growing shell. The core (raw material) continues shrinking until it is completely converted. Here, the particle volume remains constant. The gases that are possibly formed escape. In the second case (the lower route b), the nucleus shrinks and disappears during the reaction either by formation of gases or liquids or by dissolution of the reactive particle into the liquid phase. The two mechanisms [3] are often observed in practice.

15.3
Modeling of Isothermal Solid-State Reactions

For understanding the processes in a fluid/solid reaction, there are concepts that are about 50 years old [4], which correspond approximately to reality. However, mathematical equations based on models from chemical engineering describe reactions of nonporous solids with fluids, preferentially with gases. These models work with ideas of shrinking particles [3, 5–7], in which the feedstock disappears. The derived equations link turnover with reaction time.

Two physical equations are customary, namely, assuming constant or decreasing particle volume. In the first case ("unreacted core of unchanged size" [3, 4, 8], see Figure 15.2a), gases diffuse first through a boundary layer surrounding the

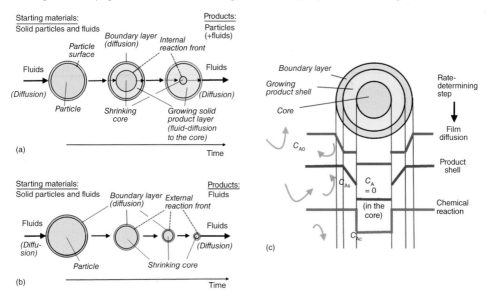

Figure 15.2 Mechanisms of different reactions from nonporous particles with a fluid media: (a) constant-volume particle with shrinking core; formation of solid and often also of fluid products; (b) shrinking/disappearing particles with formation of fluid products; and (c) concentration profiles for gas A as a function of the rate-limiting step with an internal reaction front (c_{A0}: concentration in the external phase, c_{As}: concentration at the surface of the product shell, and c_{Ac}: concentration at core surface).

particles, and thereafter react on the surface of starting materials. This shrinks the cores. Between particle surface and reaction front (core surface) emerges a shell (also called *ash*), consisting of reaction products. In addition, fluid substances must diffuse through the formed shell to the reaction front, before a further chemical reaction occurs. Consequently, diffusion paths are extended in the course of the reaction. Under certain circumstances, this may change the location of a rate-determining step.

The fluids formed in the chemical reaction must diffuse away from the core surface through the product shell and boundary layer. It is only then that the fluids can enter the surrounding continuous phase and flow away. The cores (chemistry of the raw material) gradually disappear, usually completely. An inwardly growing shell (chemistry of the product) replaces the material. The usually coreless particles that arise show sizes and shapes comparable to the starting material. This means that it is possible to influence the particle design via the starting particles. Liquid/solid reactions [9] often exhibit this mechanism.

In the second case, fluids react with particles with consumption of the starting material ("shrinking-core" models, see Figure 15.2b). In the reaction of the particles with gases, usually gases are formed. Liquid reaction products require the use of liquid reactants. The fluid substances first diffuse through the boundary layer and then react on the particle surface of the starting material with a decrease in particle size. In contrast to the first model, the diffusion paths through the boundary layer remain short in the course of the conversion. Fluids, which are formed on the surface, diffuse through the film back into the continuous phase.

Largely or partially, a reactive solid may be encased or covered with ash and minor components. These components are liable to form on particles and hinder diffusion. They drop away as the reaction progresses, when the nuclei become small. Through progressive conversion, the solid particles shrink more and more and finally disappear completely. These "shrinking-core" models contain only

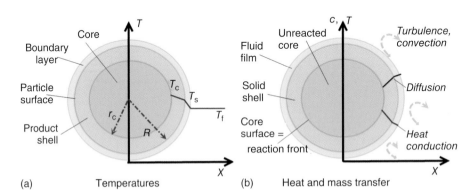

Figure 15.3 (a,b) Model concepts for temperature profiles in particles, mass and heat transport in reaction with a liquid (sprayed on), and transport resistance in the product layer: (temperatures: T_c, temperature at reaction zone on core surface; T_s, on the particle surface; T_f, in the fluid phase; and R, the radius of particles with core radius r_c).

boundary layer diffusion and chemical reaction. The slowest step determines the rate of the reaction. As this type of reaction only forms gases or liquids, there is no product design.

Different mathematical formulations result as a function of the particle shape (sphere, cylinder, and plate) and particle behavior during the reaction (constant or shrinking size) as well as location of the rate-determining step (boundary layer, product shell, core, see Figure 15.2c); these are particularly simple in the isothermal case [3, 10]. In reality, particles show shape factors from 0.65 to 0.85, which lies somewhere between a sphere, cylinder, and plate. This is confirmed by photomicrographs of sodium carbonate (see Figure 15.13).

15.4
Modeling of "Adiabatic" Solid Reactions

Numerous reactions of gases or liquids with dispersed solids do not take place isothermally, rather they are show behavior that is nearly adiabatic or in the transition area between an isothermal and adiabatic behavior. Figure 15.3 demonstrates some of these concepts. The temperature at the core surface (T_c) is significantly higher than the temperature of the surrounding gas (T_f). Therefore, undesirable side reactions may occur at the reaction zone. The heat transfer takes place via heat conduction in particle, which is rather low, depending on the fluid medium and the material of the solid and its porosity. Calculation of gas/solid and liquid/solid reactions ensues by nonisothermal "shrinking-core" models [11, 12]. As a result, the excess temperature is inversely proportional to the particle diameter as well as proportional to fluid concentration and to reaction heat. Furthermore, it depends on the constants for the diffusion, heat conduction and reaction. Mixing of the fluid and convection occur around the particles.

15.5
Reactions of Disperse Particles in the Industry

The tables in the followings sections contain typical examples of industrial gas/solid and liquid/solid reactions. These occur in fixed or in liquid phases (suspension). The different examples illustrate the multiple uses of these technologies.

15.5.1
Conversions with Gases

Many reactions between particles and gases occur in fluidized bed reactors, frequently at temperatures between 600 and 1100 °C. The reactors are of standard design or constructed as circulating fluidized beds, which show predominantly round bottom plates. They are built with large dimensions (Figure 15.4), especially

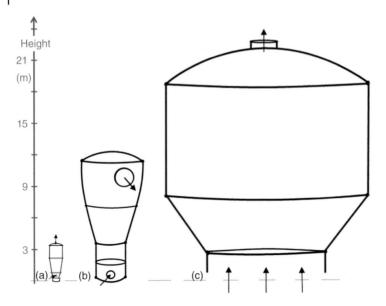

Figure 15.4 Round fluidized bed apparatuses for various industries: (a) batch apparatus for the pharmaceutical industry, diameter of the bottom plate d_B about 1 m; (b) continuous system for the chemical industry, d_B maximum about 3 m; and (c) continuous reactor for fluidized-bed roasting of ores, d_B 8–14 m, partly also with rectangular base and recirculation.

in the case of roasting plants. The bottom plates, which act as gas distributors, have an area up to 150 m². In the past, Lurgi Metallurgy built a number of such plants. After acquisition of this in 2001, Outotec (Finland) now provide these facilities.

Table 15.1 illustrates some interesting reactions in large fluidized beds. The reaction of sphalerite (ZnS) and pyrite (FeS_2) with oxygen from air forms the corresponding oxides and sulfur dioxide. Oxidic particles are similar in size to the sulfidic raw material (constant-volume particle), so the particle size distribution remains stable. After catalytic conversion to sulfur trioxide, the arising gas dissolves in sulfuric acid and after dilution with water yields commercial sulfuric acid. Sulfuric acid plants are an integral part of the production sites wherever sulfur dioxide occurs as a reaction product.

For removal of sulfur and carbon in the roasting of gold ores, some fluidized bed reactors run in large-scale plants with capacities of more than $1\,000\,000\,t\,a^{-1}$ for one unit. The gold-bearing, crushed ore rock does not change either in shape or in size. For the production of copper, the mineral chalcopyrite ($CuFeS_2$) is first ground. Enrichment is carried out by flotation before preroasting in a fluidized bed. After further reaction steps and electrolytic refining, an alloying and shaping takes place in downstream melting processes (Equations 15.3–15.6). The gas/solid reaction of first roasting, can in no way affect the design of the final product.

Table 15.1 Reaction of disperse solids with gases.

No.	Industrial processes involved

(a)   Roasting and baking of zinc sulfide (sphalerite) and iron sulfide (pyrite; FeS_2)

$$2ZnS(\text{ore}) + 3O_2(g) \rightarrow 2ZnO(s) + 2SO_2(g) \tag{15.1}$$

$$\{ZnCO_3(s) \rightarrow ZnO(s) + CO_2(g)\} \tag{15.2}$$

(b)   Roasting and desulfurization of ores, containing copper, nickel, lead, or gold

$$6CuFeS_2(\text{ore}) + 10O_2(g) \rightarrow 3Cu_2S(s) + 2FeS(s) + 2Fe_2O_3(s) + 7SO_2(g) \tag{15.3}$$

$$\{Fe_2O_3(s) + C(s) + SiO_2(s) \rightarrow Fe_2SiO_4(s) + CO(g)\} \tag{15.4}$$

$$2Cu_2S(s) + 3O_2(g) \rightarrow 2Cu_2O(s) + 2SO_2(g) \tag{15.5}$$

$$\{Cu_2S(s) + 2Cu_2O(s) \rightarrow 6Cu(s) + SO_2(g)\} \tag{15.6}$$

(c)   Reduction of iron oxide pellets with carbon/carbon monoxide (blast furnace process)

$$C(s) + O_2(g) \rightarrow CO_2(g) \tag{15.7}$$

$$C(s) + CO_2(g) \rightarrow 2CO(g) \tag{15.8}$$

$$Fe_2O_3(s) + 3CO(g) \rightarrow 2Fe(s, l) + 3CO_2(g) \tag{15.9}$$

(d)   Coal gasification with oxygen/steam for production of synthesis gas (carbon monoxide and hydrogen)

$$C(s) + H_2O(g) \rightarrow CO(g) + H_2(g) \tag{15.10}$$

(e)   Calcination and sulfation of limestone $CaCO_3$; (cement $= CaSiO_3$, CaO, $CaSO_4$, etc.)

$$\{CaCO_3(s) \rightarrow CaO(s) + CO_2(g)\} \tag{15.11}$$

$$\{CaCO_3(s) + SiO_2(s) \rightarrow CaSiO_3(s) + CO_2(g)\} \tag{15.12}$$

$$CaO(s) + \frac{1}{2}O_2(g) + SO_2(g) \rightarrow CaSO_4(s) \tag{15.13}$$

continued overleaf

Table 15.1 (*Continued*)

No.	Industrial processes involved

(f) Purification of silicon (99%) in production of solar silicon (Siemens process)
First step: reduction of silica with carbon at 2000 °C to silicon
Second step: conversion of silicon into trichlorosilane (followed by repeated distillation)
Third step: decomposition of purified trichlorosilane on pure silicon pellets (reverse reaction)

$$Si(s) + 3HCl(g) \rightleftarrows HSiCl_3(g) + H_2(g) \tag{15.14}$$

(g) Purification of titania by chloride process
First step: ground ore (rutile), mixed with fine coal, reacts with chlorine gas at 1000 °C in a fluidized bed reactor to form titanium tetrachloride
Second step: cooling, condensation, and distillation of the tetrachloride
Third step: decomposition (burning) of titanium tetrachloride with pure oxygen

$$TiO_{2(s)} + C_{(s)} + 2Cl_{(g)} \rightarrow TiCl_{4(g,l)} + CO_{2(g)} \tag{15.15}$$

$$TiCl_{4(l,g)} + O_{2(g)} \rightarrow TiO_{2(s)} + 2Cl_{2(g)} \tag{15.16}$$

(h) Conversion of solid uranium tetrafluoride with fluorine gas to the hexafluoride (a step in the enrichment of uranium)

$$UF_4(s) + F_2(g) \rightarrow UF_6(g) \tag{15.17}$$

(i) Production of metallic catalysts from the oxides
For example, iron- or nickel-based catalysts for Fischer–Tropsch synthesis

$$Fe_2O_3(s) + 2H_2(g) + CO(g) \rightarrow 2Fe(s) + 2H_2O(g) + CO_2(g) \tag{15.18}$$

(j) 1. Electrolytic refining of crude nickel
2. Preparation of pure nickel from Ni powder by reaction with carbon monoxide (50–80 °C) and subsequent decomposition of the carbonyl at 180–200 °C

$$Ni(s) + 4CO(g) \rightleftarrows Ni(CO)_4(g) \tag{15.19}$$

Some metals, metal oxides, and metal halogens form carbonyls at high temperatures and pressures, decomposing at higher temperatures to the metal, Fe, $CrCl_3$, VCl_3, $TiCl_4$, WCl_6, $MoCl_6$, Re_2O_7

Except for a small amount of slag, the particles disappear in the process of coal gasification. In this reaction, carbon from coal is predominantly converted to carbon monoxide (Equations 15.7 and 15.8). In a blast furnace, carbon monoxide reduces iron oxides, whereby the resulting iron occurs as melt because of high temperatures (Equation 15.9). Different starting materials and surcharges allow innumerable variations in product quality that are set by a metallurgist. Obviously, the melting processes with the possibilities of shaping affect the product design. This represents a very positive example of product design.

Partial oxidation of carbon to carbon monoxide and production of synthesis gas (Equation 15.10) from ground coal are examples for shrinking and disappearing of starting materials by forming gaseous products. Gases possess no product design, but only a possible enclosure. In the course of cement production, calcination, and sulfation of limestone (Equation 15.13) proceed in parallel. Some lime particles from the raw material react partly with oxygen and sulfur dioxide to form gypsum. By further sintering, this material forms the appropriate cement with lime and adjuvants such as clay, sand, and iron ore.

The preparation of pure silicon to solar silicon according to the Siemens process [13] starts with a conversion. First, crushed or powdered silicon reacts with hydrogen chloride gas to form gaseous trichlorosilane in a fluidized bed at 350 °C. Thereby, the solid starting materials shrink and disappear, while the impurities remain. Liquefied and repeatedly distilled trichlorosilane separates itself at about 1000 °C on pure silicon rods. This occurs in an endothermic batch reaction of the gas phase. In a newly developed method, deposition of Si pellets (polysilicon) is performed in a continuous fluidized bed [14] with hydrogen as fluidizing gas. The heat of decomposition originates convectively from the hot walls and from intensive radiation.

Purification of titania ores represents a reaction between two solids and a gas (see Figure 15.5d). Rutile or finely ground ilmenite, enriched by a flotation, in a mixture with coke reacts in a fluidized bed with chlorine gas at 1000 °C (Equation 15.15) for purification of titania. In this procedure, titanium tetrachloride and carbon dioxide occur. After cooling, titanium tetrachloride is condensed. Purification ensues by repeated distillation. In the process stage that follows, pure titania arises from titanium chloride at high temperatures ($\sim$1000 °C) in the presence of pure oxygen. The resulting chlorine gas flows back into the previous stage. Owing to the transfer into the gas phase, the starting material does not influence the product design, but the conditions of the decomposition determine the structure of the product. The same applies to production nickel and iron via the carbonyls (Equation 15.19).

It is important to note that in the preparation of metallic catalysts from metal oxides (Equation 15.18), the particle shape of the starting material does not change. During tempering and reduction with hydrogen, the particles (pellets, balls, or ring) remain stable in form.

15.5.2
Reactions of Particles with Liquids

Reactions of solid particles with sprayed on liquids are carried out preferably in flameproof mixers and in fluidized beds at moderate temperatures of 50–250 °C under normal pressure or pressures up to 200 bar. The reactors are relatively small (up to 25 m^3 mixer; diameter circular fluidized bed up to 2.7 m). In these reactions, the chemistry changes crucially, while usually the particle size distributions remain.

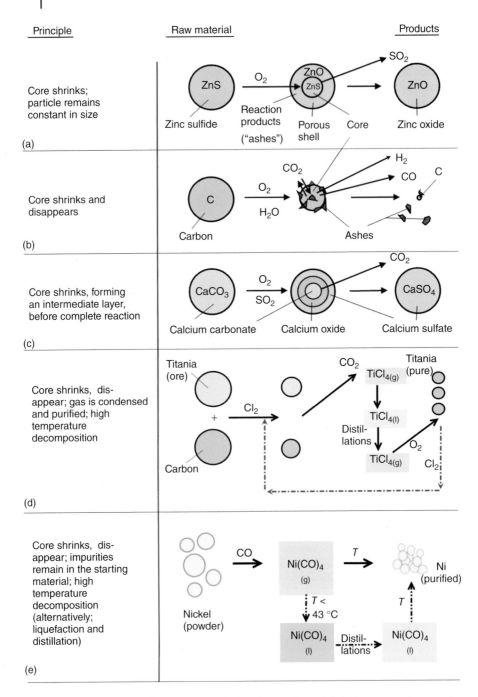

Figure 15.5 (a–e) Different conceptual ideas for reactions of shrinking cores with gases to solid and/or gaseous products.

Table 15.2 Reactions of particles with liquids in solid state (solids, suspensions).

No.	Liquid/solid reactions in the industry

(m) {Calcination of limestone

$$CaCO_3(s) \rightarrow CaO(s) + CO_2(g)\}$$ (15.11)

Extinguishing of caustic lime with water under high heat of reaction

$$CaO(s) + H_2O(l) \rightarrow Ca(OH)_2(s)$$ (15.20)

(n) Hydrothermal decomposition of silica

$$2SiO_2(s) + 2NaOH(l) \rightarrow Na_2O{\cdot}2SiO_2(s) + H_2O(l)$$ (15.21)

(o) Neutralization reactions (liquid acids, solid alkalizing, anhydrous, or aqueous)

$$R\text{-}COOH(l) + Na_2CO_3(s) \rightarrow R\text{-}COONa(s) + NaHCO_3(s)$$ (15.22)

(p) Manufacture of cellulose from wood or cotton; etherification of the alkalized cellulose fibers

$$Cell\text{-}(OH)_3(s) + 1.5NaOH(l) + 1.5CH_3Cl(l) \rightarrow$$
$$Cell\text{-}(OH)_{1.5}(s) + 1.5\ NaCl(l, s) + 1.5\ H_2O(l)$$
$$\backslash(OCH_3)_{1.5}$$ (15.23)

(q) Producing hydrogen fluoride from fluorite and sulfuric acid

$$CaF_2(s) + H_2SO_4(l) \rightarrow 2HF(g) + CaSO_4(s)$$ (15.24)

(r) Preparation of sodium percarbonate from sodium carbonate with 30% hydrogen peroxide solution

$$2Na_2CO_3(s) + 3H_2O_2(l) \rightarrow 2Na_2CO_3{\cdot}3H_2O_2(s)$$ (15.25)

Other disperse properties may differ from those of the source material. Control of product design is possible with restrictions.

Extreme changes in material properties due to the chemistry demonstrate two practice-relevant examples (see Table 15.2). Water-insoluble, solid, starting materials, namely, sand (silica) and wood (cellulose) convert themselves by a solid/liquid reaction into water-soluble products. By hydrothermal decomposition, preparation of the water-soluble, hydrated sodium disilicate (Equation 15.21 in Table 15.2) takes place. Cold water-soluble cellulose ether (Equation 15.23) is formed in the reaction of alkalized cellulose with methyl chloride. Other interesting solid reactions represent the conversion of liquid organic sulfonic acids with solid sodium carbonate, resulting in well-soluble anionic surfactants, which are solid at room temperature (Equation 15.22). Products retain the "design" of the starting materials.

Production of gaseous hydrogen fluoride results on "stripping" of calcium fluoride with concentrated sulfuric acid (Equation 15.24). The resulting gypsum is a by-product, formed out of dispersed calcium fluoride. Heavy-metal-free sodium carbonate reacts after spraying with hydrogen peroxide under drying conditions to form sodium percarbonate (Equation 15.25). This material represents a strong oxidizing agent with a high reactivity [15]. To prevent decomposition by contact with humidity, the particles get a coating layer. These intercalation compounds are characterized by a constant shape during reaction. Most of the physical properties of the raw material remain. The particles show no difference from the starting material.

15.5.3
Conversions in Liquids

Many reactions of particles occur in the liquid phase. These are conversions of solids with liquids or of suspensions with gases (see Table 15.3), which run largely isothermally and are easily controllable owing to the large amount of liquid present. For purification of calcium carbonate, precipitation out of a calcium hydroxide solution with carbon dioxide takes place (Equation 15.26). In a plurality of fluid/solid reactions, first the solid starting material dissolves. In this case, impurities and gangue of ore are left behind as solid components. In the second step under different operating conditions, dissolved valuable substances are precipitated again. This sequence takes place during the removal of impurities from bauxite (Equation 15.27; see also Figure 15.7) in production of alumina. The same mechanism allows (Equation 15.29) the extraction and purification of precious metals [16] using cyanides. In this way, gold is easy separated from the finely ground, suspended ore.

The gangue usually retains its the size, shape, and quantity, because very small quantities of gold are contained in the ores. In the subsequent step, dissolved gold can be precipitated. As in many cleaning methods, the process proceeds via dissolving, separating, and reprecipitating. In this manner, the valuable substance is accumulated and concentrated. In contrast to most gas/solids and liquid/solid reactions, in these examples, the formation of particles is independent of the disperse characteristics of the starting material. This is important, because valuable material dissolves in an intermediate stage.

Conversion of wood into cellulose takes place via removal of lignin and subsequent washing, bleaching, and drying. In this procedure, the byproduct "lignin" dissolves, while the main product retains its particle size until it is pressurized and transferred into pulp sheets.

Table 15.3   Reactions of disperse solids with liquids or gases in liquid phase.

No.	Reactions of suspended solids

(v)   Precipitation of calcium carbonate from aqueous lime milk with carbon dioxide

$$Ca(OH)_2(l, s) + CO_2(g) \rightarrow CaCO_3(s) + H_2O(l) \qquad (15.26)$$

(w)   Separating iron- and silicon compounds from bauxite for the production of aluminum oxide (Bayer process)

$$Al(OH)_3(s) + NaOH(l) \rightleftarrows Na[Al(OH)_4](l) \qquad (15.27)$$

(x)   Production of sodium thiosulfate from sodium sulfite solution by mixing in hot sulfur

$$Na_2SO_3(l) + S(s) \rightarrow Na_2S_2O_3(l) \qquad (15.28)$$

(y)   Leaching of metal ores with acids and complexing agents (cyanide)
       Gold, silver, copper, nickel

$$2Au(ore) + \frac{1}{2}O_2(g) + H_2O(l) + 4NaCN(l) \rightarrow 2Na[Au(CN)_2](l) + 2NaOH(l) \quad (15.29)$$

$$2Na[Au(CN)_2](l) + Zn(s) \rightarrow Na_2[Zn(CN)_4](l) + 2Au(s) \qquad (15.30)$$

(z)   Manufacture of chemical pulp from comminuted wood by dissolution of lignin with sodium hydroxide/sodium sulfate

15.5.4
Reactive Suspension Precipitations

In some cases, reactive precipitations or crystallizations out of suspension or paste occur. The hardening of calcium hydroxide-containing mortar after application is an example. The mortar dries in air and forms calcium carbonate during the reaction with carbon dioxide. This reaction results in hardened plasters. Gold or silver release from dissolved cyanide takes place by means of less noble zinc dust (Equation 15.30 in Table 15.3) in a reactive precipitation reaction.

Dosed addition of calcium hydroxide may reduce soluble water hardness (calcium bicarbonate) in very hard water. Waterworks filter off the precipitated calcium carbonate, which was the cause of the hardness. Another example illustrates the preparation of a chlorinating agent. When introducing chlorine into a calcium hydroxide suspension, solid bleaching powder [CaCl(OCl)] is formed in a precipitation reaction.

15.6
Formation of Products

Reaction mechanisms takes are discussed in this section in greater detail on the basis of relevant examples.

15.6.1
Product Formation in Fluid/Solid Reactions

Figure 15.5 illustrates some reactions between *gases and solids* and shows the resulting products. Such reactions are carried out industrially. Judging by the final product, there are three types of reactions: the first type form particles that correspond to the shape and size of the starting material. An industrially executed reaction of this type is the oxidation of sulfide ores with air to the metal oxides, retaining the powdered form, but changing the structures and some disperse properties.

Secondly, solids convert completely into gases. These are reactions, such as the gasification of ground coal, purification of silicon, or synthesis of uranium hexafluoride from solid uranium tetrafluoride. In coal gasification with steam/oxygen, gaseous products of value (syngas) arise. During the reaction, the ground coal particles shrink continuously and eventually disappear, leaving behind ash. Almost the entire carbon is now present as gaseous carbon monoxide (Figure 15.5b).

In the third case, in gas/solid reactions, initially two solid shells are formed, with an inner one as intermediate product. This happens as part of cement production (see Section 15.9). The important raw material is limestone, available in nature as sedimentary rocks of the minerals calcite and aragonite, which consist mainly of calcium carbonate. In cement production, presintering of milled limestone (lime powder as the main component) expires as calcination takes place at 800–900 °C. Then in a rotary kiln follows the so-called clinker formation at temperatures up to 1450 °C. After cooling, the gray-brown granules obtained are ground in a ball mill and used as a part of cement. By releasing carbon dioxide during calcination, lime core shrinks and forms calcium sulfate first in the inner shell burnt lime, and then in the outer shell under influence of reactive gases (sulfur dioxide and oxygen). The sulfation of the formed calcium oxide with sulfur dioxide, originating from minerals runs practically parallel to this. Two shells grow and the inner core disappears. Then, the outer shell can continue growing, until the product includes the entire core. An adapted "shrinking-core" model [17] is valid for this reaction. The transition state with a core and double shell is a special feature, especially because some minor constituents of lime remain in the vanishing core.

The purification of minerals may occur via a chemical gasification of ground ores (Figures 15.5d,e). The first examples are solar-silicon and titania. In both cases, at higher temperature, a chloride is formed from elementary mineral with hydrogen chloride, respectively from oxide with chlorine gas. The liquefied chloride compound can be purified by repeated distillations. During heat supply, the clean compounds decompose at high temperatures (about 1000 °C), resulting in the purified element or oxide, depending on the reactants.

In the absence of air at 40–100 °C and atmospheric pressure, nickel can be converted with carbon monoxide into the tetracarbonyl [18]. The reaction is highly exothermic. Impurities remain in the unreacted starting material. Nickel carbonyl decomposes at 150–300 °C, forming pure nickel (the Mond process). After carbonyl is condensed ($T_s = 43$ °C), the nickel can be purified further through multiple distillations. The carbonyl formation succeeds similarly for many other metals (W, Mo, and Ti). As starting material, the corresponding halogen compound is used, preferably the iodine compound (the van Arkel–de Boer method).

In most reactions of *solids with liquids in air*, which take place in the solid phase, particles largely retain their size and shape. It does not matter if the starting materials are spherical or fibrous (Figure 15.6). These reactions have this common that around the particles, a liquid film forms and the liquid diffuses inside to the reaction zone. Some of these reactions generate much heat. For example, the reaction of quicklime with water to form solid calcium hydroxide produces so much heat that some of the water evaporates, thus contributing to cooling. The water

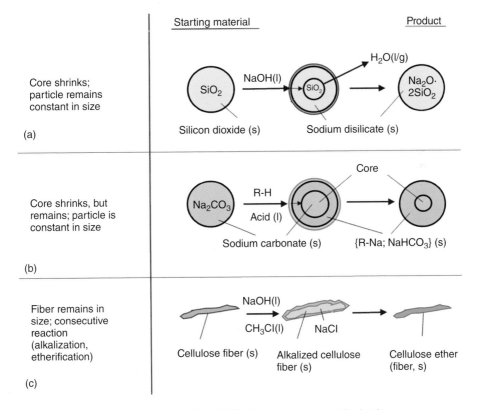

Figure 15.6 (a–c) Conceptual models for solid/liquid reactions in air with shrinking cores with shape retention of the starting materials.

forms a liquid/gas ring around the particles and diffuses through the existing layer of calcium hydroxide on the core to react with the calcium oxide until complete consumption takes place.

Again, in the thermal conversion of silicon dioxide, particles retain their original form. With liquid sodium hydroxide, they react under pressure (40–60 bar) at elevated temperature (about 260 °C), forming crystalline sodium disilicate. Drying of the product occurs after it is filtered off from the reaction mixture. As in similar cases, the drying apparatus and drying conditions affect particle design.

15.6.2
Product Formation through Reactions in Suspension

Isothermal reactions, usually of suspended inorganic solids with added gases or liquids (Figure 15.7), take place in the liquid phase. In most cases, the generated products are fine solids. The first example is that of slaking of quicklime with an excess of water (Equation 15.20) to form calcium hydroxide in suspension. The "limy paste" is suitable for production of lime mortars. In this case, particles partially dissolve, so that a significant reduction of the particles size may occur. Another example is the production of purified calcium carbonate via introduction of carbon dioxide into a calcium hydroxide suspension. By ongoing addition of

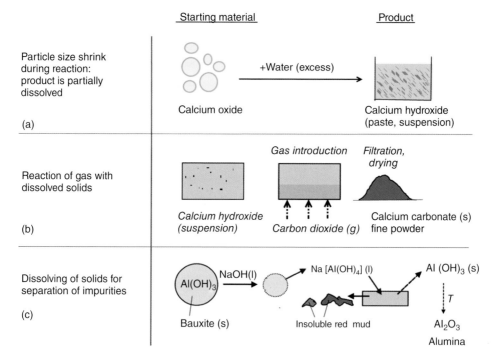

Figure 15.7 (a–c) Conceptual ideas for the reactions of suspended/dissolved solids with gases or liquids.

carbon dioxide, the system exceeds the saturation solubility of the product. Calcium carbonate precipitates from solution owing to supersaturation. Precipitation ensues similarly to controlled crystallization, but the particle design depends especially on the drying procedure.

Finely ground bauxite dissolves in sodium hydroxide solution at 175 °C and a pressure of 7 bar (the Bayer process [19]). This step is a solid/liquid reaction, which leads to a liquid. Insoluble iron ores (hematite, boehmite) can be filtered off as red mud. After cooling, dilution, and seeding, aluminum hydroxide crystallizes from the solution. Burning in a circulating fluidized bed at 1250 °C leads to alumina. One line of such a facility has a capacity of more than $1\,000\,000\,\text{t a}^{-1}$. Here, the crystallization process allows the control of product formation within limits. From the reactions between solids and fluids arise chemically and physically changed particles. They may differ in phase or in some disperse and aesthetic characteristics with respect to the starting materials. Therefore, it is interesting to see how product design can be influenced experimentally and what influences are exerted on the disperse properties of starting material. These aspects have been extensively studied for two reactions: one in an isothermal reaction in paste phase (cellulose ethers) and the other in an almost "adiabatic" reaction of acids on dry particles (dry neutralization).

15.6.3
Influencing the Product Design

The source, quality (nature and quantity of impurities), pretreatments, and disperse properties of the raw materials determine product design, arising during reaction. Pretreatment steps may be concentrations, separations, or purifications as well as grinding and screening. Depending on the type of grinding (colliding, pressing, and cutting) a specific grain size distribution is obtained (grinding/sieving). This allows influencing the reactivity of particles. In addition, the concentration of the fluids represents a setscrew for controlling the reaction. In fluidized beds,

Table 15.4 Typical reactors for reactions of gases with disperse solids.

No.	Gas/solid reactors
1	Fluidized beds, round, elongated, direct passage, circulating
2	Conveying reactor
3	Fixed bed
4	Cyclone cascade
5	Mixer
6	Spraying tower
7	Rotary kiln
8	Stage furnace
9	Mill

particle-based numbers (Reynolds, Froude, and Archimedes) are other screws for adjusting quality. For these numbers, optimal working areas exist, which should be used. Calculation of heat transfer coefficients to the walls and to the internal tube heat exchanger can be determined by means of relations [20], which have been tested in practice. Working temperature and residence time are additional design parameters. Table 15.4 displays some gas/solid reactors. For processing of solids, fluidized bed reactors are preferably used. They run in direct passage and in circulation, often as better substitutes for rotary kilns (for example, in the alumina production).

Table 15.5 depicts four main steps in the processing of solid raw materials. In the case of purification, ground solids convert into a liquid (usually a chloride), with is cleaned by distillations or crystallizations. Another possibility is the conversion into a gaseous carbonyl, followed by decomposition (Ni, Fe). By controlled thermal treatment, mainly carried out in fluidized bed systems, the final solid product is formed. These are calcinations or decompositions. Then particle design originates from the thermal process.

Table 15.5 Reactions of solids with fluids.

	Roasting	Conversion	Gasification	Purification
Examples	Pyrite	Blast furnace process	Syngas from coal	Titania
	Sphalerite	Catalyst from metal oxides	Hydrogenation of coal (Bergius–Pier process)	Silicon
	Limestone	Dry neutralization	Hydrochlorination of silicon	Alumina
	Gold ores			Nickel
	Chalcopyrite and others			Iron
Type	Ground solids react with oxygen	Solids react with gases or liquids	Solids transferred into the gaseous state	Solids transferred into a liquid or gaseous state; thermal decomposition of the purified liquid or gas
Starting particles	Maintain the size	Size changes via melt or maintain the size	Particles shrink and disappear by forming a gas	Particles shrink and disappear
Product	Chemically different particles	All shapes obtained via melt or chemical different particles	Gases	New particles formed by decomposition of intermediates, purified but the same compound is used as the starting material

Table 15.6 Design of products in fluid/solid reactions.

Parameter	Properties of formed solids
Quality	Conversion and product quality depend critically on the source, the structure, purity, and the comminution (particle size distribution) of the raw material
	Minor constituents and impurities of the raw material remain in the product
	In case of incomplete conversion, small parts of the starting material (called *cores*) remain in the middle of product particles
Chemistry	The chemical and the material properties, that is, the reactivity and physicochemical parameters corresponding to the formed products; the starting material is no longer in existence
Aesthetics	Products formed from powdered, granular, or fibrous raw materials largely retain the size and shape of the starting material
	The aesthetics (whiteness, color, luster, smell, stickiness) and the quality are typical for the product
Disperse properties	The changes in disperse properties, such as bulk density, density, and flowability, are measurable; in some cases the changes are significant
	In subsequent steps (e.g., grinding, granulation, compaction, and coating), the products can be tailored to the application
Purification/form	The purification of the solids occurs via a melt, a dissolution, an extraction, or a transfer into the gas phase by the reaction
	The melting process allows adjusting of qualities and forms

Subsequent processes can improve product quality, meeting all requirements in the application. Examples are breaking up or agglomerations with/without coatings. An adjustment of the properties to application is essential [1, 2, 15]. Some fluid/solid reactions, which take place under consumption of solid phase and form only fluid products, have no "product design" based on raw material. In the case of powdered products, an upstream granulation is possible or with coarse chunks a fine grinding.

In limits it is possible to control the design of formed particles in most solid/fluid-reactions. For shaping the final particles, the best way is the adjustment of the raw materials and the operating conditions. In subsequent steps, known basic operations are available for shaping. Pilot experiments with liquid/solid reactions, in which the particle size remained unchanged, lead to the findings summarized in Table 15.6. Further considerations, gained from other processes, are incorporated.

15.7
Etherification of Cellulose

Cellulose, etherified with methyl chloride, ethylene oxide, and propylene oxide or with the monochloroacetic acid, is used as an adhesive (wallpaper glue), as a water binder in the construction industry, as a thickener in cosmetics, and as an adjuvant in the paper and textile industries. It represents a natural product derived from wood or cotton and consists of long unbranched chain molecules. A fiber contains up to 10 000 (usually <4000) connected anhydroglucose units. In cellulose, bundled filaments are present as microfibrils that are interconnected both internally and externally by hydrogen bonds. Therefore, cellulose is insoluble in water, despite the three OH groups per anhydroglucose unit. In this cross-linked form of cellulose, the etherifying agent may react only partially. The OH groups are not sufficiently accessible on filaments. A cutting of pressed cellulose layers (pulp) in granulators allows an improvement.

"Soft" grinding prevents degradation of macromolecules, triggered by fineness and temperature. The ground material, known as *cutting mill cellulose* (Figure 15.8a), consists of cellulosic fibers that are cut to a defined particle size distribution. This material reacts in flameproof discontinuous mixers (design: 25 bar) made of salt-resistant steel. In addition to mixing elements, the reactor should have several choppers for smashing lumps. In the reactor, first an alkalization with aqueous sodium hydroxide solution takes place. Depending on temperature and time, the cellulose swells, improving accessibility of the reactants. Subsequently, isothermal etherification with liquid methyl chloride ensues under pressure. After separation and purification by distillation, the by-products methanol and dimethyl ether react with hydrogen chloride gas. This catalytic hydrochlorination forms methyl chloride that flows back, resulting in methylation of cellulose. The heat of the reaction is removed away effectively through evaporative cooling and through cooled reactor walls.

The evaluation of kinetic measurements demonstrates that the rate of chemical reaction is controlled in the range of industrial interest. Higher temperatures for acceleration of the etherification are not useful, because then macromolecules

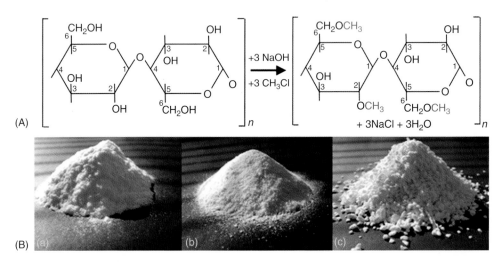

(A)

(B)

Figure 15.8 Cellulose and cellulose ethers: (A) chemical formulas for etherification of two anhydroglucose units with methyl chloride (degree of substitution: 1.5) and (B) (a) Starting material, granulator cellulose; (b) product 1: ground cellulose, mixed with polyvinyl acetate; and (c) product 2: dust-free methylcellulose after roller compaction.

degrade and change color. Further, these products show irregularities in substitution.

The fibrous cellulose ether that are formed and the by-product sodium chloride are cold water-soluble solids. Methyl cellulose and the majority of mixed ethers show flocculation points and are insoluble at higher temperatures. Therefore, the soluble salts are washed out from insoluble fibers by hot water (>80 °C). The washing occurs in continuous rotary pressure filters. Sodium chloride removal is not so easy because the salt occurs also in the interior of the fibers. A multistage countercurrent washing reduces water consumption for residual salt contents to <0.5%. Nevertheless, consumption of wash water amounts to $8-20\,t\,t^{-1}$ of cellulose ether, overall relatively high values.

The dried product shows a fibrous structure with low bulk densities and bad flowability. It cannot be used in this form. Drying takes place continuously either with hot air, or preferably with superheated steam in the circuit. For an adjustment of product design, the fibrous structure must vanish. For this purpose, there are two common paths. First, by the addition of water to the wet filter cake, the product is partially dissolved directly before being ground. This results in the elimination of the fibrousness. Therefore, a following drying and grinding [21] yields the powder. On adjusting the mill, free-flowing powder or fine granules arise. If necessary, a further grinding leads to reactive micropowders. The second way is based on dry fibers. Existing fibers from the raw material appear from the filter cake during pneumatic drying. Suitable milling processes, which mean compression, shearing, and crushing in one step, lead usually to powders, and strong grinding to micropowders (µ-powders). Because of dust explosion hazard, appropriate safety equipment for the grinding facilities must be provided. Special adhesives are formed from powders, mixed with a polymer powder (Figure 15.8b).

Figure 15.8c shows compacted material, obtained from roller compaction of powders. Granulation and tableting are also feasible, but not carried out in practice, owing to extra costs. Four commercially available alternatives for product design of cellulose ethers are illustrated in the block diagram in Figure 15.9.

The specific product design of cellulose ethers arises from raw materials, chemical reactions, and downstream processes. The pulp determines the maximum attainable viscosity in water and the average degree of polymerization of a product. Rather low viscosities result from beeches; further, medium to high from pines and high-viscosity cellulose ethers from linters (cotton). The whiteness of pulp is the result of bleaching during the production process. For etherification, the manufacturers usually use pure white pulps. In the grinding activation of pulp, the particle size of cellulose fibers exists within an optimum range First, excessive milling leads to a reduction in the maximum attainable viscosity in solution. Second, too large fibers prevent a uniform etherification, causing bittiness during dissolution.

A targeted degradation of macromolecules can occur on alkalization at intermediate temperatures (>50 °C). The duration of action or, in extreme cases, the addition of oxidation agents promotes the degradation. Comparatively low reaction temperatures prevent degradation of macromolecules and reduce discoloration.

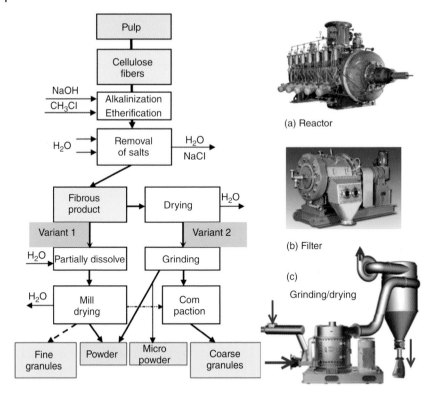

Figure 15.9 Block flow diagram of the production of free-flowing cellulose ethers, starting from cellulose fibers; four marketable qualities, powder accessible via both variants. (a) (DVT pressure (German: Druck)/vacuum/temperature) reactor; (b) rotary pressure filter; and (c) mill dryer. (Courtesy of Lödige (a), BHS (b), and Jäckering (c).)

However, this measure results in the prolongation of reaction times. From the added amounts of reactants, set according to experiments (yields), the degrees of etherification (degrees of substitution and distributions) with methyl chloride and/or ethylene oxide as well as other chemicals such as monochloroacetic acid are determined. The application determines the extent of etherification, which influences the position of the flocculation point. Temperatures at which products completely flocculate in water and the water-binding capacity at different temperatures represent further important product characteristics. In some cases (detergents), salt removal with hot water can be dispensed with fully or partially. For many applications (construction, thickeners, and wallpaper glue) salt contents of <0.5% are necessary, and in other cases a more extensive cleaning needs to take place (pharmaceuticals and eye drops). The construction industry requires a defined granulometry wherein sprayed-on additives improve wetting.

It is possible to optimize application properties by a mixing down with other ethers or polymers. Three factors, namely, raw materials, chemistry, and technology, play a role in almost all processes of raw material extraction and fluid/solid reactions. Influencing the design takes place within limits, usually via the raw material and downstream processing, as here in the refining of cellulose takes place. The product design depends on the material and process, characterized by numerous measurements (Table 15.7). It includes also aesthetic parameters (particle size distribution, shape, color, and smell).

Table 15.7 Influence of raw materials, and furthermore, of chemistry and process technology on the product design of pure cellulose ethers; small picture: granulator for cutting the pulp.

Chemistry	Technology
Sources of raw materials: Pulp (cellulose fibers from cotton and wood with different degrees of polymerization) Linters Pine Beech	*Four stages*: Grinding, reaction, washing, mill drying

Chemistry	Technology

Reactions
    Alkalinization (type, time, temperature, oxidation agents)
    Methylation (methyl chloride surplus, pressure, temperature, time, other reactants – EO, PO)	

Adjustment of the rate of dissolution
    Deceleration by linking with glyoxal
    Acceleration by auxiliaries for wetting
	1. Cutting of pulp
	2. Reactor: Mixing, breaking up, homogenization during reaction
	3. Multistage countercurrent wash
	4. Grinding and drying

Characterizing parameters for cellulose ethers
    Degrees of substitution
    Degrees of polymerization
    Solubility
    Color
    Residual salts
    Moisture
    Particle size distribution
	Special features of the process
	    Sophisticated safety devices for the reaction and grindings
	    Complex separation and purification of by-products (methanol, dimethyl ether)
	    Wastewater containing sodium chloride

EO, ethylene oxide; PO, propylene oxide.
Courtesy of Hosokawa Alpine.

15.8
Dry Neutralization

In the detergent industry, required anionic surfactants and complexing agents originate from the neutralization of the parent acids. Examples are the sodium salts of alkyl benzene sulfonic acid and of sulfates from fatty alcohols, of citric acid and 1-hydroxy-1,1-diphosphonic and polyacrylic acid. Quantitatively, the most important are the anionic surfactants with a proportion of 10–20% in detergent formulations.

The production of most detergent base powders occurs via spray towers. In these processes, all the acids react with sodium hydroxide in a tank for slurry preparation, in which the formulation is weighed together. Subsequently, the slurry flows directly to the spraying nozzles. To reduce manufacturing costs, a large part of which consists of energy costs for evaporation of water, the detergent industry has developed new methods with significantly reduced amounts of water in the recipe. The production will no longer run with an atomization of suspensions, but with granulations in mixers. The removal of excess moisture from the product happens in a downstream fluidized bed. With this modern method, the amount of water evaporated decreases by factor 6–10.

The new procedures waive neutralization with aqueous sodium hydroxide. Instead, for some anhydrous acids, the reactions are carried out with solid sodium carbonate. In these neutralizations of liquid organic acids with sodium carbonate granules, no water is available; therefore, the question of removal does not arise. Because of high heat of neutralization, the reaction of sodium carbonate with anionic surfactant acids can be seen as example for "nonisothermal" conversion of solids with liquids. These reactions are difficult to control, owing to the formation of temperature-dependent by-products. The reactions take place either in a batch or continuous mixer or in a fluidized bed.

15.8.1
Raw Materials

Sodium carbonate is found in soda lakes and as mineral (trona) in Egypt, East Africa, the United States, and Mexico. The natural raw materials differ in some parameters depending on locality, purification method (recrystallization), and storage conditions (hygroscopicity). The quality is different when the materials are obtained by synthetic methods (Solvay process). Therefore, reactions with solid sodium carbonate are (Table 15.8) resource dependent (structural properties of soda [22]).

15.8.2
Chemical Formulas

The study of dry neutralization, that is, water-free reaction of liquid sulfonic acid with solid sodium carbonate, is carried out as representative liquid/solid conversion at low temperatures. The complex reaction sequence, simplified and

Table 15.8 Hydrates of sodium carbonate as well as structural properties of sodium carbonate as a function of the manufacturing process.

$Na_2CO_3 \cdot xH_2O$ $x =$	Hydrate	Mineral	Converting temperature (°C)	Density (g cm^{-3})
0	Anhydrous	Natrit	>107	2.53
1	Monohydrate	Thermonatrit	>35.4	2.25
7	Heptahydrate	—	>32.5	1.51
10	Decahydrate	"Natron"	<32.5	1.45
$Na(HCO_3) \cdot Na_2CO_3 \cdot 2H_2O$	Dihydrate	Trona	>65 to Na-carbonate	2.14

Parameter	Na_2CO_3 calcined^a	$Na_2CO_3 \cdot H_2O$ active^b
Pore volume (cm^3 g^{-1})	0.3344	0.9718
Porosity (−)	0.4583	0.6862
Specific surface area (cm^3 g^{-1})	4.09	11.24

^{a}From sodium bicarbonate calcined at 125 °C.
^{b}From sodium carbonate decahydrate dehydrated at 25 °C.

shown in Figure 15.10, depends on the temperature ranges. Desired reactions (Equation 15.35) take place at moderate temperatures of 40–50 °C and take several minutes. In the interaction with bicarbonate, cleaved sulfuric acid generates moderate formation of carbon dioxide. At higher temperatures (from 75 °C), the decomposition of sodium bicarbonate, the intermediate formed, begins according to Equation 15.36.

15.8.3
Modeling of the Reaction Processes

The reaction starts by spraying of dodecylbenzene sulfonic acid (ABS-H) on sodium carbonate particles. The spraying should be executed similar to a coating process [15]. The acid reacts at the surface and forms sodium dodecylbenzene sulfonate (ABS-Na) and sodium bicarbonate, which also neutralizes the acid in a subsequent reaction. The products form a shell around the core. The resulting carbon dioxide and water are largely consumed in further reactions (see Figure 15.11). Localized at the reaction front, high temperatures can arise, leading to decomposition of bicarbonate. In most cases, the final product contains residues of the starting material in the core.

15.8.4
Reaction Sequence

The surface reaction of acid with sodium carbonate was examined under a microscope (see Figure 15.12). Some droplets, about 20 μm in size, are visible on the surface.

$$\text{ABS-H (l)} \quad + \quad \text{Na}_2\text{CO}_3 \text{ (s)} \quad \longrightarrow \quad \text{ABS-Na (s)} \quad + \quad \text{NaHCO}_3 \text{ (s)} \tag{15.31}$$

$$\text{ABS-H (l)} \quad + \quad \text{NaHCO}_3 \text{ (s)} \quad \longrightarrow \quad \text{ABS-Na (s)} \quad + \quad \text{CO}_2 \text{ (g)} \quad + \quad \text{H}_2\text{O (l)} \tag{15.32}$$

$$\text{Na}_2\text{CO}_3 \text{ (s)} \quad + \quad \text{H}_2\text{O (l)} \quad \longrightarrow \quad \text{NaHCO}_3 \text{ (s)} \quad + \quad \text{NaOH (s)} \tag{15.33}$$

$$\text{NaOH (s)} \quad + \quad \text{CO}_2 \text{ (g)} \quad \longrightarrow \quad \text{NaHCO}_3 \text{ (s)} \tag{15.34}$$

$$2 \text{ ABS-H (l)} \quad + \quad 2 \text{ Na}_2\text{CO}_3 \text{ (s)} \quad \longrightarrow \quad 2 \text{ ABS-Na (s)} \quad + \quad 2 \text{ NaHCO}_3 \text{ (s)} \quad \text{T} < 50\,°\text{C} \tag{15.35}$$

$$\text{T} > 75\,°\text{C} \quad \downarrow$$

$$\text{Na}_2\text{CO}_3 \text{ (s)} + \text{H}_2\text{O (l,g)} + \text{CO}_2 \text{ (g)} \tag{15.36}$$

$$2 \text{ ABS-H (l)} \quad + \quad \text{NaCO}_3 \text{ (s)} \quad \longrightarrow \quad 2 \text{ ABS-Na (s)} + \text{H}_2\text{O (l,g)} + \text{CO}_2 \text{ (g)} \tag{15.37}$$

ABS-H:

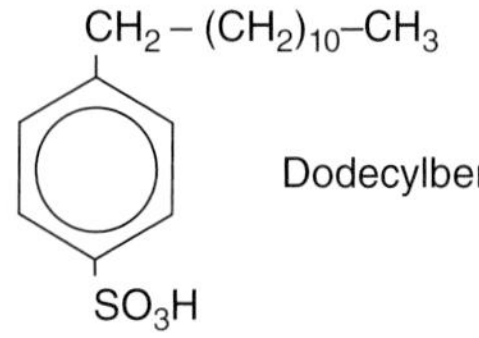

$$\text{CH}_2 - (\text{CH}_2)_{10} - \text{CH}_3$$

Dodecylbenzene sulfonic

$$\text{C}_{18}\text{H}_{30}\text{O}_3\text{S}$$

$$\text{SO}_3\text{H} \tag{15.38}$$

Figure 15.10 Reactions of anhydrous dodecylbenzene sulfonic acid (ABS-H, Equation 15.38) with solid sodium carbonate, resulting in the temperature-dependent sum Equations 15.35 and 15.37.

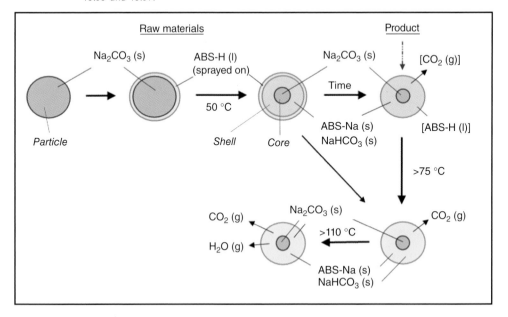

Figure 15.11 Schematic illustration of the reaction of organic liquid acids (e.g., dodecylbenzene sulfonic acid, ABS-H) with solid sodium carbonate, forming main and by-products.

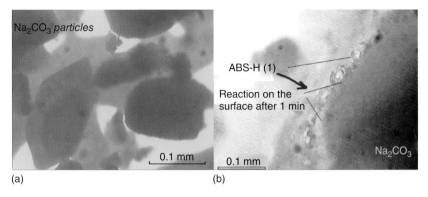

Figure 15.12 Reacting dodecylbenzene sulfonic (ABS-H) with solid sodium carbonate (light microscope views): (a) sodium carbonate particles and (b) acid on the surface of sodium carbonate.

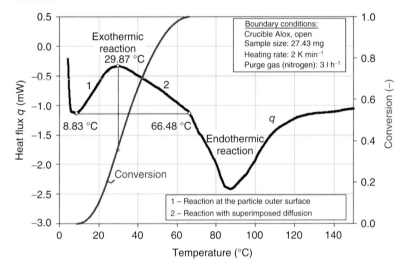

Figure 15.13 Differential scanning calorimetry: measuring the neutralization of liquid dodecylbenzene sulfonic with solid sodium carbonate (molar ratio: 1.5 Na_2CO_3/ABS-H).

The thermal analyses (DSC, differential scanning calorimetry, TGA, thermal gravimetry analysis) provide kinetic data of significant processes [23, 24]. Figure 15.13 demonstrates that in the temperature range between 10 and about 60 °C exothermic reactions take place. These are probably the neutralization of calcinated soda and later of bicarbonate. From the graph a heat of reaction of 20 kJ mol^{-1} for Equation 15.35 and a typical activation energy of 98 kJ mol^{-1} are obtained. The subsequent endothermic reaction that takes place between about 65 and 120 °C is the decomposition of sodium bicarbonate [25]. This conversion shows an activation energy of 34 kJ mol^{-1}.

The dry neutralization preferentially generates particles that can be directly admixed to detergent powders. This requires the use of suitable sifted soda qualities, because here the product sizes correspond to the starting material. Too small particles segregate in formulations. Large particles can not react completely and show after the residence time significant residual starting material inside, it remains a big core of carbonate. In these reactions, the reaction time depends on the particle sizes.

15.9
Building Materials

The hardening of many building materials happens as a fluid/solid reaction. Active solid starting materials are usually powders mixed with sandy or grainy materials. Reaction products often show large spatial structures. After addition of water, the active materials form a paste, in which the components react. Predominantly, the reactions run in the wet state. In this important area, it is possible to control the product design in many ways.

Gypsum (calcium sulfate) is present as hemihydrate or anhydride. Mixed with water, it crystallizes as dihydrate. In this more or less pasty condition, gypsum takes arbitrary shapes (spheres, plates, figures, or plaster casts). It can be preferentially used for repairing damages in walls. During curing and drying, the crystals become matted, so that after a short time, a hard mass is formed. Craftsmen use solid gypsum, which can be formed simply (stucco, plaster boards, stones) or designed by artists (sculptures). The possibilities for creating different products from very small to very large in the wet and in the dry states are unlimited.

Lime mortar (nonhydraulic mortar) is formed from burnt limestone (calcium oxide) at about 1000 °C in lime kilns. In contact with water, the material forms

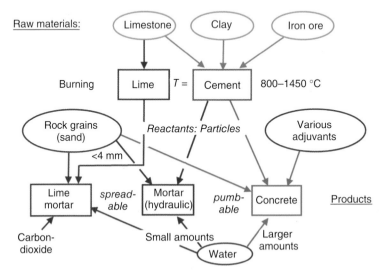

Figure 15.14 Different building materials (particles) react with water, forming plaster for large spatial structures.

calcium hydroxide, which binds with carbon dioxide from air. Solid calcium carbonate is formed in this reaction. In addition, on using cement, the material hardens in the presence of water as hydraulic mortar or as concrete. In the production of concrete (Figure 15.14), tricalcium silicates and similar compounds absorb water. They split from calcium hydroxide and crystallize into different complex compounds, which interconnect and become matted. In this way, the hardness and strength of concrete is obtained [26]. Crystallization occurs in the aqueous phase. Therefore, the concrete must be kept in wet condition, until the reaction ends.

Mortar is spreadable in layers. Pumpable concrete flows in virtually any shape. Variations in product design (sizes, shapes, colors, and strengths) are possible in many ways by including additives in concrete. Adjuvants reinforce the concrete with steel, glass, plastic, and carbon fibers, and also with silica dust (ultrastrength concrete). Asphalt, splits, and clay granules (lightweight concrete [27]) are added to obtain special properties.

15.10
Hints for Practice

In fluid/solid reactions, such as *roasting* and *conversion*, the quality of the product design depends critically on the source, structure, and purity of the raw material, also on downstream grinding. Minor constituents, impurities, and remnants from raw material remain inside the particles. These chemically modified particles retain their size and shape, while several disperse parameters (e.g., bulk density) change significantly. In some cases, the design is formed through a molten or plasticized state. Another possibility is a downstream finishing, as agglomeration and/or coating.

In *gasification*, the particles disappear, forming gases. Therefore, no design is possible. The differences show *purification* procedures. The decomposition of gases and liquids occur under controlled conditions and allows a design of purified solid substances.

The manufacturer and/or user determine the product specifications. These are results of analytical and physicochemical as well technical measurements, characterizing the product design of disperse solids. Usually, the applications laboratory constantly checks the values and adapts new findings. The importance of aesthetics depends on the use of the product. Particle size distribution (size, width), particle shape, color, smell, haptic, and flow properties must be given more attention.

As discussed in the preceding sections, product design consists of performance, convenience, and aesthetics. In general, the manufacturer should plan carefully and secure the quality by conducting pilot tests that take the cost effectiveness into account by

1) selection, targeted crushing, and purification of the raw materials;
2) specific chemical reaction (without molecular degradation, discoloration, and minimal side reactions) in an optimal designed reactor, usually fluidized beds;

3) substance isolation, concentration, purification, and shaping; the use of methods that enable the formation of the corresponding product by melting, crystallization, granulations, sintering, and coating.

In whatever manner the process runs, the most important factor is customer satisfaction.

15.11
Learnings

√ In the extraction of raw materials, the capacity of fluid/solid reactions amounts to several hundred million tons per year.

√ Reactions with gases take place in large fluidized bed reactors (bottom plate up to $200 \, m^2$), usually at high temperatures.

√ Roasting of sphalerite (ZnS) and pyrite (FeS_2) are well-known processes. During reaction, the particles retain size, while the starting material shrinks, and a product shell forms around the core.

√ The particle size distribution of the starting material determines the product design for roasting and conversion.

√ Besides roasting, conversion, gasification, and purification can also be carried out. Together, these processes are characterized by shrinking cores, described in models from the chemical engineering.

√ In gasification (burning of ground C), particles form gases and disappear. The final product arises during purification by decomposition of the gases, Examples are Ni, Si, and TiO_2.

√ Dry solid particles react usually with gases and less frequently with liquids. Some conversions run in suspension.

√ Usually, product design takes place through grinding/sieving of the raw material; in some cases, this also takes place downstream by melting, decomposition, crystallization, granulation, sintering, or coating.

√ Powdered building materials react during curing with water and/or carbon dioxide from air. These products can be well controlled in design.

References

1. Rähse, W. (2007) *Produktdesign in der chemischen Industrie*, Springer-Verlag, Berlin.

2. Rähse, W. (2004) Produktdesign – Möglichkeiten der Produktgestaltung. *Chem. Ing. Tech.*, **76** (8), 1051–1064.

3. Levenspiel, O. (2006) Fluid-particle reactions, in *Chemical Reaction Engineering*, 3rd, Chapter 12 edn, Wiley India Pvt. Ltd.

4. Yagi, S. and Kunii, D. (1961) Fluidized-solids reactors with continuous solids feed–I: residence time of particles in fluidized beds. *Chem. Eng. Sci.*, **16** (3–4), 364–371, and: Fluidized-solids reactors with continuous solids feed–II: conversion for overflow and carryover particles. 372–379, and: Fluidized-solids reactors with continuous solids feed–III: conversion in

experimental fluidized-solids reactors. 380–391.

5. Gbor, P.K. and Jia, C.Q. (2004) Critical evaluation of coupling particle size distribution with the shrinking core model. *Chem. Eng. Sci.*, **59** (10), 1979–1987.

6. Müller-Erlwein, E. (2007) *Chemische Reaktionstechnik, "Porendiffusion und Reaktion"*, Chapter 9.2.5, Vieweg+ Teubner Verlag (now: Springer Vieweg), Wiesbaden, p. 248 ff.

7. Wen, C.Y. (1968) Noncatalytic heterogeneous solid–fluid reaction models. *Ind. Eng. Chem.*, **60** (9), 34–54.

8. Moon, J. and Sahajwalla, V. (2001) Derivation of shrinking core model for finite cylinder. *ISIJ INT*, **41** (1), 1–9.

9. Pritzker, M.D. (1995) The role of migration in shrinking-core leaching reactions controlled by pore transport. *Metall. Mater. Trans. B*, **26B** (5), 901.

10. Emig, G. and Klemm, E. (2005) *Technische Chemie*, 5th edn Chapter 14, p. 402–418; Springer-Verlag, Berlin.

11. Foo Lee, Q., *Chem. Bio. Eng.*, University of British Columbia; Internet: Shrinking Core: Non Isothermal. www.chmltech.com/reactors/scm.pps

12. Adrover, A. and Giona, M. (1997) Solution of unsteady-state shrinking-core models by means of spectral/fixed-point methods: nonuniform reactant distribution and nonlinear kinetics. *Ind. Eng. Chem. Res.*, **36** (6), 2452–2465.

13. Parkinson, G. (2008) Polysilicon business shines brightly. *Chem. Eng. Prog.*, **104**, 8–11.

14. Hertlein, H. and Rednitzhembach, R. (2007) Wacker Chemie AG, DE 10 2005 042 753A1, Mar. 15, 2007.

15. Rähse, W. (2009) Produktdesign disperser Stoffe: Industrielles Partikelcoating. *Chem. Ing. Tech.*, **81** (3), 225–240.

16. Dicinoski, G., Gahan, L., Lawson, P., and Rideout, J. (2000) Application of the shrinking core model to the kinetics of extraction of gold(I), silver(I) and nickel(II) cyanide complexes by novel

anion exchange resins. *Hydrometallurgy*, **56** (3), 323–336.

17. Hillers, M. (2008) Modellierung der Turbulenzmodulation einer hochbeladenen reaktiven Zweiphasenströmung am Beispiel des Zementherstellungsprozesses. Dissertation Universität Duisburg, Essen.

18. Neikov, O.D., Naboychenko, S., Mourachova, I.B., Gopienko, V.G., Frishberg, I.V., and Lotsko, D.V. (2009) *Handbook of Non-Ferrous Metal Powders: Technologies and Applications*, Elsevier, Amsterdam.

19. Kollenberg, W. (2004) *Technische Keramik*, Kapitel 3.2.2.2, Vulkan-Verlag, Essen, S. 200 ff.

20. Reh, L. (1974) Strömungs- und Austauschverhalten von Wirbelschichten. *Chem. Ing. Tech.*, **46** (5), 180–189.

21. Andreae-Jäckering, M. and Werry, G. (1999) Altenburger Maschinen Jäckering GmbH, EP 1 082 174 B1 vom 5, Mai 1999.

22. Hartman, M., Veselý, V., Svoboda, K., and Trnka, O. (2001) Dehydration of sodium carbonate decahydrate to monohydrate in a fluidized bed. *AIChE J.*, **47** (10), 2333–2340.

23. Borchardt, H.J. and Daniels, F.J. (1956) Arrhenius kinetic constants for thermally unstable materials. *J. Am. Chem. Soc.*, **79**, 41.

24. Haines, P.J. (1995) Thermal methods of analysis – principles, applications and problems, in *Blackie Academic and Professional*, 1st edn, Chapman & Hall, London.

25. Tanaka, H. and Takemoto, H. (1992) Significance of the kinetics of thermal decomposition of $NaHCO_3$ evaluated by thermal analysis. *J. Therm. Anal.*, **38** (3), 429–435.

26. Wesche, K. (1993) *Erhärtung und Festigkeitseigenschaften*, 3 Aufl., Kapitel 3.5, Bd. **2**, Bauverlag GmbH, S. 62 ff.

27. Rähse, W. (2009) Produktdesign disperser Stoffe: Industrielle Granulation. *Chem. Ing. Tech.*, **81** (3), 241–253.

16
Materials for the Machinery

Summary

The materials for the production plant must be chemically resistant under operating conditions for a long time. Typical materials are metals, especially stainless steel, superalloys, and nonmetals (such as graphite, borosilicate glass, and ceramics), as well as plastics. The product manufacturers prefer highly alloyed stainless steel with ~17–18% Cr and 12% Ni (AISI 304 and 316 grades). The danger of pitting and stress corrosion decreases with increasing amounts of molybdenum. The pharmaceutical and food industries utilize machinery in stainless steel with surfaces in mirror finish to avoid infection with microorganisms in the entire plant. The surface smoothing is performed by mechanical methods or by electropolishing. Besides stainless steel, the chemical industry takes advantage of nonferrous metals, alloys, metallic oxides, and carbides as well as plastics for processing and storage of aggressive substances. These materials are mainly used for linings and coatings. In many cases, they are applied by thermal spraying.

16.1
Motivation

The product design describes the development of a new product, which solves a customer's needs by involving the customer. Product design includes product performance, handling, and product esthetics (see Chapter 1). In addition, for food, pharmaceutical, biotechnology, and cosmetic preparations, and also for some other products (e.g., liquid detergents), product hygiene plays an essential role with regard to quality. Product hygiene means that microorganisms are absent or their numbers are below fixed limits. Certain microorganisms must always be absent. Product hygiene requires that the production is carried out in an appropriate facility (plant hygiene; see Section 13.10). The plant must be designed dead space free to avoid the formation of biofilms. On the other hand, a firm adhesion of microorganisms on the walls of equipment must be prevented. Therefore, the materials and their processing are of great importance for hygiene and product

Industrial Product Design of Solids and Liquids: A Practical Guide, First Edition. Wilfried Rähse.
© 2014 Wiley-VCH Verlag GmbH & Co. KGaA. Published 2014 by Wiley-VCH Verlag GmbH & Co. KGaA.

quality. Furthermore, a suitable plant design with appropriate materials simplifies the system's cleaning and reduces the wastewater pollution.

While some very large companies have specialists in the field of materials who develop optimal solutions for each process, most of the others are dependent on the knowledge and experience of their employees from process development, project management, and production. During the purchase of new machinery and equipment, the material and its processing is usually defined in a joint discussion with the manufacturer. This chapter is intended to give the developers and operators of plants some evidence on the properties of materials. This knowledge is important when ordering new machinery and for trouble-free production of high-quality products. The remarks come from own experience and background in the biotechnology and consumer goods industry, supported by reading the literature.

16.2
Relationship between Material and Product Design

Chemical engineers who work in the field of product and process development for the food, chemical, or pharmaceutical industry, must not only specify the materials when ordering equipment but, in addition, often the finish of wetted surfaces and welds. In many cases, specifications exist because the company prescribes for some or all plants a standard stainless steel (such as AISI 316Ti, Din 1.4571); optionally, it also specifies a surface finish. Therefore, the storage in the company's workshops is reduced; furthermore, the processing in the plant construction is standardized. If a device requires a better material grade against chemical corrosion and abrasive wear, the decision must be made between alternatives. Howsoever, the economy (cost, repair, resistance, and cleaning cycles) of the material stands against the product quality (metal contents and hygiene).

Quality includes product safety and product hygiene, which require extensive plant hygiene. Especially in the pharmaceutical and biotechnology industries (Section 10.5), as well as for food and cosmetics, stainless steels are used with ground and polished surfaces. On the one hand, the polishing of the product's wetted surfaces significantly increases corrosion resistance. On the other hand, a residue-free cleaning can be performed, avoiding cross-contamination and infection by microorganisms in the entire production plant; when, as a prerequisite, a dead space-free, fully drainable plant is available. Examples of avoiding dead spaces can be found in DIN EN 1672-2:2005 and the European Hygienic Engineering and Design Group (EHEDG) (3-A and NSF International) brochures [1].

The adhesion of microorganisms, the formation of biofilms, the adhesion of product residues and deposits, and the development of crusts depend on the material and surface properties [2]. The same applies to the cleaning and disinfection of the surfaces and the removal of residues with appropriate solutions. These parameters worsen over the years on account of mechanical wear and chemical attack. Typical aging processes affect mainly plastics (bellows, seals). Because rough surfaces, incipient corrosion, and the gradual formation of microcracks offer numerous opportunities for the microorganisms to adhere, these are hygienically questionable [3] and not allowed. The DIN standard 1672-2 specifies that in the food industry, product contact surfaces must have an average surface roughness Ra $\leq 0.8\,\mu$m. The required processing of stainless steel by grinding/polishing or pickling/electropolishing clearly reflects in the purchase price.

The cleanability of the plant is not only an essential quality but also a cost factor, because the system's availability depends on the duration of cleaning. The cleaning method is a function of the material, its surface smoothness, the distance between roughness mountains and the composition of cleaning agents (surfactants, oxidants, pH-value), as well as the cleaning conditions (flow, temperature). The pharmaceutical industry often produces in electropolished equipment. The electropolishing of the surfaces ensures the necessary hygiene and a rather easy cleaning. This allows reducing the duration and consumption of cleaning solutions. In addition, it is possible to disinfect the production plant including the filling lines with 2-propanol, which regularly takes place in the cosmetic industry after cleaning. Even higher demands on sterility meet with chemical and steam sterilization.

The choice of material depends on the chemistry and operating conditions as well as the raw materials and product phases (solid, liquid, gas, multiphase systems). Theoretically, for each part of the production line (tank farms, silos, main plant, filling line, and product storage) another material might come into question. This means that the term *plant design* includes not just the machinery and equipment of the core facility. If the standard grade of stainless steel is not sufficient, there are alternative materials for each part of the production plant. Examples are materials for silos made of glass-fiber-reinforced plastic (GRP) or stainless steel, for acid baths made of steel with rubber or with an acid-resistant plastic liner (such as polyvinylidene fluoride, PVDF), for agitator vessels of borosilicate glass or enamel, for heat exchangers made of graphite or Hastelloy®.

As an integral part of product design, the quality depends crucially on the materials and surface finish in the entire plant. Further, they also affect the production costs.

16.3
Choice of Material

The materials are chosen on the basis of chemical, thermal, and mechanical stresses. Materials for the production plant can be divided into metals (stainless steel, nickel, titanium, etc.) and inorganic nonmetals (such as graphite, glass, and ceramics) as well as organic polymers (plastics). For machinery, the chemical, pharmaceutical, biotechnology, and food industry utilizes mainly alloyed stainless steel of high quality. Typically, they work with standard grades, mostly AISI 304 or 316 qualities. If the following critical substances are involved in the process, another grade or material must be chosen [4]:

- Strong organic and inorganic acids;
- Acidic gases (including chlorine and bromine);
- Chloride, bromide, and fluoride ions;
- Alkalis (especially at high temperatures);
- Molten salts;
- Hydrogen.

Several chemical processes on the material surface can trigger corrosion. This ensues in various forms. The best known are probably surface corrosion, crack, contact, pitting, and stress corrosion. Further, fatigue cracking and intergranular corrosion occur. The corrosion resistance of the material depends on its chemical composition (usually Fe alloy) and the processing conditions. From the outside, chemical substances attack and abrade the surfaces by raw and auxiliary materials, intermediate, secondary, and end products. These attacks determine the lifetime of the machinery in detail via:

- chemistry of substances present in the process;
- application time and temperature;
- concentration of problematic substances, gases, liquids, suspensions, melting, and solids;
- pH;
- presence of oxidizing or reducing substances;
- presence of abrasives;
- presence of salts, particularly of chlorides (Na, K, heavy metals).

For many substances, corrosion data are available in the literature [5]. For high-temperature and/or high-pressure reactions as well as for operations in the petrochemical industry special grades of steel are used. These steels are tailored to the application, and represent particularly economical solutions. Without guidelines of the company, the choice of material usually takes place after the price–performance ratio.

16.4
Stainless Steel

In the chemical apparatus, stainless steel is by far the most common material (list of stainless steels: EN 10088.1). In general, stainless steels have a chromium content (Cr) of $> 10.5\%$ and contents of sulfur (S) and phosphorus (P) of $< 0.025\%$, and carbon (C) $< 1.2\%$. They are characterized by forming a thin protective layer on the surface in contact with air. After minor damage, the layer regenerates by itself, when the steel is exposed to air and moisture [2]. The weldability increases with decreasing carbon content. Stainless steel can be improved in some properties by further alloy constituents. For this purpose, nickel (Ni) and molybdenum (Mo) are particularly suitable.

16.4.1
Standard Grade

Chemical, pharmaceutical, and food industries require mainly high-alloy steels containing at least 17% Cr (up to 30%) and 12% Ni (up to 26%) for the plant and machinery in production, besides C < 0.1%, Si < 1%, and Mn < 2%. These form in contact with oxygen from the air a thin, invisible passivation layer of chromium (III) oxide (Cr_2O_3), which protects the underlying metal. The addition of Ni improves the corrosion resistance of stainless steel, because it oxidizes very slowly (in contrast to iron). But the standard stainless steels cannot meet all the requirements of the chemical apparatus. For example, the steels are additionally alloyed with 2–5% molybdenum (Mo) to avoid pitting and stress corrosion cracking. Other requirements lead to the addition of titanium (Ti), copper (Cu), aluminum (Al), niobium (Nb), and others. The compositions of frequently used steels are listed in Table 16.1.

The standard grade AISI 304 (1.4301) is probably the first commercial stainless steel and currently most often processed. This austenitic, nonmagnetic material, and other AISI 304L qualities with less carbon (1.4306 and 1.4307), which are easier to weld and polish, show good resistance to water, steam, moisture, fruit acids, and weak organic and inorganic acids. The food, beverage, pharmaceutical, and cosmetic industries use these steel grades commonly in the processing of liquids. For improved cleaning and corrosion resistance, the product contact surfaces are usually ground and/or polished.

The stainless steel AISI 316Ti (1.4571) also includes titanium, to improve the stability and weldability. Together with the added molybdenum, this steel shows increased chemical resistance, especially against intergranular corrosion, pitting, and stress corrosion cracking. As this alloy covers most requirements, it is widely used in chemical apparatuses, providing the standard material in many companies (chemicals, consumer goods). Owing to the stabilization of the stainless steel with titanium, the electropolishing is difficult. Mechanical grinding of the surfaces cause many scratches by the very hard particles from titanium carbide, torn from the surface. This concerns the Ti, Nb-stabilized steels (AISI 316Ti, 316Nb, and 321) that are less suitable for use in the pharmaceutical, food, biotechnology, and cosmetic industries. In these fields, the machinery must be polished to mirror finish. The engineers use for these applications some low-carbon grades, such as AISI 316L (1.4404 and 1.4435) or for lower requirements AISI 304L (1.4306 and 1.4307).

Most equipment is made in standard stainless steels (all AISI 304/316 grades). In some cases, another material, as mentioned in Table 16.1, is required. The standard qualities show a medium, mechanical stress tolerance (e.g., as material in mills), and a mean resistance to chemical agents. They represent an economic compromise, and are widely used in process industries for processing of solids, suspensions, pastes, liquids, and liquids/gases at moderate temperatures.

Table 16.1 Comparison of material numbers (DIN-AISI) of high-alloy stainless steels, their composition, and relative prices.

Example: Material number according to DIN: 1.4301; composition according to European standards EN: X5CrNi18-10, number after X = carbon content multiplied by factor 100, 5 = 0.05%, CrNi18-10 = 18% Cr, 10% Ni; American Iron and Steel Institute Standard: AISI 304

Material number (Din EN 10088-1; formerly: DIN 17440 and 17445)	DIN EN 10088-1 Steel composition	AISI standard, (and ASTM-International)	Estimated relative prices (V2A = 100%); May 2013, Germany[a]		
			Steel bar 100% = 1.53 (%)	Sheet 100% = 1.23 (%)	Seamless pipe 100% = 2.32€/kg (%)
1.4006	X12Cr13	410	30	—	26
1.4301 (V2A)	X5CrNi18-10	304	100	100	100
1.4307	X2CrNi18-9	304L	100	100	100
1.4401	X5CrNiMo17-12-2	316	148	154	135
1.4404	X2CrNiMo17-12-2	316L	148	154	135
1.4435	X2CrNiMo18-14-3	316L	166	173	151
1.4439	X2CrNiMoN17-13-5	317LNM	184	193	172
1.4462	X2CrNiMoN22-5-3	—	117	126	100
1.4539	X1NiCrMoCu25-20-5	904L	274	282	247
1.4541	X6CrNiTi18-10	321	114	115	100
1.4563	X1NiCrMoCuN 31-27-4	—	311	—	281
1.4571 (V4A)	X6CrNiMoTi17-12-2	316Ti	150	157	135
1.4580	X6CrNiMoNb17-12-2	316Nb	156	—	135
1.4762	X10CrAlSi25	446	41	48	34

[a]Calculated by alloy surcharges (according to a table from Stappert/Düsseldorf for wholesalers).

In contrast, the mentioned aluminum-containing steel AISI 446 (Table 16.1) is a specialist. The inexpensive high-temperature steel is used up to 1100 °C and shows a high resistance to sulfurous gases.

16.4.2
Corrosion

The choice of steel quality is based on the chemical and mechanical resistance and processability, availability, and costs. Steel producers and distributors have published a series of chemical resistance tables, mainly for pure liquids at ambient temperature. In the literature [4–6], extensive data collections exist. It is widely known that stainless steels are attacked by many acids. Less known, but a big problem, is the pitting and stress corrosion cracking, caused by halide ions, preferably chloride ions [2, 6]. These effects occur in particular at elevated ion concentrations (>50 ppm) and temperatures, increased in combination with low pH values. The corrosion starts at a flaw in the passivation layer. In addition, movements of the liquid have an impact. The extent of damage increases in stagnant aqueous fluids (formation of local elements). Therefore, it is advisable after cleaning, to drain off the water, and then to dry the plant for a longer standstill.

Accompanying substances affect corrosion, either positively or negatively. Therefore, in complex cases, some simple measurements of the product mixture must be carried out in the laboratory. However, the corrosion resistance against chloride can be determined by measuring the electric potential in chloride solutions as a function of temperature (Figure 16.1). Measurements showed that AISI 304 stainless steels are not suitable for processing chloride-containing solutions. Depending on the concentration, at elevated temperatures all AISI 316 stainless steels are not sufficiently stable. Higher levels of alloyed molybdenum provide the required chemical resistance; especially the grades 1.4439, 1.4539, and 1.4563 (Table 16.1). By progressive alloying, the stability increases significantly against a variety of chemicals. Unfortunately, by this alloying the commodity prices rise. In many cases, the material AISI 317LNM (1.4439) represents a good compromise between the required resistance and costs.

From the chemical perspective, both materials 1.4539 and 1.4563 are best suitable against chloride-induced pitting. The resistance must be ensured throughout the facility, taking into account the temperatures. In addition to the machinery and equipment, this concerns all piping, valves, sensors, and welds. The welds should be derived from a similar material, and be performed (automatically) in high quality. In all cases, the weld must be post-treated by careful pickling and optionally grinding/polishing. By carbide formation, the chromium content can be lowered at the weld, leading in extreme cases to intergranular corrosion. Properly executed welds form in air a closed, continuous passivation layer in one or two days.

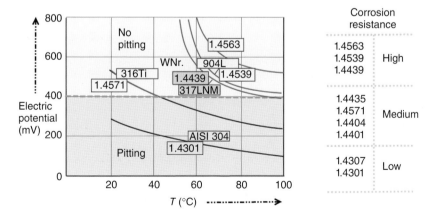

Figure 16.1 Measurements of the electric potential as a function of the temperature in 3% sodium chloride solution, using various stainless steels (WNr. = material number Din EN).

16.4.3
Smoothing the Metal Surfaces

16.4.3.1 Preparations

The manufacture of machinery and equipment requires the processing of stainless steel in the form of sheets, seamless pipes, and partly steel bars (for shafts). These materials are available in different qualities with surfaces that are smooth, polished, brushed, or shiny polished [7]. The average surface roughness (Ra) of cold-rolled sheets is in the range of 0.2–0.5 μm [1], depending on the pretreatment. This starting material facilitates further operations for the production of machinery with surfaces in mirror finish.

Table 16.2 Improving stainless steel surfaces.

Procedure	Action	Execution	Achievable roughness, mean value, Ra (μm)
Cleaning as upstream process step	Removing coarse dirt	High pressure cleaning, steam cleaning (140 °C)	
	Removing dirt and grease, possibly supported by ultrasound	Dipping into a surfactant-alkaline cleaning solution, or: spraying with a mixture of phosphoric acid/surfactant	
	Removing tough dirt	Stainless steel brush, electrolytic processes	
	Removal of caking	Ultrasound	
Mechanical smoothing	Making of a clean, matte layer	Blasting of surfaces with quartz sand or beads of glass or of stainless steel or carbon dioxide	< 1.2
	Satinized matte structure	Brushes (coarse to very fine)	
	Shiny, planar surface	Buffing (felt disc with decreasing grain sizes of the pastes)	
	Smooth surface (up to mirror finish)	Grinding or polishing (with decreasing grain size)	Up to 0.1
Chemical smoothing	Passivation layer	Accelerated formation with dilute nitric acid	
	Removal of scale layers after heat treatment, and of tarnishing after welding, further of external rust	Pickling with acids (mixture of nitric and hydrofluoric) in the bath or by spraying	0.5–1.0
	Smooth surface (up to mirror finish)	Electropolishing, Plasma polishing	0.2–0.5 < 0.01

For cleaning and smoothing the steel surfaces, several methods (Table 16.2) exist. The procedures are usually applied individually or in combination. Absolutely necessary steps represent the precleaning of the machinery and the pickling of all welds. A smoothing of the surfaces may be carried out chemically or mechanically. If subsequently an electropolishing follows, the equipment always must be completely pickled.

For a perfect result of the pickling, a previous cleaning of the surfaces is required. In the procedure, the dirt and residue from the stickers are removed and a pretreatment for degreasing executed. Alkaline solutions at elevated temperatures provide good results. Clinging dirt, slag residues, or scale layers can be removed with stainless steel brushes or by electrolytic processes or ultrasound. Only completely purified surfaces are able to form a continuous passivation layer.

16.4.3.2 Mechanical Procedures

Through blasting, brushing, grinding, buffing, pickling, and polishing, esthetic structures on stainless steel surfaces arise for goods in the private and public sector (gates, railings, wall coverings, household items, and appliances). In the industrial sector, however, the surfaces must be corrosion resistant and easily cleanable. Size, accessibility, and application of the metal surface determine the method for smoothing. On large areas, such as the interior walls of a stirred tank, the smoothing is performed mechanically by hand with the aid of grinding and polishing machines. Fissured surfaces are preground in a first step with coarse sandpaper (grade 60 or 80). Each subsequent step requires a finer grain size in the gradation up to the maximum factor of 2, that is, with pastes grit 120, 240–320 and with sanding belts grit P120, P 240, P 320–600 (Table 16.3) until the desired target (approximately Ra 0.2–0.4) is reached.

Table 16.3 Classification of the grade for abrasive pastes and abrasive papers/sanding belts.

Abrasive grit for pastes according to FEPA[a]/ANSI[b]	Average grain size of pastes d_p (μm) FEPA/ANSI	Abrasive grit for sandpaper and sanding belts FEPA	Average grain sizes of sandpaper and sanding belts d_p (μm)	Achievable roughness, mean value, Ra (μm)[c]
60/–	260	P20	1000	
80/–	185	P60	269	3.5
100/100	129/125	P120	125	1.1
180/180	69/70	P240	58.5	0.2–0.5
220/220	58/58	P320	46.2	0.15–0.4
320/360	29/28	P600	25.8	0.1–0.25
360/400	23/23	P1000	18.3	—
400/500	17/18	P2000	10.3	—

[a]FEPA: Fédération Européenne des Fabricants de Produits Abrasifs.
[b]ANSI: American National Standards Institute.
[c][EH].

Figure 16.2 Large twin-shaft mixer with clean surfaces, polished to a mirror finish on walls, shafts, and tools (Courtesy of Nara Machinery Co., Ltd. Europe).

The coarse abrasives eliminate the imperfections in the metal surface, such as sanding marks, nicks, and hairlines. With progressively finer grits all scratches disappear, even the invisible ones. The high-gloss finish requires appropriate polishing pastes and sanding belts, discs, and high-speed machines. The grinding result is highly dependent on the skill of the polishers, their expertise, and equipment (grinding wheel machine, sanding pastes). During the "dry" applications in steel polishing, waxes and kerosene are used as a lubricant and a coolant. For the smoothing of stainless steel, the polisher needs in graduated grain sizes not only sanding particles, made of alumina, silica, or zirconia but also aluminum carbide and silicon carbide and sometimes diamond dust. These mechanical methods remove the scratches, but the base material is not eroded appreciably. The result of extensive polishing is shown in Figure 16.2, in the case of a large two-shaft mixer.

16.4.3.3 Pickling

Pickling [5] is the process for removing scale layers after heat treatments, residues from welding, annealing colors, sanding dust, abraded metal, external rust, and chromium carbide (which can arise during drilling without a coolant). By action of the acid (usually nitric acid mixed with hydrofluoric, HF acid), a metal erosion of $1-5\,\mu m$ occurs at $\sim 40\,(\pm 20)\,°C$. For some parts of the plant or for the entire apparatus, the acid bath is applied. Pickling ensures the subsequent formation of a complete passive layer, to protect the metal against corrosion. After rinsing with soft water, the formation of the layer begins in contact with air on the clean surface. Two to 8 h later, the passivation layer is completely present in a thickness of about 20 nm. By the addition of oxidizing agents, such as dilute nitric acid or hydrogen peroxide, the layer is formed immediately on all surfaces. In the food and pharmaceutical industries, the passivation is preferably carried out with citric acid at temperatures above 40 °C over a period of a few hours.

Pickling is advisable before a mechanical polishing; but constitutes an indispensable part of electropolishing. It takes place with acid mixtures in a dipping and spraying process, or by pumping through tubes (Table 16.4). Welds must always

Table 16.4 Acid treatment of stainless steel [8, 9].

Action	Acid (mixture) (nitric, sulfuric, phosphoric acid)	Hydro-fluoric (HF) (%)	Additives	Temperature (°C)	Duration (min)
Pickling "classic"	10–28% HNO_3 (50%)	3–8	0.1% surfactants	15–60	20–300
Pickling "nitrate-free"	10–25% H_2SO_4/H_3PO_4	3–5	1–5% H_2O_2	15–60	20–300
Elektro-polishing	1:1; H_2SO_4 (96%)/H_3PO_4 (86%)	—	—	40–75	2–20

Figure 16.3 Pickling of large equipment in acid baths (right: 90 m³) (Courtesy of Siedentop and Poligrat).

be pickled. If it is not done with the whole apparatus in the bath, the acid is often applied in the form of a paste with the brush. Careful pickling allows the formation of a passive layer on the welds to prevent corrosion. In addition, pickling improves the appearance (Figure 16.3).

To generate a complete passivation, pickling is in use for stainless steels, nickel alloys, titanium, and titanium alloys (Figure 16.4). Normal steel, copper, aluminum, and their alloys are cleaned and pickled before electropolishing, chemical deburring and polishing, anodizing and plating.

16.4.3.4 Electropolishing

Depending on the diameter and length of tubes, a mechanical smoothing of the inner surfaces is difficult to accomplish. And especially for small parts, such as pipe bends, fittings, valves, sensors, and pumps, pickling/electropolishing is the appropriate method. But large apparatuses, preferably for the pharmaceutical and biotechnology industry, are also electropolished. The method [9] begins with a cleaning of the metal surfaces with alkali followed by pickling. Then, electrolysis expires in a mixture of concentrated sulfuric and phosphoric acid at elevated temperatures (see Table 16.4). The steel parts are placed in a basket, which is connected as the anode. By applying a DC voltage (Figure 16.5), the "hills, mountains, and elevations" on the surfaces dissolve, by forming ions. Either

Figure 16.4 Untreated and pickled stainless steel fitting, along with a titanium sheet (Courtesy of Poligrat).

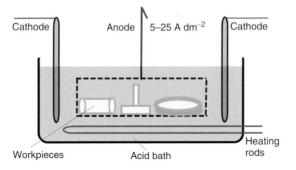

Figure 16.5 Smoothing of steel surfaces by electropolishing (2–20 min at 40–75 °C).

they remain dissolved or fall as mud at the cathode on the ground. During the electropolishing, about 10–40 µm of steel are dissolved from the entire surface, depending on the time, temperature, composition of the acids, and the applied voltage. Thus, not only the craters but also the inhomogeneous superficial layers and ridges disappear. The average roughness is halved by the process. All these advantages especially appreciate the pharmaceutical industry and produced in electropolished plants.

Microscopy images, as shown in Figure 16.6, demonstrate that edges are converted into roundness. For the achievement of the same smoothness (Ra values

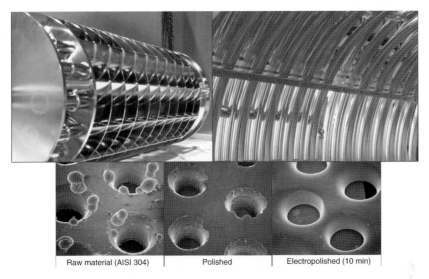

Raw material (AISI 304)   Polished   Electropolished (10 min)

Figure 16.6 Mixers and a heat exchanger, electropolished to prevent the formation of deposits, and a filter plate in 50-fold magnification (Courtesy of Poligrat).

of ~0.2 μm), electropolishing is a superior method. It neither leaves residues nor influences the surface mechanically. The method can be completely reproduced worldwide. In contrast to mechanical polishing, the electropolishing leads, besides the smoothing, to a (desired) significant erosion of inhomogeneous layers in the superficial areas. During mechanical polishing, the surfaces are (minimal) deformed by heat and pressure. Often, there remain microscopic scratches and traces of metal dust and polishing paste. Advantageously, the mechanical smoothing allows to start with raw material surfaces of much higher average roughness (> 3 μm). By using pastes and sanding belts of various grits, especially when successively applied, there is a gradual improvement of the surface. In contrast, the electropolishing is usually carried out with a material having an average surface roughness of about 0.5 μm. Therefore, some parts (welding seams) must often be mechanically preprocessed.

In the electropolishing process, iron and nickel dissolve relatively quickly. Thus, the resulting surface is rich in chromium. Chrome determines the appearance and corrosion resistance by the passivation layer formed. Electropolishing of titanium-stabilized stainless steel is difficult, because TiC does not dissolve, and the carbon content of the steel is relatively high. Therefore, similar alloys without stabilization and with reduced carbon content (AISI 316L, 1.4404, 1.4435) are preferred. After the electrolysis treatment of work pieces, a rinsing with nitric acid takes place. This converts the hardly soluble heavy metal salts (sulfates and phosphates) in slightly soluble nitrates. The metal nitrates can be rinsed off with water, leaving a completely clean metal surface.

16.4.3.5 **Plasma Polishing**

The new electrolytic polishing method for metals provides extremely smooth surfaces with low average roughness values. These are, with Ra values lower than 0.01 μm, more than a factor of 10 below electropolishing. In the procedure, the (untreated) metallic work pieces are placed in a bath and connected to the anode. The bath consists, in contrast to electropolishing, not of acids but of an aqueous salt solution (2%). According to information from Plasotec, by applying a high voltage (>300 V), a process-induced ignition and spreading of plasma occurs. This encloses the work pieces as a thin gas layer. Thermal, chemical, and electrochemical processes give a shine on the surfaces that is not achievable with other polishing methods. The temperature of the work pieces remains below 100 °C. A minimal weight loss resulted during the treatment, caused by the leveling of microroughness and the removal of ridges. Furthermore, organic and inorganic surface contaminants are removed by oxidation.

The amount of abraded metal can be controlled by the processing time. The typical metal removal is 3–30 μm at a rate of 3–10 μm min^{-1}, wherein the geometric shape is maintained. Plasma-polished metals exhibit higher corrosion resistance than the starting material. Figures 16.7 and 16.8 demonstrate the results of the plasma polishing on photo shoots in the microscope. The first example is a laser-drilled stainless steel sheet together with additional images from the scanning electron microscope (SEM). In the second example, a rasp made of a

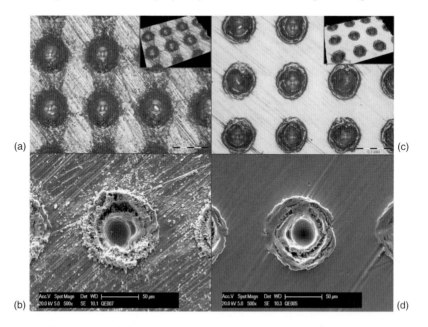

(a)

(b)

(c)

(d)

Figure 16.7 Laser boreholes spaced 150 μm in stainless steel (AISI 316L, 1.4404); microscopy and SEM images; (a, b) initial state and (c, d) after plasma polishing (Courtesy of Plasotec).

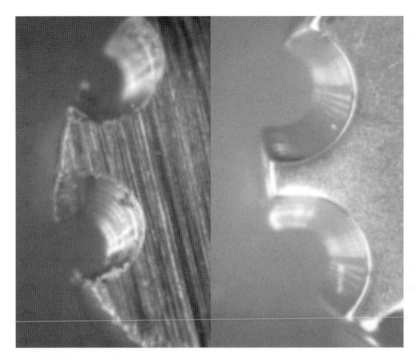

Figure 16.8 Effect of plasma polishing, demonstrated on a tool made of a titanium alloy (Courtesy of Plasotec).

titanium alloy is shown, which obtains a completely smooth surface by plasma polishing.

16.5
Nonferrous (NF) Metals and Alloys

Nickel (Ni 99.2, alloy 200, 2.4066) shows under reducing conditions (absence of air) in the entire temperature range an excellent resistance to alkalis. Furthermore, nickel is resistant to dry halogens and hydrogen halides. In weakly oxidizing media, it forms an oxide layer for protection against acids. But in highly oxidizing media such as nitric acid or bleach, nickel may corrode rapidly. Caution is advised when using other acids and/or in the presence of chloride ions from iron, nickel, and copper compounds.

For maximum mechanical and chemical requirements of the material, nickel alloys are used, such as for the construction of some chemical reactors and turbines (aircraft, power plants). They show highest stability and good resistance to pitting, crevice corrosion, and stress corrosion cracking, as well as excellent resistance to acids (e.g., HCl) and caustic solutions, even at high temperatures. Caution is advised in the presence of heavy metal chlorides, similar to pure nickel. Under

Table 16.5 Composition of important nickel alloys (in (%)), ordered by increasing Ni and Mo contents and decreasing Fe and Cr contents; materials for extreme chemical attack.

Alloy No. Brand name Material number	Ni	Fe	Co	Cr	Mo	W	Si	Mn	C
825 Incoloy 2.4858	38,0- 46,0	27,0	-	19,5- 23,5	2,5- 3,5	-	0,5	1,0	0,025
G-3 Inconel, Nicrofer, Hastelloy, 2.4619	39,0- 50,0	18,0- 21,0	5,0	21,0- 23,5	6,0- 8,0	1,5	1,5	1,0	0,015
X Hastelloy 2.4669	47,0	18,5	1,5	22,0	9,0	0,6	1,0	1,0	0,1
625 Inconel, Nicrofer 2.4856	62,0	5,0	1,0	21,5	9,0	-	0,5	0,5	0,1
C4 Nicrofer, Hastelloy 2.4610	65,0	3,0	2,0	14,0- 18,0	14,0- 17,0	-	0,08	1,0	0,015
B3/B4 Hastelloy, Nicrofer 2.4600	65,0	1,0- 6,0	2,5- 3,0	0,3- 3,0	26,0- 32,0	0- 3,0	0,01- 0,05	1,5- 3,0	0,01
B2 Hastelloy 2.4617	69,0	2,0	1,0	1,0	26,0- 30,0	-	0,1	1,0	0,02

the brand names Monel[®], Inconel[®], Hastelloy[®], Nicrofer[®], and others, several comparable nickel alloys of high corrosion resistance are available (Table 16.5). For each application, a suitable alloy exists.

Incoloy[®] (alloy 625) is used industrially for the construction of evaporators for phosphoric acid. The alloy 825 is approved as material for pickling baths, but also in plants for the production of sulfuric and phosphoric acid, and for the concentrating of sodium hydroxide solution. Hastelloy B2 and C4 are also suitable for wet chlorine and any acid, in particular, hydrochloric acid, at elevated temperatures. Monel (400, 401, and 404) is a Ni/Cu alloy with 30–53% Cu and the only resistant metal against fluorine, hydrogen fluoride, and hot concentrated hydrofluoric acid. The superalloys based on nickel represent a continuation of the stainless steel variants in which the Ni- and Mo content is continuously increased at the expense of the iron and chromium content.

Nickel alloys are more frequently used for extreme conditions. The special materials titanium (Ti), zirconium (Zr), and tantalum (Ta) are rarely encountered in chemical apparatuses. Titanium is resistant to dilute sulfuric acid, hydrochloric acid, chloride-containing solutions, cold nitric acid, and most organic acids and bases as well as against sodium hydroxide. In concentrated sulfuric acid, however, it dissolves with the formation of the purple titanium sulfate. A thin, dense zirconium oxide layer is formed on contact with atmospheric oxygen and passivates

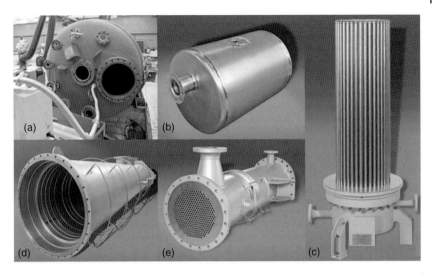

Figure 16.9 Equipment with product wetted parts made of tantalum (Ta). (a) Steel vessel with Ta lining (explosion bonded); (b) Ta-tank; (c) vertical bajonet heater, bottom plate made of steel with a tantalum lining, tantalum tubes (Ta + 2.5% W); (d) column shell of stainless steel with Ta lining ($d_i = 1.1$ m); and (e) heat exchanger made of steel (stainless steel compensator) with Ta + 2.5% W-tubes (Courtesy of Tantec).

the metal. Zirconium is therefore insoluble in acids and alkalis. Only aqua regia and hydrofluoric acid can attack the metal.

Comparable to zirconium, tantalum (Ta) is protected by a thin superficial oxide layer of tantalum (V)-oxide. Because of the strong passivation, tantalum is insoluble in acids, even in aqua regia, and can be used therefore almost universally up to 150 °C (Figure 16.9). The chemical and pharmaceutical apparatus engineering used tantalum often as alloy with 2.5% tungsten for increased stability. The metal is attacked only by hydrofluoric acid and oleum (a mixture of sulfuric acid and sulfur trioxide) and molten salts. As base metal, tantalum reacts with oxygen or halogens at temperatures above 300 °C.

The prices vary greatly. They are dependent on the quality and application shape. In 2013, the following prices per kilogram were published for the metals: Ti (ingot): €15–20.50, Zr (99.6%): €25–115, Ta (> 99.8%): €290–400, calculated with an exchange rate of 1.34 $/€. The high price of tantalum is relativized by the long life of the equipment and the high credit note after scrapping the system. An estimate from 1997 gives the following price ratios [10], if the stainless steel AISI 316Ti is set to 100%: Ti = 270%, Hastelloy = 470%, Ta = 2070%, and graphite = 145%.

Using platings, linings, or superficial coatings, the base material of the plant and machinery is protected and separated from the corrosive medium. Stainless steels, special alloys, and superalloys can be applied over rolling plating or explosion plating (about 3 mm thick plates). For this, various methods have been developed [11].

16.6
Inorganic Nonmetallic Materials

Inorganic materials such as borosilicate glass or graphite are in use because of their extreme resistance to most aggressive chemicals. The glass can be combined with superalloys for heating and vaporizing. Other inorganic compounds are suitable for the formation of a coating layer on a base material such as steel. These coatings are performed by spraying on multiple layers, and stand in competition to the plastic and plastic inliners. The layer thickness reaches 300 μm. Chemical resistances, pressure, and temperatures as well as the mechanical load capacity determine the best coating material. Known examples are enamel, aluminum oxide, and zirconium dioxide. Mixers, reactors, absorber columns, membranes, mills, pumps, and many other machinery can be coated with corrosion-resistant oxides or carbides using a spray method. In addition to the known methods (flame and high-velocity flame spraying with powders, arc and plasma spraying, and cold spraying) and its variants (vacuum arc and vacuum-plasma and high-frequency plasma spraying), there are new processes such as laser spraying and molten bath [11]. Except the sinter and melt procedures, the aluminum oxide can be processed in the form of tiles, mosaic tiles, and strips [12].

Besides the carbides and oxides, almost all metallic materials and alloys can be applied with these spray processes. The metal in the form of a wire, rod, or powder melts in the spray gun. In the electric arc spraying, the metal wire is melted at 4000 °C at the top and applied via an inert carrier gas (Ar or N_2) on the workpiece in a layer thickness of 0.2–20 mm. This is also possible for the super alloys and special metals.

16.6.1
Borosilicate Glass

Owing to its extreme resistance to aggressive chemicals and superior hygienic properties, borosilicate glass has been proven in many laboratory and pilot plants as well as in small productions. The pharmaceutical industry often uses this material, which is assigned to the highest (best) hydrolytic class 1 (a measure of the alkalinity of the glass according to ISO 719). Borosilicate glass is resistant to all chemicals except hydrofluoric acid and hot phosphoric acid, and is characterized by high thermal stability and resistance to sudden temperature changes. Possible temperatures are −50 to 200 °C, limited by the bellows and sealing materials. The glass itself withstands −196 °C (liquid nitrogen) up to 250 °C. Many parts of a production line may consist of glass in combination with other special materials, for example, tantalum or Hastelloy, for processing the corrosive chemicals (Figure 16.10). The glass limits the size of tanks, reactors, and pipelines and the height of the operating pressure. Only atmospheric, reduced pressure or minimal excess pressure can be applied, depending on the size of the components.

Also, to observe processes such as the operations in the apparatuses, the installations are built of transparent material. During the tension-free construction

Figure 16.10 Plants partly or entirely made of borosilicate glass (a) small pharmaceutical plant; (b) condenser; (c) extraction and rectification column; (d) gas absorber; and (e) chemical reactor (Courtesy of QVF®).

of the system, it is necessary, similar to using graphite, to pay attention to the fragility of small components.

16.6.2
Vitreous Enamel

When carrying out reactions *under pressure*, enameled reactors (Figure 16.11), tanks, columns, pipes, and valves are used for aggressive substances in the pressure range of −1 to 6 bar with temperatures of −25 to 200 °C. On the wetted steel surfaces of

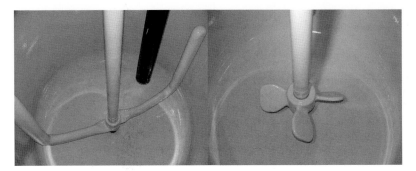

Figure 16.11 Anchor and turbostirrer in enameled chemical reactors (Jurec, Wikipedia, GNU Free Documentation).

the stirred vessel, the manufacturer applies two to five layers of a thickness from 1.4 to 2 mm, up to a volume of $80\,m^3$ (EN 15159-01:2006). The enamel layer is made of glass-forming oxides such as borax (B_2O_3), quartz (SiO_2), and feldspar (Si, Al, B, K, Na ...), besides clay (Al_2O_3) and various heavy metal oxides. Enamel shows extremely good chemical resistance in an acidic environment, as with all glasses. Because of the chemical properties of silicates, they are not suitable for the alkaline region.

16.6.3
Graphite

Graphite is used in chemical apparatus engineering for heat exchangers, coolers, and condensers as well as falling-film evaporators and absorbers, especially for the production and processing of mineral acids (such as hydrochloric, sulfuric, hydrofluoric, and phosphoric acid). For heat exchangers, large heat exchange areas exist (over $2000\,m^2$). The standard version operates up to $6\,bar/180\,°C$ (Figure 16.12), special designs up to 12 bar.

Graphite tubes are also used as support material for micro- and ultrafiltration membranes with zirconia or alumina as the filter layer.

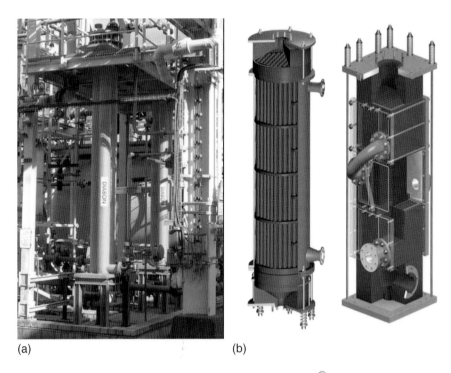

(a)   (b)

Figure 16.12 (a) Shell and tube heat exchanger with Diabon®-tubes and (b) a graphite block heat exchanger (Courtesy of SGL-Carbon).

16.7
Plastics

To ensure chemical stability, coatings are in use at present [13]. These originate not only from metals (e.g., superalloys) but also from nonmetals. As discussed, typical inorganic materials are oxides, graphite, and glasses. Widespread organic materials represent the different polymers with interesting properties regarding the processability and chemical stability [14].

In chemical plants, various plastics are used, mainly because the plastic materials offer high corrosion resistance and cost advantages in comparison to other variants. Examples (Figure 16.13) are containers, large tanks, big bags, silos, piping, pumps, valves, bellows, and sealings, in which the liquids come in contact only with plastic. The chemical resistance of steel machinery and equipment, especially of fiberglass tanks, is ensured by lining and coating with suitable plastics (Table 16.6) [15]. Alternatively, the manufacturer uses plastic inliners in containers or equipment made entirely of plastic. Inliners are hanged bags without a fixed connection to the base material.

The stability and chemical resistance of plastics is limited in temperature. All plastic materials tolerate 50 °C, several plastics 90 °C, and only a few (polytetrafluoroethylene, PTFE and similar compounds; see Table 16.6) even up to 200 °C. The chemical resistance must be verified in detail in corrosion tables and/or in experiments. High attention is always necessary for polar solvents (acetone), aromatic and chlorinated hydrocarbons, as well as for strong acids and bases (Table 16.7).

Sealless pumps of high chemical stability up to 180 °C are coated with plastics (Figure 16.14). Aggressive, hazardous liquids can be transported safely with a PFA-lined (perfluoroalkoxy) centrifugal pump, equipped with a magnetic drive (conforms to DIN EN 22858). This sealless centrifugal pump is resistant to

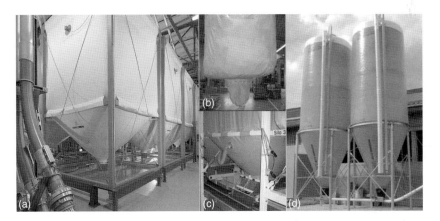

Figure 16.13 Storage silos for bulk materials: (a) silos with flexible walls for plastic pellets; (b) emptying of a big bag; (c) discharge aid for a flexible silo; and (d) glass-fiber-reinforced polyester silo (Courtesy of A.B.S. Silo and conveyor systems (a–c) and Hans Gaab (d)).

Table 16.6 A method for lining and coating of equipments and components with plastics; polymers partially reinforced with fibers, nonwovens, mica, carbon fibers, and glass slides.

Process	Principle	Plastic (examples, abbreviations see also Table 16.7)	Layer thickness (mm)
Lining with plates, tiles	Stick on and weld together of preformed panels	PE, PP, PVC, PVDF, PU, (PU)-elastomer (Vulkollan®), GFK, PFA, FEP, E-CTFE	3–20
Lining with sheets or laminates	Stick on and weld together; application and hardening of impregnated glass fiber mats	HDPE, PVC, PIB, rubber, resins (epoxy, furan, vinyl ester, polyurethane)	3–12
Smooth over, paint	Applying a ceramic reinforced epoxy resin, or other systems	Various resins (epoxy, vinyl ester, polyester)	About 2
Dipping	All-round seamless coating	PVC systems with adhesion promoter	1–8
Thermal spraying (primarily flame spraying)	Melting in the spraying unit, heat supply to the workpiece by the flame	All thermoplastics, sprayable resins (epoxy, novolac, vinyl ester, bisphenol, and mixtures thereof), and rubbers (systems)	0.03–3
Injection molding in an adapted form	Insertion and fixing of the moldings	Many thermoplastics	About 0.5–5
Powder coating	Sprinkle the powder on the workpiece by a flat spray nozzle and then the powder melts by heat	PE, PA, PU, powder coatings, epoxy, and polyester resins	0.015–0.03
Electrostatic powder coating	Negatively charged powder meets positively charged component and then the powder melts by heat	PE, PA, PVC, polyester, epoxy, and polyacrylate resins	0.04–0.6 (up to 1 mm)
Fluidized bed sintering	Placing the preheated component in a fluidized bed, and the powder melts at the surface	PE, PA, PES, epoxy resin	About 0.1–0.3
Sewer renovation	A long hose with a wound up resin/hardener soaked tissue is inflated and hardens in the old pipe	Unsaturated polyester resins (UP resins), according to DIN 18820	About 1–3

FEP, fluorinated ethylene propylene; E-CRFE, ethylene chloro trifluoro ethylene; HDPE, high density polyethylene; PIB, polyisobutylene; PA, polyamide; and PES, polyethersulfone.

Table 16.7 Plastic materials in chemical apparatus engineering for tanks, containers, big bags and silos (t), inliners and lining material (i), pipes (p), coatings (c), and sealings (s).

Material (chemical name)	Abbreviations	Application	Maximum operation temperature (°C)[a]	Characteristics of corrosion resistance
Polyvinyl chloride (without plasticizer)	PVC	p	60	Resistant to most acids, alkaline, and salt solutions, furthermore to water-miscible organic compounds Not resistant to aromatic and chlorinated hydrocarbons
Polyvinyl chloride after chlorinated	PVC-C	p	90	Same features as for PVC, but for higher temperatures
Polyethylene	PE	t, p	60 (40–90, increasing with crystallinity)	Resistant to aqueous solutions of acids, alkalis, salts, and several organic liquids Not resistant to oxidizing acids
Polypropylene (heat stabilized)	PP	t, i, p, c	90	Analog to PE, but for higher temperatures
Polyvinylidene fluoride	PVDF	i, p, c, s	130	Resistant to acids, salt solutions, aliphatic and aromatic and chlorinated hydrocarbons, alcohols, and halogens Limited resistance to ketones, esters, ethers, organic bases, and alkalis
Polytetrafluoro-ethylene	PTFE	i, p, c, s	180–200	Resistant to almost all chemicals
Perfluoroalkoxy-polymers (Teflon, and other)	PFA	i, p. c, s	180–200	Similar to PTFE, but meltable for processing
Glass-fiber reinforced polyester resin	GFK-UP	t, p	80	Resistant to acids and aqueous salt solutions, oils, and glycerol Not resistant to many organic hydrocarbons
Glass-fiber reinforced epoxy resin	GFK-EP	t, p	100	Analog to GFK-UP
Nitrile-butadiene rubber	NBR	i, c, s	−30 up to 70 in aqueous, 90 in gaseous systems	Resistant to oil, gasoline, and aliphatic hydrocarbons Not resistant to oxidizing substances, benzene, esters
Fluorinated hydrocarbon (e.g., Viton®)	FKM	s	−25 up to 200	Resistant to ozone, oils, and many organic solvents, best chemical stability of all elastomers Not resistant to polar solvent (acetone), glycols, organic acids, and bases

(continued overleaf)

Table 16.7 (Continued)

Material (chemical name)	Abbreviations	Application	Maximum operation temperature (°C)^a	Characteristics of corrosion resistance
Ethylene propylene diene monomer	EPDM	s	90	Resistant to ozone, and to many aggressive chemicals, also diluted acids Not resistant to oils, fats, and petroleum products
Polyacrylate rubber	ACM	i, c, s	−10 up to 150	Resistant to ozone, hot oil, and oxidizing environment Not resistant to moisture, acids, and bases
Silicone rubber	VMQ (ASTM)/ MVQ (Din, Iso)	s	−55 up to 100 (oxygen, water), 210 for hot air	Resistant to oil and fats, glycols, ozone
				Not resistant to aromatic and chlorinated hydrocarbons, superheated steam, acids, and bases
Polyurethane	PU	i, c, s	−50 up to 130	Lining and coating material; PU forms resistant casted seals, single use only Resistant to oils and fats, diluted acids Not resistant to bases

^{a}Depending on the application.

corrosion. The liquids come in contact only with PFA (polymer similar to Teflon®). The polymer is a further development of PTFE and can be processed as a typical thermoplastic. In another type of pumps, the housing and the impeller are made completely of plastics, accordingly to the chemical and thermal requirement, out of PP (polypropylene) or PVDF.

GRP represents a fiber-plastic composite made of a plastic and fiberglass [16]. As base, thermosets (e.g., polyester or epoxy resin) and thermoplastics (PA, polyamide) are used. GRP (also called glass-fiber-reinforced, GFK) shows an excellent corrosion resistance in aggressive environments, especially in mineral acids (Figure 16.15). In addition, in special cases, a lining improves the chemical resistance. Therefore, GRP is not only a suitable material at normal and moderate temperatures (up to 80 °C) for the production equipment and container in the chemical industry but also an economical solution in comparison to stainless steel. Pure GRP is problematic for use in the pharmaceutical, biotechnology, and cosmetic industries, because microorganisms can penetrate into the material, unattainable for the cleaning solutions. In these and similar applications, polished stainless steel represents the better material.

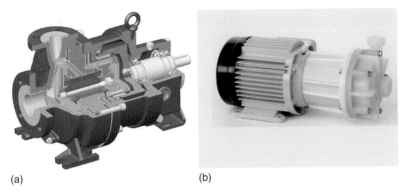

(a)   (b)

Figure 16.14 Two different sealless centrifu- CP Pumps). (b) Impeller and housing com-
gal pumps with magnetic drive for aggres- pletely made of plastic (Courtesy of Schmitt
sive liquids: (a) PFA-lined pump (Courtesy of centrifugal pumps).

(a)   (b)

Figure 16.15 GFK storage tanks for liquids: (a) demineralized water tank PVDF/GRP and
(b) two acid tanks for 33% hydrochloric, each 1000 m^3 in size (Courtesy of Christen &
Laudon).

16.8
Learnings

√ Under the operating conditions, the materials must be chemically resistant.

√ Mainly, the manufacturers prefer highly alloyed CrNi-stainless steels as standard material (AISI 304 and 316 grades).

√ For avoiding pitting and stress corrosion, molybdenum is alloyed, especially
in stainless steels with high resistance (AISI 904L = 1.4539).

√ Food, pharma, and biotech industries use stainless steel machinery in mirror
finish. This can be reached by mechanical polishing or electropolishing.

√ After the processing of stainless steel, the welds must be pickled, to achieve
the corrosion protection of the chrome.

√ For aggressive chemicals, such as mineral acids at higher temperatures, Ni-superalloys have proven themselves.

√ Stainless steel, superalloys, special metals, oxides, carbides, and vitreous enamel can be applied on a base material as lining by thermal spraying in multiple layers.

√ The chemical industry uses different plastics as linings and coatings for equipment in steel (pumps, tubes, containers, and tanks) and in GRP (tanks and silos).

References

1. EHEDG (2004) *Hygienic Equipment Design Criteria*, 2nd edn, EHEDG publication. *http://www.ehedg.org/uploads/DOC_08_E_2004.pdf* (accessed 06 December 2013).
2. Hauser, G. (2008) *Hygienegerechte Apparate und Anlagen*, Wiley-VCH Verlag GmbH, Weinheim.
3. Bode, U. and Wildbrett, G. (2006) Anforderungen an Werkstoffe und Werkstoffoberflächen bezüglich Reinigbarkeit und Beständigkeit. *Chem. Ing. Tech.*, **78** (11), 1615–1622.
4. Behrens, D., Kreysa, G., and Eckermann, R. (1988–1992) *DECHEMA Corrosion Handbook – Corrosive Agents and Their Interaction*, Vols. 1–12, Wiley-VCH Verlag GmbH.
5. Craig, B. and Anderson, D. (1995) *Handbook of Corrosion Data*, Materials Data Series, ASM International.
6. Davis, J.R. (2000) *Corrosion, Understanding the Basics*, ASM International.
7. Hess, D. *et al.* (2006) *Beizen von Edelstahl rostfrei*, Merkblatt 826, 3 Aufl., Informationsstelle Edelstahl Rostfrei, Düsseldorf.
8. Bettenworth, E. *et al.* (2012) *Die Verarbeitung von Edelstahl rostfrei*, Merkblatt 822, 4 Aufl., Informationsstelle Edelstahl Rostfrei, Düsseldorf.
9. Kosmac, A. (2010) *Elektropolieren nichtrostender Stähle, Erste Ausgabe*, Reihe Werkstoff und Anwendungen, Bd 11, Euro Inox, Merkblatt 974, Informationsstelle Edelstahl rostfrei, Düsseldorf.
10. Blass, E. (1997) *Entwicklung verfahrenstechnischer Prozesse*, 2. Aufl., Springer, Berlin.
11. Kuron, D. (1997) in *Apparate: Technik - Bau – Anwendung* (ed. B. Thier), 2. Aufl., Vulkan-Verlag, Essen, S. 246–270.
12. Kalweit, A., Paul, C., Peters, S., and Wallbaum, R. (eds) (2006) *Handbuch für Technisches Produktdesign; Material und Fertigung, Entscheidungsgrundlage für Designer und Ingenieure*, Springer, Berlin.
13. Schweitzer, P.A. (2006) *Paint and Coatings: Applications and Corrosion Resistance*, CRC Press.
14. Schweitzer, P.A. (2007) *Corrosion of Polymers and Elastomers*, 2nd edn, CRC Press.
15. Revie, R.W. (ed.) (2011) *Uhlig's Corrosion Handbook*, 3rd edn, John Wiley & Sons, Inc., Hoboken, NJ.
16. Rosato, D.V. and Rosato, D.V. (2004) *Reinforced Plastics Handbook*, Elsevier Ltd.

17
Principles of Product Design

Summary

Product Design provides an earlier market entry and improved the sale. At the beginning of the development, a specification list is worked out for defining the targets. The main features are the involvement of the customer and the marketing. The development is performed from highly motivated team members with the support of the top management.

17.1
Characteristic Features

The characteristic features (1–8) of a good product design, performed in development projects, can be briefly described as follows:

1) Involvement of the customer
   a. Industrial customers participate and interact directly in the development projects.
   b. Consumers take part indirectly, by attending as focus groups or by testing of products in the household (HUT, home-use test).
2) Assessment of proposals for a novel product (elaborated on by the customer, by own employees, or inspired by the competitor)
   a. Feasible product ideas are evaluated according to a scoring table.
   b. With positive outcome, the development starts.
3) Involvement of marketing
   a. Important, large projects are managed by a team consisting of a chemist and a chemical engineer as well as an employee in marketing.
   b. For small projects, the project manager (chemist/chemical engineer) keeps close contact with the marketing team.
4) Support of the top management
   a. The top management provides the necessary resources (personnel, money, rooms, and equipment).

Industrial Product Design of Solids and Liquids: A Practical Guide, First Edition. Wilfried Rähse.
© 2014 Wiley-VCH Verlag GmbH & Co. KGaA. Published 2014 by Wiley-VCH Verlag GmbH & Co. KGaA.

    b. One member of the top management steers the project and checks the progress and the timetable from time to time.

5) Engaged project team
    a. Motivated by the top management and the team leaders (core team).
    b. Members take responsibility for subprojects.

6) Elaboration of a specification list
    a. First, the specifications of the new product are defined. The successful completion of a development project depends on the strict compliance of this list. Changes lead to project delays or termination.
    b. The specification list includes several parameters concerning the product performance in the application, and for the handling and esthetics.

7) Expiration of the development
    a. The core team assigns the work to different groups and experts, which is performed inside and outside of the company, for example, in cooperation with a customer, a raw material supplier, and machinery manufacturers.
    b. The core team meets every week (or every 10 days) for closely controlling the progress.
    c. All works run parallel, from laboratory up to scale-up preparation and pilot trials as well as the development of packaging and market research.
    d. The customers can examine and evaluate the products out of the pilot plant.
    e. After intensive tests in the laboratory that is responsible for these applications, the product is ready for production.

8) Speedy implementation
    a. The basic design of the responsible chemical engineer serves as the basis for adapting an existing facility or for building up a new production facility.
    b. A rapid market entry leads to immediate revenue, so that possible improvements in the production can be paid without any problems (speed before accuracy).

17.2
Targeted Production of Particles and Fluids from Different Raw Materials

The kind of raw materials that must be processed for reaching the product particles in the corresponding unit operations is shown in Figure 17.1. Each process also includes a grinding-sieve procedure (not shown) to adjust the particle size distribution. If necessary, an additional improvement of the shape ensues in the spheronizer. Individual processes are described in detail from Chapters 5–8. The product design of liquids ensues on the one hand via solid or fluid renewable resources, as shown in Figure 17.2 with several examples. Biomass, milk, plant parts, fruits, and vegetables as well as seeds, and others serve as starting materials (Chapter 12). A similar view in Figure 17.3 starts predominantly with synthetic substances and allows the representation of the broad spectrum of liquid supply forms. Figure 17.4 shows the variants when using a liquid as starting material, and Figure 17.5 when using a melt.

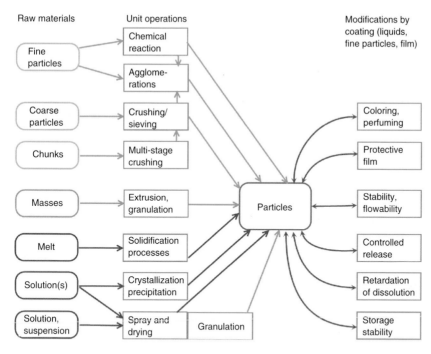

Figure 17.1 Processes in product design for the manufacturing of tailor-made particles.

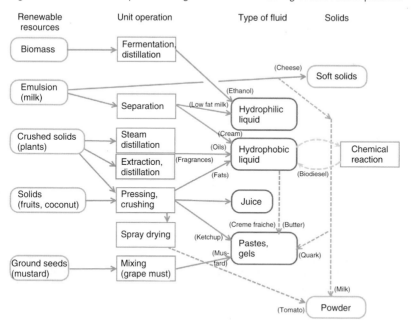

Figure 17.2 Different types of liquids from renewable resources with examples (see also Figures 12.2, 12.3, and 12.9).

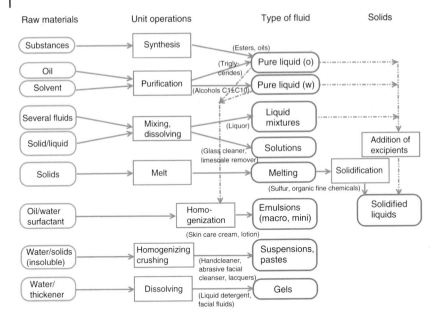

Figure 17.3 Different types of liquids with examples.

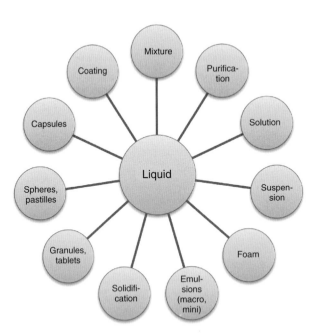

Figure 17.4 Possibilities for the product design of a liquid (see also Figures 12.11 and 12.12).

Raw material   Unit operation   Soild shape

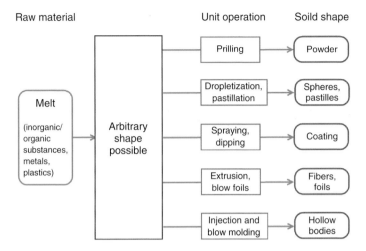

Figure 17.5 Arbitrary solid shapes by solidification of melts (see also Table 9.6).

17.3
Learnings

Finally, dealing with the product design gives many excitations and leads to innovations:

√   Make the development of new products and processes to your passion. Do dare to take on difficult tasks. Much can be learned and implemented quickly, also in cooperation with colleagues and employees.

√   Consider the processes without the usual boundaries, regardless of whether it is mechanical or thermal process engineering, organic or inorganic chemistry or plastics, biotechnology, or pharmaceutical.

√   Learn from other disciplines and people and from literature. This leads to broadening of views and knowledge.

√   Stay in contact with all team members, especially with the marketing people.

√   Focus the development on the needs of your customers.

List of Companies

These are the companies providing the images used in this book.

A.B.S. Silo- und Förderanlagen GmbH
Industriepark 100
D-74706 Osterburken
Phone: +49 6291 6422-0
E-mail: info@ABS-Silos.de
Internet: www.abs-Silos.de

ATS License GmbH
Orchideenstr. 13
D-76437 Rastatt
Phone: +49 7222 988 074
E-mail: info@ats-license.com
Internet: www.ats-license.com

BAYER AG
Kaiser-Wilhelm-Allee 1
D-51368 Leverkusen, Germany
Phone: +49 214 30-1
Internet: www.bayer.com

BHS-Sonthofen GmbH
An der Eisenschmelze 47
D-87527 Sonthofen
Deutschland
Phone: +49 8321 6099-0
E-mail: info@bhs-sonthofen.de
Internet: www.bhs-sonthofen.de

BIOZYM, Kundl, Austria
(production closed 2011)

CHRISTEN & LAUDON GmbH
D-54634 Bitburg-Staffelstein
Phone: +49 6563 51-0
E-mail: info@christen-laudon.de
Internet: www.christen-laudon.de

COGNIS (now part of the BASF Group) BASF Personal Care and Nutrition GmbH
Rheinpromenade 1
D-40789 Monheim
Phone: +49 2173 4995-0
Internet: www.cognis.com

CP Pumpen AG
Im Brühl
CH-4800 Zofingen/Switzerland
Phone: +41 62 746 85 85
E-Mail: info@cp-pumps.com
Internet: www.cp-pumps.com

Parfümerie DOUGLAS GmbH
Kabeler Straße 4
D-58099 Hagen
Phone: +49 800 - 690 690 5
E-Mail: service@douglas.de
Internet: www.douglas.de

Industrial Product Design of Solids and Liquids: A Practical Guide, First Edition. Wilfried Rähse.
© 2014 Wiley-VCH Verlag GmbH & Co. KGaA. Published 2014 by Wiley-VCH Verlag GmbH & Co. KGaA.

Maschinenfabrik Gustav EIRICH GmbH & Co KG
Walldürner Str. 50
D-74736 Hardheim
Tel.: +49 6283 510
E-mail: eirich@eirich.de
Internet: www.eirich.de

EKATO HOLDING GmbH
P. O. Box 10 70
D-79010 Freiburg
Phone: +49 7622 29-0
E-mail: info@ekato.com
Internet. www.ekato.com

The FITZPATRICK Co. Europe:
Entrepotstraat 8
B-9100 Sint-Niklaas, Belgium
Phone: +32 3780 6200
E-mail: info@fitzpatrick.be
Internet: www.Fitzpatrick.be
Corporate Headquarters:
832 Industrial Drive
Elmhurst, IL 60126 USA
Phone: +1 630 530-3333
E-mail: info@fitzmill.com

FLEXXOLUTIONS BV
Faraday Weg 4
NL-7591 HD Denekamp
Phone: +31 541 760400
E-mail: info@flexxolutions.nl
Internet: www.flexxolutions.com

FrymaKoruma GmbH
Fischerstrasse 10
D-79395 Neuenburg
Phone: +49 7631 7067-0
E-Mail: frymakoruma
   @romaco.com
Internet: www.frymakoruma.com

Hans GAAB
Ammonschönbronn 8
D-91632 Wieseth
Phone: +49 9855-567
E-mail: gaab-hans@t-online.de
Internet: www.hans-gaab.de

GEA NIRO
Mauchener Str. 5
D-79379 Müllheim
Phone: +49 7631 70539-0
E-Mail: gea-niro.info@gea.com
Internet: www.niro.com

GEA Niro Soavi
Via A. M. Da Erba Edoari 29
I-43123 Parma/Italien
Phone: +39 0521 -965411
E-mail: info.GeaNiroSoavi
   @geagroup.com
Internet: www.niro-soavi.com

GEA TUCHENHAGEN GmbH
Am Industriepark 2-10
D-21514 StadtBüchen
Phone: +49 4155-49-0
E-Mail: geatuchenhagen
   @gea.com
Internet: www.tuchenhagen.com

GLATT Ingenieurtechnik GmbH
Nordstrasse 12
D-99427 Weimar
Phone: +49 3643 47-0
E-mail: info@glatt-weimar.de
Internet: www.glatt.com

HENKEL AG & Co. KGaA
Henkelstraße 67
D-40589 Düsseldorf
Phone: +49-211-797-0
Internet: www.henkel.com

HOSOKAWA ALPINE AG
Peter-Dörfler-Str. 13-25
81699 Augsburg
Phone. +49 821 5906-0
E-mail: marketing
    @alpine.hosokawa.com
Internet:
www.hosokawa-alpine.com

**JÄCKERING Mühlen- und
Nährmittelwerke GmbH**
Vorsterhauser Weg 46
59067 Hamm
Phone: +49 2381 4220
E-mail: info@jaeckering.de
Internet: www.jaeckering.de

Amandus KAHL GmbH & Co. KG
Dieselstraße 5 - 9
D-21465 Reinbek
Phone: +49 40 72 771 -0
E-mail:
info@amandus-kahl-group.de
Internet:
www.amandus-kahl-group.de

**KORZILIUS (production closed
2013)**

KURARAY Europe GmbH
Philipp-Reis-Straße 4
D-65795 Hattersheim am Main
Phone: +49 69 305 85 300
E-Mail: info@kuraray.eu
Internet: www.kuraray.eu

**LIHOTZKY (acquired by
HB-Feinmechanik)
HB-Feinmechanik GmbH & Co.
KG**
Finsinger Str. 1
D-94526 Metten
Phone: +49 991-91 07-0
E-mail: sales@hb-fein.de
Internet: www.hb-fein.de

**Gebrüder LÖDIGE Maschinenbau
GmbH**
Elsener Straße 7 - 9
D-33102 Paderborn
Phone: +49 5251 309-0
E-Mail: info@loedige.de
Internet: www.loedige.de

**MENKE Industrieverpackungen
GmbH & Co. KG**
Beckedorfer Bogen 7
D-21218 Seevetal
Phone: +49 410558 522-0
E-mail: info@menke-
industrieverpackungen.de
Internet: www.menke-
industrieverpackungen.de

NARA Machinery Co., Ltd.
Zweigniederlassung Europa
Europaallee 46
D-50226 Frechen
Phone: +49 2234-2776-0
E-mail: contact@nara-e.de
Internet: www.nara-e.de

Erich NETZSCH GmbH & Co. Holding KG
Gebrüder-Netzsch-Straße 19
D-95100 Selb
Phone: +49 9287 75-0
E-mail: info@netzsch.com
Internet: www.netzsch.com

PALLMANN Maschinenfabrik GmbH & Co. KG
Wolfslochstrasse 51
D-66482 Zweibrücken
Tel.: + 49 6332 802 0
E-Mail: process@pallmann.eu
Internet: www.pallmann.eu

PFEIFER & LANGEN GmbH & Co. KG
Aachener Straße 1042 a
D-50858 Köln
Phone: +49 221 4980 - 0
E-mail: info@pfeifer-langen.de
Internet: www.pfeifer-langen.com

PLASOTEC
Grünauer Fenn 42,
D-14712 Rathenow
Phone: +49 3385 51 98 97 - 0
E-mail: info@plasotec.de
Internet: www.plasotec.de

A. POHLI GmbH & Co. KG
Hölker Feld 10 - 12
D-42279 Wuppertal
Phone: +49 202 648 24 - 0
E-Mail: info@pohli.de
Internet: www.pohli.de

POLIGRAT GmbH
Valentin-Linhof-Strasse 19
D-81829 München
Phone: +49 89 42778-0
E-mail: info@poligrat.de
Internet: www.poligrat.de

QVF® DE DIETRICH Process Systems GmbH
Hattenbergstr.36
D-55122 Mainz
Phone: +49 6131 9704 -0
E-mail: info@qvf.de
Internet: www.qvf.de

Alfred RITTER GmbH & Co. KG
Alfred-Ritter-Str. 25
D-71111 Waldenbuch
Phone: +49-7157-97 0
E-Mail: info@ritter-sport.de
Internet: www.ritter-sport.de

SAMSUNG
Samsung Electronics GmbH
Am Kronberger Hang 6
D-65824 Schwalbach / Ts
Internet: www.samsung.com

SANDVIK Holding GmbH
Heerdter Landstrasse 243
D-40549 Düsseldorf
Phone: +49 211 5027-0
Internet: www.sandvik.de
Headquarters: Sandvik AB
Kungsbron 1
SE-101 30 Stockholm
Phone: +46 8 456 11 00

SCHMITT-Kreiselpumpen GmbH & Co. KG
D-76262 Ettlingen
PO-box: Postfach 842
Tel: +49 7243-5453-0
E-mail: info@schmitt-pumpen.de
Internet:
www.schmitt-pumpen.de

SGL CARBON SE
Söhnleinstrasse 8
D-65201 Wiesbaden
Phone: +49 611 6029-0
E-mail: cpc@sglgroup.com
Internet: www.sglgroup.com

SIEDENTOP GmbH
Neckarstr. 9
D-38120 Braunschweig
Phone: +49 531 889393-0
E-mail: siedentop@top-beizen.de
Internet: www.top-beizen.de

TANTEC GmbH
Rodenbacher Chaussee 6
Gebäude: 801 / TGZ
D-63457 Hanau
Phone: +49 6181 90669-0
E-Mail: info@tantec-group.com
Internet: www.tantec-group.com

Apollo VREDESTEIN GmbH
Rheinstraße 103
D-56173 Vallendar
Phone: +49 261 807 6600
E-mail: customer.de
@apollovredestein.com
Internet: www.vredestein.com
Headquarters: P.O. Box 27
NL-7500 AA Enschede
Netherlands

Index

Industrial Product Design of Solids and Liquids: A Practical Guide, First Edition. Wilfried Rähse.
© 2014 Wiley-VCH Verlag GmbH & Co. KGaA. Published 2014 by Wiley-VCH Verlag GmbH & Co. KGaA.

YOU C

AN INTERACTIVE ADVENTURE

CAN YOU
BECOME A
PRO ATHLETE?

BY MATT DOEDEN

CAPSTONE PRESS
a capstone imprint

Published by Capstone Press, an imprint of Capstone.
1710 Roe Crest Drive
North Mankato, Minnesota 56003
capstonepub.com

Library of Congress Cataloging-in-Publication Data
Names: Doeden, Matt, author.
Title: Can you become a pro athlete? : an interactive adventure / by Matt Doeden.
Description: North Mankato, Minnesota : Capstone Press, [2022] | Series: You choose
: chasing fame and fortune | Includes bibliographical references. | Audience: Ages 8-12
| Audience: Grades 4-6 | Summary: "Do you have what it takes to compete against the
world's most elite athletes? Be ready to put in the time, sweat, and tears that it takes.
Choose which path to take on your journey to athletic excellence. Some choices lead to
the big leagues, while others introduce other opportunities, or even a fall from grace"—
Provided by publisher.
Identifiers: LCCN 2021029952 (print) | LCCN 2021029953 (ebook) | ISBN
9781663958983 (hardcover) | ISBN 9781666323856 (paperback) | ISBN
9781666323863 (pdf)
Subjects: LCSH: Sports—Vocational guidance—Juvenile literature. | Professional
athletes—Juvenile literature. | Professional sports—Juvenile literature.
Classification: LCC GV734.3 .D64 2022d (print) | LCC GV734.3 (ebook) | DDC
796.023—dc23
LC record available at https://lccn.loc.gov/2021029952
LC ebook record available at https://lccn.loc.gov/2021029953

Editorial Credits
Editor: Mandy Robbins; Designer: Heidi Thompson; Media Researcher: Jo Miller;
Production Specialist: Tori Abraham

Photo Credits
Getty Images: JAVIER SORIANO, 107 (right), Randy Faris, 53, simonkr, 75,
South_agency, 69, vm, 86; Shutterstock: Air Images, 31, cctm, 37, Cynthia Farmer,
25, Dmitry Molchanov, 82, Flystock, 99, IfH, Cover (left), lzf, 10, Milos Kontic, Cover
(right), PhotoProCorp, 61, 65, Scott Meivogel, 107 (left), tammykayphoto, 18, Vasyl
Shulga, 43
Design Elements: Shutterstock: koltsovserezha, Mia Stendal, Nina_FOX, Southern
Wind, ZaZa Studio

TABLE OF CONTENTS

MAKING IT BIG!

YOU have big dreams. You want to get paid to play the sport you love. And you're willing to do whatever it takes to make your dream come true. But it's never a straight shot to the top, especially in the world of competitive sports. Find out if you have what it takes to find fame and fortune as a pro athlete.

The first chapter sets the scene. Then you choose which path to read. Follow the directions at the bottom of each page. The choices you make will change your outcome. After you finish one path, go back and read the others for more adventures.

• Turn the page to begin your adventure.

YOU CHOOSE
THE PATH YOU TAKE TO
BECOME A
PRO ATHLETE.

HITTING THE PAVEMENT

It's 6 a.m. The sun is just rising. A big part of you wants to crawl back into bed.

But you don't. Instead, you lace up your running shoes, throw on a hoodie, and head outside. The air is cool and damp. No one else is around. After a few quick stretches, you start your morning run. Within minutes, your heart is pumping. You feel energized. Thoughts of going back to bed fade away.

• Turn the page.

A quick check of your smart watch tells you that you're coming up on one mile. You've got four to go. Thump, thump, thump. The sound of your shoes against the pavement settles into a smooth rhythm. You're in the zone now.

For the next hour, it's just you and the road. You'll still have time to get home, shower, and be ready for school. Then it's off to practice, a little weight training, then home for homework and bed.

It doesn't leave you a lot of time for doing anything else. Your friends are out having a great time, while you're spending every free moment working on being in the best shape possible.

You're missing out on some fun. But that's a price you're willing to pay. Becoming a pro athlete takes hard work and dedication. Lots of people want to become pro athletes. But only a select few are willing to do everything it takes. It's your dream, and you'll do everything you can to make it a reality.

- To pursue a career in professional baseball, turn to page 13.
- To set your sights on basketball and the Women's National Basketball Association (WNBA), turn to page 47.
- To see if you have what it takes to become a professional snowboarder, turn to page 73.

ON THE
DIAMOND

"Back up," shouts Eduardo, your team's second baseman.

You move your feet back to the edge of the infield dirt. From your position at shortstop, you look on as the opposing team's best hitter steps to the plate. It's the last inning, and your high school team—George Washington Carver High—is clinging to a one-run lead.

• Turn the page.

Your team has one out, and there's a runner on first. This is a pivotal moment.

With a conference championship on the line, the stakes are high. And you know the stakes are even higher for you, personally. As one of the state's top players, you've caught the attention of college and pro scouts. You can see a few of them sitting in the bleachers behind home plate, taking notes.

The pitcher winds and delivers. The batter pounces on a low fastball, driving it on the ground to your right. Your instincts kick in. You dart toward the ball, diving to stop it before it reaches the outfield. In one swift move, you rise to your feet, pivot, and fire the ball to Eduardo just in time to get the runner at second. Eduardo turns and relays it to first base. Double play!

The crowd is on its feet as you and your teammates celebrate the biggest win so far.

"Conference champs, yeah!" Eduardo shouts as he wraps you up in a bear hug.

It's a clutch play! Not long after the game, you get a scholarship offer to join one of the nation's top baseball programs. A few months later, the Texas Rangers select you in the fifth round of the Major League Baseball (MLB) Draft. They offer you a big signing bonus if you'll sign on to their minor-league system.

Your coach calls you to his office. "You have a decision to make," he says. "What are you thinking?"

"Well," you answer, scratching your head, "my mom really wants me to go to college. She says that an education is a sure thing, and that if I'm good enough, I'll get another shot at the pros after college."

• Turn the page.

Coach nods. "There are two ways to look at it. This shot is in front of you right now. What if you get hurt in college? You might never get another shot. On the other hand, your mom is right about an education. It's something no one can ever take away from you. It's really about what you want."

You shake your head. These same thoughts have been running through your mind all day. "My dream is to be a professional. The Rangers are offering me a big signing bonus if I join one of their minor-league teams. I'd be playing full time, making money, and it's the fastest way to my dream of reaching the big leagues. That's pretty hard to turn down. What should I do, Coach?"

Coach drums his fingers on the desk. He smiles at you and shakes his head. "I can't tell you that. It's your choice. What's best for you?"

That night, you take a run to clear your head. That's when it clicks. Suddenly, you know exactly what you're going to do.

- To accept the scholarship and play college baseball, turn to page 19.
- To sign a deal with the Rangers, turn to page 21.

Your mom is right, and you don't want to let her down by not going to college. You learn pretty quickly that the level of competition there is a big step up from high school. The players are bigger, stronger, and faster. In your first game, you strike out three times and make an error in the field.

It's just the beginning of a rough season. You're not hitting well. Your fielding improves, but you don't feel like you're helping the team. One day at practice, you're playing catch with some of the pitchers to help them stretch out their arms. That's when you notice Steve, the team's pitching coach, staring at you.

"What's up?" you ask.

"How much pitching have you done?" Steve asks. "I'm watching you throw. You've got a really strong arm."

• Turn the page.

You shrug. "I pitched a little in high school. But mostly I played shortstop."

Steve asks if he can work with you. The two of you go to the bullpen area, and you show him your stuff.

"I think you're playing the wrong position here," he says. "You've got a big-league quality arm. Your control isn't there, but I could work with you on that. What's your interest in switching positions and joining our pitching staff?"

A switch to pitching? You've never really thought about it. All your life, you've imagined being a big-league shortstop. Are you ready to abandon that plan so quickly?

- To switch to pitching, turn to page 24.
- To stick at shortstop, turn to page 27.

College is great in theory, but you've always wanted to be a pro. Signing with the Rangers gives you a nice payday. Now it's time to earn it.

You report to the minor leagues to start your career. "Hey kid," booms a deep voice when you walk into the clubhouse. It's another player—a big, strong first baseman named James. "Welcome to the show."

This is James's second year in the minors. He's a slugger and your roommate. He also quickly becomes your best friend. He shows you the ropes, helping you learn to study film on opposing pitchers so you can hit off of them better. He teaches you the ins and outs of being a minor leaguer.

• Turn the page.

It's a big help. But you feel overwhelmed on the field. The pitchers are so much better than you ever saw in high school. While James is belting long balls, you're popping out and hitting weak ground balls.

One day, James pulls you aside and shoves a bottle of pills into your hand. "Listen, kid. It's tough making it to the big leagues. We need every advantage."

"What is this?" you ask, your voice shaking. You already know the answer.

"It'll make you stronger. Faster. Better," James says. "Trust me."

You stare at the unmarked bottle. It's a performance-enhancing drug (PED). Taking it will help your game. But it's at a terrible risk to your body—and your career. PEDs are illegal and can get you suspended. Worse, it can lead to all sorts of health problems, including heart and liver trouble.

- To refuse the PEDs, turn to page 29.
- To swallow a pill, turn to page 40.

It's an tough choice. But if becoming a pitcher is your best chance to get on a big-league track, you've got to take it.

It's not an easy switch to make. You don't play anymore that first year. Instead, you spend long afternoons working on bullpen sessions. You learn how to improve your footwork and increase the spin rate on your pitches. Your fastball soon reaches 95-miles-per-hour. You learn to throw a nasty slider that breaks away from the plate.

When you finally take the mound for the first time in a game, you're ready. You strike out the first batter you face. You start out as a relief pitcher, throwing just one inning at a time. But by your third year, you're a starter. And as you grow more and more comfortable on the mound, you become one of college's top pitching prospects.

• Turn the page.

In your final year, your team goes to the College World Series. You're on the mound with everything on the line. But during warm-ups, you feel something strange in your shoulder. It doesn't hurt ... it just feels ... off.

Your team is counting on you. But do you trust your arm?

- To pitch anyway, turn to page 36.
- To pull yourself out of the game, turn to page 38.

"Thanks, but I think I'll keep working at it as a shortstop," you answer.

Steve shrugs his shoulders. "Your call, kid. Let me know if you change your mind."

Over the next three years, you work to balance baseball with your studies. You discover a passion for astronomy and decide to major in it. When you're not doing homework, you're in the weight room, batting cage, or taking ground balls in the infield.

You make slow but steady progress. By your third year, you're a valuable part of a team that wins the conference championship. And in your fourth and final year, you're the best hitter on the team.

• Turn the page.

When the MLB Draft rolls around again, you wait. No one selects you. It's not a big surprise, but you hoped it might happen.

But now it's decision time. You could keep chasing your dream. There's always smaller independent leagues where you can make a living playing baseball, even if it's far from the majors. And who knows, maybe you could catch some scout's eye and get another crack at the big leagues? Or maybe baseball just isn't meant to be. Maybe it's time to turn your attention to a new dream. Maybe instead of becoming a star, you should focus on studying the stars.

- To keep playing, turn to page 33.
- To give up on baseball and continue your education in astronomy, turn to page 35.

Taking a shortcut to the big leagues is tempting. But after a moment of hesitation, your mind is made up. "No, I can't," you tell James. He shrugs. "Okay. If you change your mind, you know where to find me."

Within a week, James gets the call-up to the majors. He belts a home run in his first at bat. But his success doesn't last. He gets caught using PEDs, and his career is quickly over. You're sad for your friend, but you know that he made his own choices. He'll have to live with them.

As for you, life in the minors is tough. There are long bus rides from city to city. You play in front of mostly empty stadiums. You live in hotels. It would all be easier to take if your game was going well. But it's not. You struggle at the plate, barely hitting above .200 in your first year.

• Turn the page.

Yet, as the seasons pass, you slowly make progress. You get bigger and stronger—the natural way. Still, the majors feel like they're a long way away. After five years in the minor leagues, you're 23 years old and starting to get frustrated.

That's when you get a surprising offer. A team in Japan, the Yomiuri Giants, wants to sign you. You'll be the starting shortstop on their team. They offer you a contract worth a lot of money. You're tired of minor-league life. This is a chance to play for a big-time Japanese team. But playing in Japan was never your goal. Your home is here, and you're not sure you're ready to pack up and leave everything and everyone you know.

- To stay in the Rangers' minor league system, go to page 31.
- To sign with the Giants and go to Japan, turn to page 41.

It's a good offer. But you belong here. You won't give up on your goal.

In your fifth season, a new batting coach joins the team. Kent was an all-star in his big-league days, and he has an idea. "What if we change your batting stance?" he suggests. "An open stance will give you a longer look at the pitch. You'll lose some power, but I think you'll make better contact."

• Turn the page.

Changing a batting stance is a big adjustment—and a risk. You've used your stance all your life. It's all you know. Trying something different could change your results—for good or bad. You've heard stories of players who changed their stance and could never get their swing back. Or it could be just the thing to break you out of your slump. Is it worth trying?

- To change your batting stance, turn to page 42.
- To stick with the stance you have, turn to page 44.

You can't just give up. Dreams are worth working for. So you keep going. For the next three years, you bounce from team to team in the independent leagues. It's not a glamorous life. You have to endure long bus rides and games in front of barely any fans. You play well, but not well enough to get noticed by the big leagues. You don't make much money.

In your fourth season of independent ball, you injure your knee sliding into second base. "I'm sorry," says your doctor, shaking his head. "You're looking at a year of recovery, and you'll probably never get your speed back."

• Turn the page.

As you sit in the doctor's office, you know it's the end. Your dream is over. It's time to look for a new future. Maybe you'll go back to school to continue your studies. Or maybe you'll just settle down and find a job. It's a new world of possibilities—baseball just isn't one of them.

--- THE END ---

To follow another path, turn to page 11.
To learn more about becoming a pro athlete, turn to page 101.

All your life, playing baseball has been your dream. But maybe it's not anymore. People change. You have changed. When you look out into the night sky, your mind is filled with questions about stars, planets, comets, and galaxies. Where does it all come from? How will it all end? You want to spend your life finding out.

So it's on to graduate school to become an astronomer. Your love of baseball never leaves you. In the summer, you play for a local town ball team. Later in life, when you have kids of your own, you coach their Little League teams.

You're happy with your life and your choices. But every now and again, you wonder if you would have made it if you'd have done things differently.

--- THE END ---

To follow another path, turn to page 11.
To learn more about becoming a pro athlete, turn to page 101.

"You alright?" asks DeShawn, your catcher. "Your slider isn't breaking today."

You wave him off. "It's nothing. Let's go do this."

You start off shaky. The first two hitters reach base. But then you settle in and start mowing down batters.

By the fourth inning, your shoulder feels tight. You keep going. Your team is ahead 3–0 and you don't want to quit now. But then something pops. You can hear it and feel it as you deliver a fastball. You double over in pain and grab your shoulder. The training staff rushes you back into the locker room and puts ice on your shoulder. From there it's on to the hospital for scans.

The news is not good. You have a badly torn ligament in your shoulder. Surgeons do their best to repair it. But you're never the same again. Your fastball has lost its zip. Your slider doesn't break like it used to. You had a big-league arm, but you weren't careful with it. Now your career is over.

--- THE END ---

To follow another path, turn to page 11.
To learn more about becoming a pro athlete, turn to page 101.

"Something's not right," you call out to your catcher, DeShawn. Within minutes, the team's coaches and trainers are gathered around you. Your team looks dejected when they hear the news. Their star pitcher won't be pitching in the biggest game of the season.

You watch from the dugout as your team loses, 7–4. It hurts to know that you can't do anything to help. But pulling yourself from the game proves to be a wise move. Doctors find a very small tear in a ligament in your shoulder. It will heal without surgery. But you got lucky. Who knows what might have happened if you had tried to pitch?

That summer, the Boston Red Sox make you their first-round pick. You sign a multimillion-dollar contract and report to their minor-league team. Within a year, you're on the big-league roster, living your dream. You're on your way to becoming one of the best pitchers in the league, and you know that every little decision you made along the way helped you to get there.

--- THE END ---

To follow another path, turn to page 11.
To learn more about becoming a pro athlete, turn to page 101.

With a deep breath, you pop one of the pills into your mouth. For the next month, you take one every day. Sure enough, you notice a change. Your pop flies turn into home runs. Your ground balls turn into line drives. You're tearing up the league, on the fast track to the majors.

Then a random drug test brings it all crashing down. Your secret is out. Just like that, the team cuts you. And with your PED bust, nobody is willing to sign you again—even when you're clean.

You had your shot, and you blew it. You only wish you could go back and make a different choice.

--- THE END ---

To follow another path, turn to page 11.
To learn more about becoming a pro athlete, turn to page 101.

It was never your dream to play overseas. But baseball is big in Japan. And you're not making much progress here. So you take the offer. You thank the Rangers for the opportunity and board a flight to Tokyo.

It's quite an experience. In the United States, you were a career minor leaguer. In Japan, you're a star for the Yomiuri Giants. You make a life for yourself in Japan. You learn the language, fall in love, and get married.

You play 14 years for the Giants, winning three championships. You start a family. Japan becomes your home, and you love it there. You never would have imagined your dreams coming true in this way. It may not be exactly what you had hoped. But it's definitely a happy ending.

--- THE END ---

To follow another path, turn to page 11.
To learn more about becoming a pro athlete, turn to page 101.

You need a spark, and maybe Kent can give it to you. The two of you spend hours in the batting cage, perfecting a new stance. At first, it doesn't go well. But after a month, you start to feel comfortable. You're lacing line drives all over the park.

The Rangers call you up in August. They're in a pennant race, and their shortstop is out with an injury. They slot you in as the starting shortstop—and you never give the job back. You become one of the league's best contact hitters. You don't have much power, but you get hits in bunches, use your speed on the bases, and play great defense. You help the team reach the playoffs in your first season, and you're a World Series hero in your second.

It's the start of a long and brilliant career.
All-Star games, batting titles, and three World
Series titles later, you finally retire as one of the
most beloved Rangers in history. You did it!
You chased your dream, and it was everything
you ever hoped it would be.

--- THE END ---

To follow another path, turn to page 11.
To learn more about becoming a pro athlete, turn to page 101.

You shake your head. "Sorry, I don't think that's for me."

At first, it looks like a good decision. You start out with your best season yet. By July, you're batting .330. Finally, the Rangers take notice. They call you up to the big leagues.

It's finally happening! After all your hard work, you're about to make your major league debut. You can hardly believe it.

When you step up to the plate for the first time, you can feel the crowd buzzing. There's a runner on second late in a tie game. You need a hit.

The pitcher looks in and fires a blazing fastball. You swing and miss. He throws another. This time, you're sure you're ready. You swing with all your might—only to realize it wasn't a fastball at all. It was a change-up—a slower pitch designed to throw off a batter's timing. One more fastball and it's over—you've struck out.

You thought making the big leagues would be the greatest time of your life. But it's miserable. You don't get a single hit in 19 plate appearances. The Rangers send you back down to the minors. And you never get another chance. You become a career minor leaguer—with just a tiny taste of the big show.

You gave it your all. Sadly, you came up just a bit short.

--- THE END ---

To follow another path, turn to page 11.
To learn more about becoming a pro athlete, turn to page 101.

HOOP DREAMS

"Yo! I'm open," shouts Keisha, sticking her hand up. You're at the top of the key, dribbling as the game clock ticks down. You fake a drive, throwing your defender off balance. Then you step back and thread a perfect pass to Keisha in the paint—the area near the rim. She rises up, banks the ball off the backboard, and drops it through the net. It's a game winner! The buzzer sounds, and your teammates swarm you and Keisha, as you celebrate a big victory.

• Turn the page.

Keisha crashes at your place that night, and the two of you stay up for hours reliving the victory. You're two of the best high school players in the state, and you spend most of the night talking about the next step—college ball. You've been best friends since sixth grade, and you dream of playing on the same college team.

"Let's do it," Keisha says, for the fifteenth time. "Come with me." Keisha has just one scholarship offer, to a local state school. She's already committed to the university. It's a decent team but not a championship contender. You, on the other hand, have a lot more options. You're ranked in the top 100 in the country, and you've got your choice of schools. You glance over to your desk, where a stack of recruiting letters are piled up.

There are letters from Tennessee, Notre Dame, Connecticut, Stanford, and more—the best basketball programs in the country. And they've all offered you scholarships.

"Think about it," Keisha presses on. "We can run our famous pick-and-roll for another four years. You and me—we can build the team into a winner. And we'll be close to home. Your family will be able to come to all of the games."

You close your eyes. It sounds great. It's all you ever wanted. But that was before the best schools in the country came calling. Playing with your best friend is appealing. But playing in the National Collegiate Athletic Association (NCAA) Tournament, competing for national titles . . . that's a hard offer to refuse.

• Turn the page.

You finally drift off to sleep. Your dreams are filled with images of knocking down step-back jumpers and slashing to the rim. The dream feels so real—and when you wake up, you know exactly what you want. Your mind is made up. You're ready to commit to a school.

- To go with Keisha to the state school, go to page 51.
- To accept a scholarship to one of the nation's best basketball programs, turn to page 52.

What could be more important than playing ball with your best friend and staying close to the people you love? Keisha squeals with delight when you tell her the news. Your family is happy. Your teammates are happy. And most importantly, you're happy with the decision.

It's an easy transition to college life. You and Keisha live together in a dorm room, studying, working out, and honing your skills. Over your first three years, you can feel your game growing. Your ball handling is smoother. You focus on playing lock-down defense. And your three-point shot is becoming deadly accurate. All the while, you're studying hard and earning a valuable education. You love college. Everything is working out great.

• Turn to page 58.

"I'm sorry," you tell Keisha after you announce your decision. "I know you'll do great. I just couldn't pass up this chance."

Keisha admits she's disappointed. But then she flashes you a smile and gives you a hug. "I get it. You have a chance to go pro. You have to do what's right for you. I'll be rooting for you."

The next year is a whirlwind. In the fall, you dig into your classes. Once basketball season starts, you spend more and more time in the gym. You hit the weight room. You shoot what seems like a thousand free throws a day. You run. You're ready.

The crowd is rocking for your first game. You don't get to start, but the coach puts you in late in the first half. You launch your first shot—an air ball. But you slowly improve.

• Turn the page.

Your numbers aren't great, but you have your first taste of college ball. And you like it.

Your team is a powerhouse. Getting playing time is tough. You spend your first two years mostly on the bench. You're eager to be a starter in your third season—but then a new recruit takes the spot. Frustrated, you storm out of the gym. Another year on the bench? Are you ever going to get your chance to shine?

You call Keisha to vent. "You know, if you're not happy there, you could always transfer. You'd be a starter right away here," she informs you.

- To stick it out where you are, go to page 55.
- To leave the program and join Keisha, turn to page 57.

You're no quitter. You'll put in the work. You'll earn your minutes. It's late, but you feel charged. You head to the gym for a serious workout. You've always worked hard. Now you double your efforts.

Your work pays off. In your team's opening game, you score 20 points. You become the team's spark plug. You put up points night after night while playing lock-down defense. In your final year, you earn the starting job and help your team reach the Final Four. It took some time, but you finally got the chance to show what you can do.

You eagerly wait for the WNBA Draft. You know you won't be a first-round pick. You just didn't put up enough numbers in college. But your hard work pays off. The Minnesota Lynx choose you in the second round. It's a dream come true. You're headed to the WNBA!

• Turn the page.

That's when disaster strikes. You injure your ankle while working out. You'll be out for the season! The Lynx cut you before their new season begins. Just like that, you're out of a job. You could always play in Europe. Lots of players do it. Or you could stay here and hope to get another shot at the WNBA.

- To try to get another shot in the WNBA, turn to page 60.
- To sign with a European team, turn to page 64.

You're not happy here. You can barely get onto the court to show what you can do. Pro scouts won't be watching you on the bench. So what's the point of staying? It's time to try something new. So you transfer. You're heading home to be close to the people you love. Keisha and the rest of the team welcome you with open arms.

"Now we're ready to take on the world," Keisha says as she helps you unpack your things.

It's a great start to the season. You're filling up the stat sheet and winning games. Being out there from tip-off to crunch time feels so good. After two years of riding the bench, you know this is where you belong. Whether you end up getting drafted to the WNBA or not, you'll get to play the sport you love for at least two more years.

• Turn the page.

Your fourth and final season comes with big expectations. You and Keisha are seniors, and the heart of the team. At times, you feel unstoppable. Keisha dominates the paint. You are a sharpshooter from the outside. Opposing defenses scramble to cover both of you.

Your hard work pays off—the team earns its first bid to the NCAA Tournament in more than 10 years. Your first game is against Ohio State—a powerhouse. You're heavy underdogs. But you believe in yourselves.

The game is a nail-biter. With just 10 seconds to play, you're behind by 2 points. You've got the ball in your hands, with a defender all over you. You're the best player on the team, and the coach wants you taking the final shot. So you fake left, then back off to the right. Five seconds remain. There's a small opening to your right.

You could drive into the lane and try to get an easy layup to tie it. Or you could step back and launch a three-pointer for the win. It will be a harder shot, but you will have the chance to end it right here. With the roar of the crowd growing louder by the second, you make your move.

- To drive for the tie, turn to page 62.
- To shoot a three-pointer for the win, turn to page 70.

Playing overseas isn't a bad option. But it's not what you want. So you stay home, spending your days working on your skills. You put in hours at the gym. You eat right. You keep in peak shape.

Then you get your break. The New York Liberty give you a tryout. You don't waste the chance, knocking down shots and putting on a dribbling clinic. The Liberty sign you.

It's a great opportunity. New York is in the playoff hunt. Their star guard suffered an injury that knocked her out for the season. You're helping to fill the role she left behind.

In the final game of the season, you need a win to make the playoffs. You're in a tight game with the team that drafted you—the Lynx. In the final seconds, the ball is in your hands. You're behind by one point. You need a basket.

You fake a move to your left, then spin to the right. The lane opens up before you. To one side, you see your team's center, Anna, break free of her defender. Decision time!

- To pass to your open teammate, turn to page 66.
- To take the outside shot, turn to page 67.

You've got an open lane. You have to take it. So you put your head down and drive into the paint. The defense begins to collapse in on you. But you're quick. You rise up and lob the ball up toward the rim.

CLANK! No good!

You fall to your knees as the final buzzer sounds. Your season is over. Keisha helps you off the court. "It's okay," she assures you. "We came further than anyone thought we would."

It's true. But it doesn't feel like enough. The WNBA Draft comes and goes. No one selects you. You aren't even invited to a tryout. You have offers from a few European pro teams. But that means leaving your home and your country. You're not sure that's what you want.

"There's always a place for you here," your coach tells you. "Join my coaching staff. You've got a great mind for the game. What do you say?"

- To go to Europe to play, turn to page 64.
- To take on a new challenge in coaching, turn to page 68.

You still want to play. So you sign with a team in Italy. You pack your things and board a flight.

Life overseas is a big adjustment. You work on your Italian, but you're not very good. At first, you feel completely out of place. But once you hit the court with your team, you're comfortable. Most of your teammates speak some English. And one of your teammates, Aliyah, is from Ohio.

Italy never quite feels like home. But you make it work. Your combination of strength and quickness makes you a nearly unstoppable force in the Italian league. You lead your team to the finals in your first year. In your second, your team wins the championship, and you're the league's Most Valuable Player (MVP).

You're making decent money. You're playing at a high level. And you're experiencing a different culture and way of life. You've got a long career in front of you. It's a good life, even if it's not exactly the one you had in mind.

--- THE END --

To follow another path, turn to page 11.
To learn more about becoming a pro athlete, turn to page 101.

The moment takes you back to your days in high school. You remember the game where you hit Keisha down low with a pass for the game winner. So that's what you do. You rifle a chest pass down into the paint. Anna is ready. She grabs the pass. In one smooth move, she pivots and lays the ball off the glass. It falls through the net—game winner! You're in the playoffs!

That's where you really shine. You're on fire, knocking down three-pointers and running the offense. Your great play helps your team advance all the way to the finals. The Liberty rewards you with a new long term contract.

You've done it! You're a WNBA player with a bright future. You never stopped believing in yourself, and it's paying off big time.

--- THE END ---

To follow another path, turn to page 11.
To learn more about becoming a pro athlete, turn to page 101.

Anna is open, but you're feeling hot. You're sure you can make the shot. So you stop and pop, launching a shot from just outside the three-point arc. It feels good coming out of your hands.

The ball sails through the air. It drops down, right onto the rim. *CLANK!* The ball rattles off, and a Lynx player scoops up the rebound.

The game is over. You missed the playoffs, and you feel like it was your fault. The Liberty don't offer you a contract for the next year, and you never get another shot at the WNBA.

You had your shot at glory, and you missed it. But that's still more than most people can say. At least for a few bright, shining weeks, you were a professional basketball player.

--- THE END ---

To follow another path, turn to page 11.
To learn more about becoming a pro athlete, turn to page 101.

You love the game, and a big part of you wants to keep playing. But you also love your home, and the idea of leaving is too hard. You take the offer to become a coach. It's a big adjustment. Now, instead of hitting the gym, you spend your time studying game film. As time passes, you come to love the job, and you're really good at it. You find that helping the team succeed is every bit as rewarding as hitting a big shot yourself.

After a few years as an assistant coach, you get an offer to become the head coach at a nearby college. You take it—becoming the youngest head coach in the NCAA. It's not quite the career in basketball you'd always imagined. But you're happy. If you could go back in time, you wouldn't change a thing.

--- THE END ---

To follow another path, turn to page 11.
To learn more about becoming a pro athlete, turn to page 101.

You don't hesitate. You give your defender a head fake, buying just a little space. Then you step back, raise up, and launch a long three-point shot. The crowd collectively gasps as the ball sails through the air. You hold your breath, knowing everything rests on this one shot.

SWISH! It's good! The final buzzer sounds as the ball drops through the net. The game is over, and you've knocked off one of the nation's best teams. It's a highlight that makes every sports news program in the country. You ride that success all the way to the Elite Eight—by far the deepest NCAA run your school has ever made.

Your great season and magical shot have made you a top prospect in the WNBA. The Atlanta Dream selects you with the third pick in the WNBA Draft. "You're the future of this team," the coach tells you as you meet on the draft stage.

You've made it! You know that life as a WNBA player won't be easy. It will take hard work and lots of travel—all without the huge salaries that the top men make in the NBA. But right now, you're not thinking about that. You just can't wait to show the league what you've got.

--- THE END ---

To follow another path, turn to page 11.
To learn more about becoming a pro athlete, turn to page 101.

GOING **BIG**

"Good snow today," says your sister Tonya, as she steps into her board.

You nod. The weather is perfect, hovering just around freezing. But then, the two of you are out on the halfpipe pretty much every day, no matter the weather.

"You going to try it today?" she asks. She means the 1440 tail grab—a super-difficult trick. You've been working on it for a year, but you've still never landed it. And you've paid the price for failure with a few hard falls.

• Turn the page.

"Not sure yet," you answer. "I'll see how the run goes. If I'm feeling it, I might try it."

"Well, Samir is down there filming. You really should give him a show."

You grin. Tonya is always pushing you to go bigger and bigger. "See you at the bottom," you say, as you drop into the pipe.

Right away, you can feel that it's going to be a good run. You grab some air on your first two jumps, or passes. Then you do a full spin in the air for a simple 360 nose grab. You come down clean, and you know the time is right. You build up speed on the next pass. Then comes the moment of truth. You rise and twist your body, rotating rapidly while reaching back to grab the back of your board.

It's a whirlwind of motion as you spin around four full rotations. As you come down, you need to time your landing perfectly.

• Turn the page.

You've got it! Everything feels perfect. But when your board touches down, you've rotated just a bit too much. It's a messy landing, but you manage not to fall. Samir is filming at the bottom. He hoots and hollers so loud that you're sure everyone on the mountain can hear him.

A minute later, Tonya comes down, sticking a 720 on her final pass. "You did it!" she shouts. "Samir, you gotta post that on social media!"

You smile. You did it. But it wasn't perfect. It feels good, but you're going to keep working on it until it's just right.

That night, the three of you are chowing down on pizza when Samir's phone buzzes. "Whoa," he gasps. "Check out who just liked your video! He left a comment too!"

You look over Samir's shoulder. One of the best snowboarders in the world has complimented your video. Watching his videos on social media was a big part of what made you take up snowboarding in the first place. Your heart is racing just thinking that he watched your trick.

"Nice air," reads the post. "Landing needs work, but you show a lot of potential. Sent you a pm."

"No way!" Tonya shouts. You scramble to pull out your phone and read the message. Your jaw drops as you stare at the screen. "Any interest in coming to Colorado to train with my crew? LMK."

• Turn the page.

You're half in shock. The training facility in Colorado has almost legendary status in the snowboarding world. Some of the world's top boarders work out there. It's like an exclusive snowboarding club, and you just got invited. But as exciting as the idea is, it's also terrifying. How can you possibly belong in that kind of company? You're afraid you'll make a fool of yourself. The thought of looking like a noob in front of professionals fills you with dread. Maybe if you keep working on your skills here, you'll be ready in a year or two.

- To accept the offer to train with the pro crew, go to page 79.
- To continue training at home, turn to page 81.

You grin. Who knows if you have what it takes? There's only one way to find out. You reply with a quick message. "I'm in!" Within a week, you're on your way to Colorado.

A young woman named Ling picks you up at the airport. She introduces herself and tells you that she's one of the up-and-coming snowboarders training with the team. Ling drives you to the facility, which is complete with a private halfpipe. It's like a dream come true. You get there just in time to watch your idol drop in and throw down some serious tricks.

"Welcome!" he greets you at the bottom. "Loved your tail grab. I think we can help you stick that landing. Ready to show us what you've got?"

• Turn the page.

There's no turning back now. As you head up to the top of the pipe, your head is spinning. What should you do? A basic run with some big airs and rotations? It'd be the safe choice. You've never ridden this pipe and you're feeling pretty nervous. Or should you just go big right away and try the 1440 tail grab?

- To do a safe, basic run, turn to page 83.
- To try the 1440 tail grab, turn to page 85.

"Nah," you say with a nervous laugh. You don't want to tell Samir and Tonya about your doubts. Instead, you make up an excuse. "It's probably not even for real. For all we know, that's just someone posing as a famous snowboarder to try to scam me. Let's go back out tomorrow. I'll work more on the 1440."

They look at you like you're crazy. But then they just shrug. "Okay, man," Samir says.

You get to work, really pushing yourself to be better. You make slow progress. You add more tricks to your bag, and crowds gather at the local halfpipe to watch your runs.

A few months later, when a low-level pro tournament comes to town, everyone wants you to enter. "Come on," says Samir. "You have what it takes."

• Turn the page.

"Do it," Tonya chimes in. "If not now, then when?"

The idea makes you nervous. What if you wipe out? Will you ever be able to show your face on the slopes again?

- To enter the tournament, turn to page 88.
- To skip the tournament, turn to page 98.

You're surrounded by some of the best snowboarders in the world. They know that you're going to need to build up to big tricks. So you keep it simple. You drop into the pipe and pull off some really big airs. Then you finish with some 720s, sticking every landing perfectly.

Your hero nods. "Okay. I can see that you've got your basics mastered. That's good. Now let's get to work on the tricks you'll need if you want to go pro."

You spend the next few months training. It's amazing to learn from so many of the best. You're adding new tricks and improving the ones you already know.

• Turn the page.

It's time to test yourself. Two big snowboarding competitions are coming up. The best of the best are going to the World Snowboarding Championships. A few of the other newer up-and-comers are headed to a smaller competition just for local riders.

"You're ready," Ling says. "It's time for you to show the world what you can do."

• To go to the World Snowboarding Championships, turn to page 87.
• To try out the smaller competition, turn to page 88.

You've come all this way to prove that you belong. What better way than to bust out your best trick? You drop into the pipe and pick up speed. You set up the big moment by catching huge air. Then you jet across the pipe and scream up the other side. You launch and rotate. The world spins around you, but you're in complete control. At the last moment, you straighten up and land your board perfectly.

The others are hooting and hollering at the bottom of the pipe. "You're legit, kid!" Ling says. Several of the other competitors pat you on the back. And when your idol follows with a great run of his own, he's all smiles.

You become his top student. The two of you spend hours working on technique. With his help, your skills grow by leaps and bounds.

• Turn the page.

Soon, you're winning pro competitions. A national snowboarding magazine even lists you as one of the sport's up-and-coming stars. Everything is happening so fast, you can barely believe it.

• Turn to page 90.

The championships are one of the highlights of the snowboarding year. When you get there, the crowd is electric.

Your first run is flawless, but you stumble in your second. That leaves you about halfway down the leaderboard, well behind the leader.

You'll have to go big in the final run to have any shot at the podium. You pull out your signature 1440 tail grab. But you'll need something more. So on your last pass, you bust out a 900 McTwist—a front flip with two-and-a-half spins. It's not perfect, but you manage to land it. The score shoots you all the way up to third place—good enough for the bronze medal! Tonya and Samir are there in the crowd, cheering you on. You can't believe how quickly you've reached your pro snowboarding dream!

--- THE END ---

To follow another path, turn to page 11.
To learn more about becoming a pro athlete, turn to page 101.

It's a small tournament, so that means less pressure. But as you watch some of your competition, you realize it's going to be tough. They're landing hard tricks with ease. You're going to need to be perfect to keep up.

Your heart is pounding as you make your first run. It's very clean but had easy tricks. In your second run, you go for it. You catch big air and try to pull off a double cork 1080. It's a complicated trick that has you spinning and flipping all at once. It requires precise timing to land just as you complete both a spin and a flip—and yours is off.

You don't get enough rotation. Your board comes down at a bad angle, sending you sprawling down hard onto the packed snow. Your confidence is shot, and you fizzle out near the bottom of the standings.

If you bombed that badly at a small tournament, how could you ever compete on a bigger stage?

Dejected, you head home. Once you get there, you don't even want to look at a snowboard. So you take up skateboarding instead. You and Samir hit the local skate parks, and you discover that you're really good at street style skating. But when winter returns, you have a decision to make. Are you ready to give up snowboarding? Or is skateboarding your new future?

- To chase a new dream of being a pro skateboarder, turn to page 91.
- To head back to the slopes, turn to page 92.

A year later, you give Tonya a call that makes her shriek with joy. "I made it. I'm on the Olympic team!" you tell her. It's a dream come true, and you can't wait.

You've got a big choice to make, with spots available in both the halfpipe and big air competitions. You can do one or the other, but not both. Halfpipe is your comfort zone. But you've also been working on your big air, where you get one trick off a huge ramp. You've gotten really good at it, and you think the field of opponents might not be as strong as in the halfpipe.

You make your choice and get to training. Then it's off to Beijing, China, for the Winter Olympics. Which event is your path to Olympic glory?

- To test yourself in the halfpipe event, turn to page 94.
- To try the big air event, turn to page 96.

It's no surprise to you that you're also good at skateboarding. It uses a lot of the same skills as snowboarding, and many of the tricks are similar. That's why superstars like Shaun White have won gold medals in both sports.

You spend your days perfecting new grabs, grinds, and other tricks. You enter a few small competitions and do well. You're never going to be a world-class skater. But you're good enough to earn a little money and even get a few small sponsorships.

It's not quite the life you had imagined. But you get to spend your time doing something you love. And there's nothing wrong with that.

--- THE END ---

To follow another path, turn to page 11.
To learn more about becoming a pro athlete, turn to page 101.

You're not quite ready to give up on snowboarding, even if your attempt at going pro was a disaster. Yet even as you get ready to drop in for your first run of the year, you flash back to your big wipeout. It's messing with your mind. The snow is good and your heart is pumping. But no matter how hard you work, you can't land the tricks that you used to. Your confidence isn't there.

You're still good enough to wow all the kids that gather around the pipe. They watch and cheer, and you're happy to give them tips. One day, the manager of the slopes offers you a job teaching snowboarding classes.

"You should do it," Tonya says. "You'd be a great teacher."

You decide to give it a shot—and you love it. You get to spend your days on the slopes and in the pipe. You have the chance to share your love of the sport with a new generation of boarders. You always imagined a career filled with Olympic glory. But now you know that this was the life for you all along.

--- THE END ---

To follow another path, turn to page 11.
To learn more about becoming a pro athlete, turn to page 101.

The halfpipe has always been your first love. The field may be tougher, but you really want to test yourself. You give it everything you've got, and you manage to advance to the finals.

After two runs, you sit in the middle of the pack. If you want to win a medal, you'll have to go huge. Your friends help you map out a routine filled with tricks that will push you to your limits.

Your heart is thumping when the time finally comes for your last ride. A cold wind blasts you in the face as you drop in. You don't waste any time, throwing down a backside rodeo 720, flipping over backward two full times before landing. It's a trick you've struggled with in the past. But with your blood pumping, you stick it. You can hear the crowd roaring as you grab air and set up your next big trick.

Then, on your final pass, you do your signature 1440 tail grab. Success! The crowd roars! It's a 98.00—the best score of the games! Your fellow snowboarders mob you at the bottom of the pipe. You can barely believe it. It's a gold-medal run! You've just cemented your place as one of the greatest of all time.

You've already done more than you could ever have dreamed. And your career is just getting started. It's time to work on becoming the greatest snowboarder who ever lived. You won't stop working until you've achieved your new goal.

--- THE END ---

To follow another path, turn to page 11.
To learn more about becoming a pro athlete, turn to page 101.

You love the halfpipe. But you're sure big air is your best shot at Olympic glory. In big air, you get a much bigger launch than in the halfpipe. So there's lots more time for twists, flips, and grabs. Each run of this event requires just one outrageous trick. You consider a bunch of different options. You finally settle on a backside quad cork 1800 indy grab. You've pulled off the trick in practice a few times. As you stand at the top of the slope, you try to forget the dozens of times you wiped out.

"Positive thoughts," you tell yourself as you start your run. You crouch low to build up speed. Then you shoot up the huge, curved ramp and soar into the air. The world spins around you as you try to pull it off. As you combine a series of high-speed spins and twists, you can feel that your flip rotation is just a little behind.

You tuck in tight, desperately trying to save the trick. But you just don't have the speed that you need. There's nothing you can do except brace for impact.

It's a complete disaster. The ground rushes at you, and you can't straighten up in time. You crash down onto the packed snow. Bones crunch and snap. Your head slams into the ground hard, and the world goes black.

It's a terrible fall. An ambulance rushes you to the hospital, where you're treated for broken bones and a bruise to your brain.

You went big, and you failed. Now the question is whether you'll heal well enough to ever ride again.

--- THE END ---

To follow another path, turn to page 11.
To learn more about becoming a pro athlete, turn to page 101.

You shake your head. "Nah," you say. "I really just like doing this stuff for fun. Maybe someday."

Tonya storms off. Samir just looks disappointed. Later that night, when you're by yourself, you feel like it was the right choice.

Time after time, you've had a chance to chase your dream. But every time, it hasn't felt quite right. Your hopes to be a pro someday were just fantasies. You don't really want the fame and pressure of pro competitions. It would take all the fun out of the sport for you.

Tonya and Samir are mad at you for a few days, but they get over it. The three of you end up starting a new social media channel.

Unlike most snowboarding channels, it's not based on landing big tricks. It shows all of your spectacular wipeouts. People love it. Within a year, you've got a million subscribers. Who knew that failure could be so successful?

--- THE END ---

To follow another path, turn to page 11.
To learn more about becoming a pro athlete, turn to page 101.

A **LONG** ROAD

Becoming a professional athlete takes natural talent, hard work, and maybe even a little bit of luck. Whether it's baseball, basketball, snowboarding, or another sport, you'll have to be among the best in the world to make it pro.

So what does it take? Pro athletes need to train hard. They have to perfect their skills and techniques with countless hours of practice. They need to be in peak physical condition. That means working out—running, lifting, and doing cardio.

But training isn't the only ingredient to success when it comes to being a pro athlete. Elite athletes need to eat right, avoid harmful substances such as drugs, and be at their best when the stakes are highest. High-level athletes often see therapists to help them deal with the intense pressure they face.

The road from amateur to pro is different for every sport. Football players usually advance through high school and college before they get their shot at the National Football League (NFL). Golfers work their way up through amateur or college ranks. Gymnasts and swimmers may build their skills in clubs, starting when they're extremely young. There's no one way to get to the top.

It's also important to remember that careers in pro sports don't end with athletes. Teams need coaches. They need trainers to keep athletes healthy. Equipment managers keep everything in top shape. Front-office executives make the big decisions on which players teams draft and sign. Scouts look for new talent and give reports on upcoming opponents. Referees and officials enforce the rules and keep games moving.

And it doesn't end there! Security workers keep players and fans safe. Ushers help fans to their seats. Vendors sell food, drinks, and souvenirs. Reporters, camera operators, and broadcast crews help bring the action to fans at home. So even if you don't make it as a pro athlete, the dream of working in sports doesn't end!

Most athletes who try to make it to the pros fall short. But a select few have what it takes. Their talent and hard work pay off, and they get to compete on the biggest stage. Do you have what it takes? The only way to find out is to try.

SPORTS
FAME OR FORTUNE
REALITY CHECK

EDUCATION OR EXPERIENCE

In general, there's no specific education or experience needed to become a pro athlete. Some athletes spend years working their way up through amateur levels or minor leagues. Others burst onto the scene and take their sport by storm at a young age. What all successful pro athletes share in common is hard work, dedication, and natural talent.

SALARY RANGE

The sky is the limit when it comes to what a pro athlete can earn. In 2020, tennis star Roger Federer earned $106 million! Fellow tennis player Naomi Osaka was the highest paid female athlete, at $37.4 million. Most pro athletes earn much less than these amounts, of course. In Major League Baseball, the minimum salary is a little over $550,000 per year. In the WNBA, the average is about $130,000. Unfortunately, even at the highest levels, men tend to make more money than women.

WHAT THE PROFESSION ENTAILS

Fans just see the games and matches. That's a big part of a pro athlete's life. But it's only a small part of the job. Countless hours of practice and workouts make it possible. A pro athlete's entire life revolves around keeping in peak physical condition. That's what it takes to perform when everyone is watching.

WELL-KNOWN PEOPLE IN THE PROFESSION

LeBron James (NBA)

Breanna Stewart (WNBA)

Mike Trout (MLB)

Serena Williams (tennis)

Alex Morgan (soccer)

Jenny Jones (snowboarding)

Patrick Mahomes (football)

LeBron James

Jenny Jones

OTHER PATHS TO EXPLORE

Making it as a pro athlete is hard. What about people who don't make it or pursue other opportunities?

Imagine your dream is to be a sports executive. They're the people who build teams, sign players, and run the day-to-day operations. What might it be like to go after a career as a sports executive?

Injuries are a big part of sports. Some are minor. Others are career-ending. Imagine you're a star athlete on your way to becoming a pro. But then a terrible injury ends your dream. In a heartbeat, everything you ever wanted is gone. How would you react? Would you still want to be a part of the sport? Or would you turn away from it altogether to take on new life challenges?

For every athlete who makes it to the big time, there are dozens who fail. Some are prepared for a different career. Others are not. What would you do if you failed in your dream? Would you be ready to find happiness in another line of work?

GLOSSARY

amateur (AM-uh-chur)—a sports league that athletes take part in for pleasure rather than for money

debut (day-BYOO)—a player's first game

draft (DRAFT)—the process by which teams in a professional league select new players

performance-enhancing drug (PED) (pur-FOR-muhnss en-HANS-ing DRUG)—a substance that helps an athlete build muscle, recover faster, or improve their performance—often putting their body at risk; most pro leagues ban PEDs

professional (pruh-FESH-uhn-nuhl)—a person who is paid to perform a sport as their career

recruit (ri-KROOT)—to ask someone to join a college or professional team; also the name of a person who has been recruited

salary (SAL-uh-ree)—money paid to a person for performing a job

scholarship (SKOL-ur-ship)—money given to help pay for a student's expenses, including tuition, books, and living expenses

transfer (trans-FUR)—to change to a new college

SELECTED BIBLIOGRAPHY

Burton, Rick et. al. *20 Secrets to Success for NCAA Student-Athletes Who Won't Go Pro*. Athens, Ohio. Ohio University Press, 2018.

Epstein, David. *The Sports Gene: Inside the Science of Extraordinary Athletic Performance*. New York: Penguin, 2014.

Forbes—Highest-Paid Athletes 2021
forbes.com/athletes/#553d366c55ae

NCAA: Research
ncaa.org/research

Professional Athlete · Job Description, Salary & Benefits
allaboutcareers.com/job-profile/professional-athlete/

READ MORE

Coleman, Ted. *Great Careers in Sports*. Lake Elmo, MN: Focus Readers, 2022.

Doeden, Matt. *The Paths to Pro Basketball*. North Mankato, MN: Capstone Press, 2022.

Kerstetter, Greg. *Work in the Professional Sports Industry*. San Diego: ReferencePoint Press, Inc., 2020.

Mattern, Joanne. *What It Takes to Be a Pro Basketball Player*. Mankato, MN: 12-Story Library, 2020.

INTERNET SITES

Become an Athlete
bigfuture.collegeboard.org/careers/sports-fitness-athletes

How to Become a Professional Athlete
allaboutcareers.com/job-profile/professional-athlete/

Professional Athlete
careerkids.com/pages/professional-athlete

ABOUT THE AUTHOR

Matt Doeden is a freelance author and editor from
Minnesota. He's written numerous children's books
on sports, music, current events, the military, extreme
survival, and much more. His books *Sandy Koufax*
(Twenty-First Century Books, 2006) and *Tom Brady:
Unlikely Champion* (Twenty-First Century Books, 2011)
were Junior Library Guild selections. Doeden began his
career as a sportswriter before turning to publishing.
He lives in Minnesota with his wife and two children.